PLANT LIFE OF WESTERN AUSTRALIA

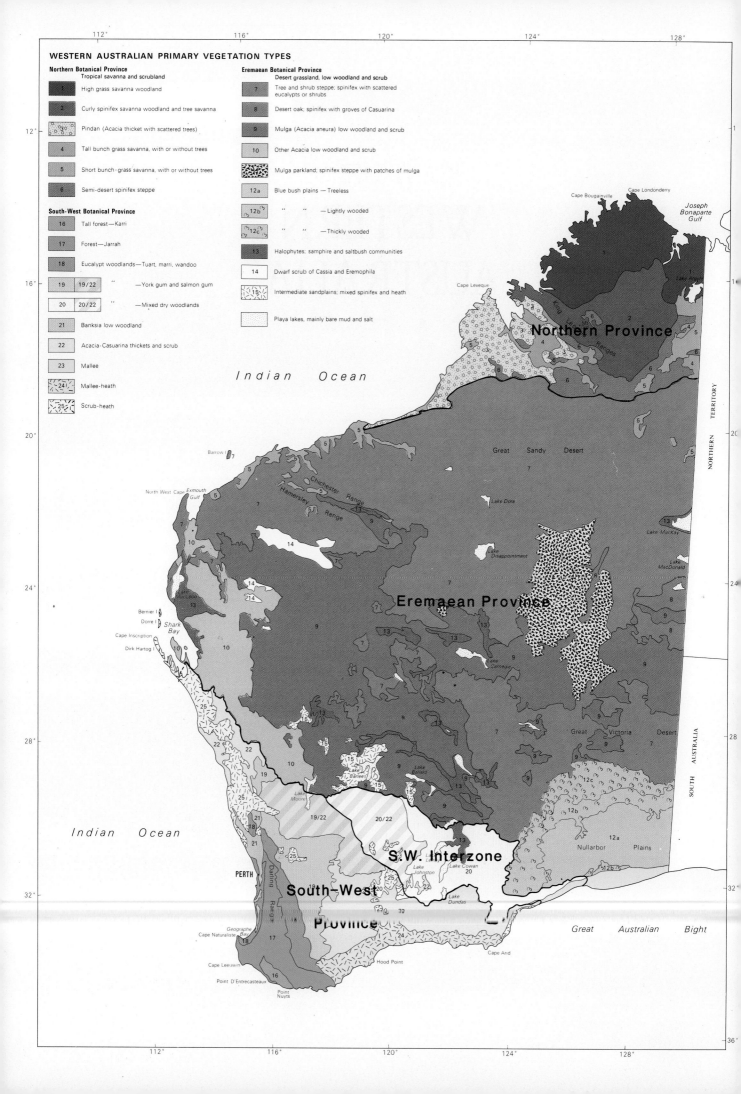

PLANT LIFE OF WESTERN AUSTRALIA

J.S. Beard

Kangaroo Press

Acknowledgments

Thanks are due to Alcoa of Australia Limited and to Westralian Sands Limited for financial contributions to the cost of producing this book.

The front cover shows a view of the Weld Range near Cue in the Western Australian Eremaea or Arid Zone; scattered shrubs of mulga *(Acacia aneura)* cover the hills, with the mulla-mulla *Ptilotus obovatus* flowering white in the foreground. Jacket design by Darian Causby.

The frontispiece map shows the natural vegetation of Western Australia simplified to 25 'primary vegetation types' at the scale of 1:10 000 000, based on a map by the author in *Western Australia: an Atlas of Human Endeavour*, ed. N.T. Jarvis 1979.
    Scale determines the detail in which vegetation can be depicted. The author's larger 1:3 000 000 map of the State (1981) showed 50 types, and the 1:1 000 000 Vegetation Survey (Fig. 1.12) showed 130.

© John S. Beard 1990

Beard, J. S. (John Stanley), 1916–
Plant life of Western Australia.

Bibliography.
Includes index.
ISBN 0 86417 279 6.

1. Botany – Western Australia. I. Title.

581.9941

First published in 1990 by Kangaroo Press Pty Ltd
3 Whitehall Road (P.O. Box 75) Kenthurst NSW 2156
Typeset by G.T. Setters Pty Limited
Printed in Hong Kong through Global Com Pte Ltd

ISBN 0 86417 279 6

Contents

Preface 7

CHAPTER 1
Introduction 9

CHAPTER 2
The Environment 29
1 The Natural Regions 29
2 Climate 37
3 Geology 40
4 Landforms 43
5 Soils 49
6 Fire 55

CHAPTER 3
The Southwest Province
and Southwestern Interzone 59

- 3.1 Southwest Forest Region
  —Darling Botanical District 59
- 3.1A Karri Forest Subregion
  —Warren Botanical Subdistrict 59
- 3.1B Southern Jarrah Forest Subregion
  —Menzies Botanical Subdistrict 77
- 3.1C Northern Jarrah Forest Subregion
  —Dale Botanical Subdistrict 80
- 3.1D Swan Coastal Plain Subregion
  —Drummond Botanical Subdistrict 87
- 3.2 Northern Sandplains Region
  —Irwin Botanical District 98
- 3.3 Wheatbelt Region
  —Avon Botanical District 115
- 3.4 Mallee Region
  —Roe Botanical District 127
- 3.5 Esperance Plains Region
  —Eyre Botanical District 139
- 3.6 Southwestern Interzone
  —Coolgardie Botanical District 155

CHAPTER 4
The Eremaean Province 168

- 4.1 Nullarbor Region
  —Eucla Botanical District 171
- 4.2 Great Victoria Desert
  —Helms Botanical District 178
- 4.3 Murchison Region
  —Austin Botanical District 186
- 4.4 Carnarvon Region
  —Carnarvon Botanical District 202
- 4.5 Gascoyne Region
  —Ashburton Botanical District 213
- 4.6 Little Sandy Desert Region
  —Keartland Botanical District 223
- 4.7 Gibson Desert
  —Carnegie Botanical District 226
- 4.8 Central Ranges Region
  —Giles Botanical District 234

4.9 Pilbara Region
—Fortescue Botanical District *240*

4.10 Great Sandy Desert
—Canning Botanical District *252*

4.11 Tanami Desert
—Mueller Botanical District *259*

CHAPTER 5
The Northern Province *264*

5.1 Dampierland Region
—Dampier Botanical District *267*

5.2 East Kimberley Region
—Hall Botanical District *277*

5.3 Central Kimberley Region
—Fitzgerald Botanical District *285*

5.4 North Kimberley Region
—Gardner Botanical District *292*

Glossary *311*

Index *313*

PREFACE

This book has been planned essentially as a photographic record of 25 years of work in Western Australia. Hitherto I have published much written information on the plant life of the State and have mapped it overall. Illustration, however, had always to be confined to a relatively few black and white pictures although naturally I had been taking numerous colour slides in the course of field work. For years I have wished to see the best of these published also, serving both as a scientific record of what was there at the time and as a source of pleasure to the reader. It has now been possible to realise this ambition, thanks to the generosity of my sponsors Alcoa of Australia Ltd and Westralian Sands Limited.

All photographs reproduced in this book were taken by myself except where acknowledged to others in the text. I am most grateful to those who loaned slides to be included, particularly Dr Byron Lamont, Dr Ken McNamara and Professor John Pate, who furnished slides of particular technical subjects. Thanks are due to the Western Australian Department of Land Administration for the provision of cartographic material for Fig 3.1. I also wish to record my thanks to those many people, staff of King's Park or of the Western Australian Herbarium and others, who accompanied me on expeditions, some lengthy and into the remote outback. Most of them are immortalised by having served as scale objects in the photographs which appear in the book. Forty-five different people lent their services in this way, chief among them my wife Pamela who appears 38 times, King's Park staff members Herb Demarz (21) and Fred Lullfitz (14), and botanist Alex George (13). Mere objects were pressed into service at other times and include a Land Rover (17 times), my car (6), a boat (1) and a 5 cent piece (3)! They bring back for me pleasant recollections of days spent in the field.

I am no less grateful to all those who took part in or helped with the Vegetation Survey of which this book is the outcome: my collaborator Professor Martyn Webb, Geoffrey Ward in his department, the cartographers who drew maps, all the staff of King's Park and Botanic Garden, and the numerous taxonomists in various herbaria who identified botanical specimens and gave advice on nomenclature, especially Pauline Fairall and Grosvenor Selk of the King's Park herbarium. All these contributed in a major way to this long-lasting project.

Perhaps the most stimulating experience of those early days was the feeling of exploration, of being the first in the field. In 1964 I was, to my knowledge, the first professional botanist to enter the Great Sandy Desert. In 1966 in company with Alex George we made the first botanical north-south crossing of the Great Victoria Desert and explored the Nullarbor Plain. In 1967 I traversed the Gibson Desert. In the far north I was the first to report on the vine thickets of the Mitchell Plateau, in 1976. Even in the southwest where the vegetation was already better known, new and undescribed species were continually brought in by our collections.

The 1960s were the years of the great surge of agricultural development when the Brand government committed itself to the release of one million acres of Crown Land each year, without environmental assessment other than rudimentary soil surveys. In vain I urged a botanical survey of such land to determine what was there before it was ploughed out. This was not politically acceptable as it was feared that the discovery of rare species would lead to demands for reserves. We shall never know how many species have been rendered extinct in consequence. Throughout the farming areas at that time relentless destruction was the order of the day, and I had to count myself privileged to have been actively involved at all with the famous southwestern flora before it finally disappeared.

Today, of course, the picture is quite different. Environmental awareness has spread widely, our flora still survives and is no longer under direct assault, while government departments are much better organised to protect it. Threats remain from the competitive spread of agricultural weeds and *Phytophthora* root disease, we have little long-term hope of saving roadside flora, but in general we may feel that the crisis has passed. Today the pendulum has even swung too far the other way—too many people calling themselves conservationists are swayed by emotion, lack real understanding of the facts and are hotly in pursuit of spurious issues. No doubt, too, this is a crisis which will also pass.

John Beard
Perth
December 1989

1
INTRODUCTION

The world has been aware that the plant life of southwestern Australia possessed special and unique features ever since the famous botanist Ferdinand von Mueller drew attention to them in 1867[1]. An exceptionally high number of the plant species is endemic, that is, they occur nowhere else. Many of them exhibit strange and bizarre plant forms and many of them flower brilliantly in season so that the wildflowers of Western Australia have become a byword and a major tourist attraction.

It is therefore strange that only one comprehensive book dealing with the plant life in Western Australia in one volume has ever been published before, and that a long time ago and in the German language! Two German botanists, Ludwig Diels and Ernst Pritzel, arrived in Western Australia on 30 October 1900 and travelled widely until their departure at the end of December 1901, covering virtually all the country that was accessible at that time, mainly by rail, through agricultural and mining settlements, and making an additional trip to the northwest by sea (Fig 1.4). After their return to Germany Diels wrote a book of 400 pages dealing with the southern half of the State which he entitled *Die Pflanzenwelt von West-Australien südlich des Wendekreises*[2]—literally 'The plant world of Western Australia south of the tropic'. However a more idiomatic translation would be 'The plant life of Western Australia' and this has inspired the title of this present book.

Diels' book was translated bit by bit at the Botany Department of the University of Western Australia until it was completed in 1950, but only typed copies were made and it has never been published in English.

In 1914 J.T. Jutson produced a vegetation map of Western Australia to accompany an authoritative work on physiography, now a classic[3], but descriptive text (on vegetation) was confined to a few notes. In 1942 the Government Botanist Charles Gardner gave a presidential address to the Royal Society of W.A. entitled 'The Vegetation of Western Australia with special reference to climate and soils' which was published in the Society's journal[4] in 1944 presumably in an expanded form as it runs to 76 pages. This paper covered the whole State and materially extended Diels' coverage. Data given for the north were new and original, particularly for the Kimberley where Gardner had taken part in exploration. Remote parts of the south such as the Nullarbor Plain, of which Diels had no knowledge, were also described, but the various deserts of the interior remained largely unknown.

In 1961 the State government funded the King's Park Board in a project to create a botanical garden in King's Park for the cultivation and display of Western Australian native plants. The present writer was appointed Director in that year, to take charge of the project. Planting commenced in 1963 and the garden was officially opened in 1965. Considerable difficulty

Pioneer Ecological Botanists of Western Australia

It is possible to look at the plant cover in the wild in two ways: as species to be studied and classified individually, or as vegetation in the form of plant communities. The former is the field of the systematic botanist, otherwise called the taxonomist. The latter is the field of the plant ecologist. Our three notable pioneers who laid the foundations of our knowledge of vegetation in Western Australia were distinguished taxonomists as well as ecologists.

This book is about vegetation as it gives character to the landscape. We shall speak of vegetation types, plant communities, plant forms, and plant species too.

State Library of Victoria

Fig 1.1 Ferdinand Jakob Heinrich Mueller
1825–1896

Botanischer Garten und Museum, Berlin-Dahlem

Fig 1.2 Friedrich Ludwig Emil Diels
1874–1945

Western Australian Herbarium

Fig 1.3 Charles Austin Gardner
1896–1970

Ferdinand Mueller was born at Tönning in Schleswig-Holstein in 1825 and as a boy was apprenticed to an apothecary where botany formed part of his studies. He took a degree in pharmacy from the University of Kiel in 1846, and a PhD in botany in the same year. To escape the progressive destruction of his family by tuberculosis he emigrated to South Australia with his sister in 1847, where he went to work in a pharmacy but privately carried out botanical studies and began to publish papers. These secured him the position of government botanist of Victoria in 1853, a post which he held for the rest of his life, becoming the most active, effective and famous botanist in Australian history. Early in his career he took part in major exploring expeditions such as the North Australia Expedition under Augustus Gregory in 1855–56 which discovered the Sturt Creek and awarded the name Mt Mueller to a hill near Billiluna. Later he helped to promote expeditions and the collecting of specimens for him, such as the Elder Exploring Expedition which crossed the Great Victoria Desert in 1891. Mueller himself visited Western Australia in 1877 at government invitation to report on the forests of the Colony, which he traversed, as well as travelling up to Shark Bay.

Mueller had a fad for soliciting honours and died as Baron Sir Ferdinand von Mueller, KCMG, MD, PhD, FRS.

F.L.E. Diels was born in Hamburg, Germany, to a family of distinguished academics. He took early to botany and studied taxonomy under Engler, geography under von Richthofen, at Berlin. His doctoral thesis was on 'Vegetation–biology of New Zealand'. He had not, at that time, visited that country, but in company with Ernst Pritzel he spent 14 months travelling in Western Australia in 1900–01, and in 1906 published his classic book *Die Pflanzenwelt von West-Australien südlich des Wendekreises* (The plant world of W.A. south of the tropic) which remains the only comprehensive account of the vegetation of the State in one volume but has never been published in English. Diels joined the staff of the Berlin University in 1900 and was called to the chair of Plant Geography and Systematics at Marburg in 1906. Later he became the second director of the Botanical Museum at Berlin-Dahlem in succession to Urban and worked under Engler as director-general. He succeeded to the latter post himself in 1928 and held it till his death in 1945, saddened by the wartime destruction of the herbarium and whole institution in Berlin.

C.A. Gardner was born in Lancaster, England, and brought to Western Australia as a boy of 13 when his family emigrated to farm at Yorkrakine. He early showed an interest in botany and a talent for art employed later in botanical illustration. Although he had no opportunity for academic study and training, his interests secured an appointment to the Forests Department as collector in 1920 and this led to his participation in the Easton Exploring Expedition in the north Kimberley in 1921. His subsequent report made his name as a botanist. He transferred to the Dept of Agriculture in 1924 and was appointed government botanist in 1929, continuing till his retirement in 1960. Throughout his life Gardner travelled widely in the State, collecting and recording. He placed the W.A. Herbarium on a firm basis and greatly expanded its collections. He is credited with 320 published items including a Census of W.A. Plants (1930) and a series on the eucalypts illustrated by himself. He described eight new genera of plants and some 200 new species. An ecologist as well as a taxonomist he helped to define the State's phytogeographic regions and published a classic paper on the State's vegetation in 1944.

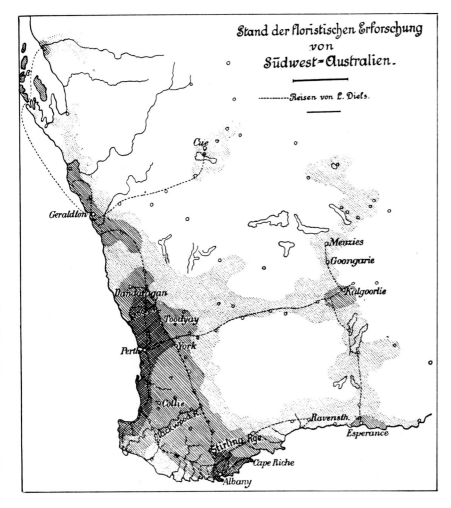

FIG 1.4
Original diagram from Diels' *Pflanzenwelt* p.68 showing the journeys of L. Diels as dotted lines, and the extent of botanical exploration in the southwest of the State in 1905, indicated by hatching in four degrees of intensity. Diels and Pritzel travelled largely by train, making excursions from the railway. In more remote areas, expeditions accompanied by collectors for von Mueller had provided some general knowledge.

was encountered due to the lack of detailed botanical knowledge at that time. If the project was to be competently carried out, we needed to know precisely what species of plants made up the flora, where they were to be found and under what conditions they grew. Field expeditions were mounted with the dual object of acquiring this information and of collecting propagating material for the garden. It was quickly seen to be desirable to systematise the collection of information into formal inventories, one of plant species and one of plant communities. The former, compiled into a 'Descriptive Catalogue of West Australian Plants' was completed with the assistance of Mr Charles Gardner in his retirement and of the W.A. Herbarium and published in 1965[5]. It listed all known species of flowering plants for the State (with the exception of grasses and sedges in the first edition), summarising for each the information of greatest horticultural interest i.e. the type of plant, height at maturity, flower colour, season of flowering and districts of occurrence. A previous census of the flora published by Charles Gardner in 1930 listed only species, with references to taxonomic literature.

The inventory of plant communities, intended to give the *ecological** information as to conditions under which plants grow in the wild was much more difficult, requiring extensive field surveys and detailed mapping from aerial photography, and in fact took 17 years to complete. A project for the purpose called 'The Vegetation Survey of Western Australia' was established in 1964 by the writer jointly with Prof. M.J. Webb, then Professor of Geography at the University of Western Australia. It was funded initially by the King's Park Board, later by the Australian Commonwealth. Maps were produced in the University Department of Geography and published with explanatory texts by the University of W.A. Press[6]. At the mapping scale of 1:1 000 000 seven sheets were required to cover the huge area of the whole State (2½ million km²), so that the accompanying texts appeared in seven volumes (Fig. 1.12).

---

*Ecology is the study of forms of life in relation to environment.

Herbert Demarz

Fred Lullfitz

Fig 1.5
The author in the field with fellow botanist Paul Wilson (right). In the course of botanical surveys over a period of 20 years, the author covered all accessible parts of the State, travelling some 200 000 kilometres.

Fig 1.6
At the end of the day in the comfort of the caravan, away from the dust and the flies, the author takes stock of the day's collection of specimens which will be catalogued, pressed and dried and later identified by taxonomist colleagues.

Fig 1.7
Hon. David Brand, MLA, Premier of Western Australia (right), talks to the author and his wife after the official opening of the King's Park Botanic Garden on 4 October 1965.

Fig 1.8
The Pioneer Women's Memorial Fountain, focal point of the King's Park Botanic Garden. It was donated by subscription of descendants of pioneer women.

The author was appointed Director of King's Park, Perth, in 1961 to take charge of the formation of a botanic garden for cultivation and display of Western Australian native plants. Extensive fieldwork had to be undertaken to collect propagating material and to study native species in their habitats, since their ecological requirements were expected to afford clues to their horticultural requirements.

Fig 1.9
The purpose of the botanic garden was to cultivate and display W.A. native plants. Here *Thryptomene elliottii* (foreground) and taller *Grevillea leucopteris*.

Fig 1.10
A section of the botanic garden. A display of *Verticordia*, a genus which provides some of the most colourful small plants in the flora, with some kangaroo paws *(Anigozanthos)* and behind, a bush of *Eucalyptus macrocarpa*.

Fig 1.11
Lambs' tails *(Lachnostachys eriobotrya)* one of a great many species found difficult to propagate owing to poor seed production and viability, and problems with vegetative reproduction. Modern techniques of tissue culture have helped to overcome such problems.

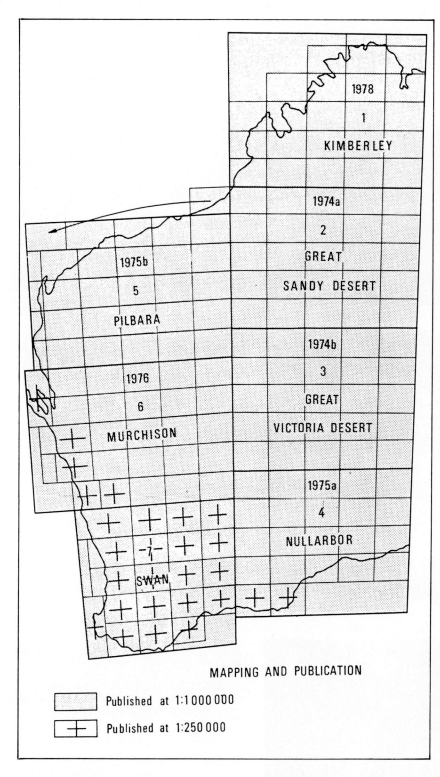

FIG 1.12
Mapping and publication of the Vegetation Survey of Western Australia.
  In order to provide a regular framework for the collection of ecological information and a programme for its publication, a project called the Vegetation Survey of Western Australia was established by the author and Professor M.J. Webb of the Department of Geography, University of W.A. Vegetation maps prepared in the Department were gradually produced, covering the whole State at the scale of 1:1 000 000, and the southwest at the larger scale of 1:250 000. An explanatory book accompanied each map.
  A portion of one of the vegetation maps is illustrated as an example in Chapter 2, Fig 2.4.

FIG 1.13 (OPPOSITE)
Vegetation is classified by the principal kinds of plants—trees, shrubs, grasses and so on, by their height and spacing, and lastly by plant species. This table shows a classification for the principal structural forms of vegetation found in Western Australia, in idealised sketches. (Based on a diagram in Volume 6, 'Vegetation', of the *Atlas of Australian Resources*, 1989.)

These describe in detail the area covered by each map sheet, its climate, geology, physiography and soils, and its plant cover in the various vegetation types concerned. The total text amounted to 821 pages and may be consulted by those seeking greater and more scientific detail. Greater detail still is available for the Southwest Province in a companion series of 24 maps at the scale of 1:250 000 with appropriate text[7].

The present book has been designed to give the general reader a more concise account in one volume, as well as to publish in full colour for the first time a selection of the numerous photographs taken by the writer in the course of field expeditions. Many have appeared before, in the volumes of the Vegetation Survey, but only in black and white, which does not do justice to them.

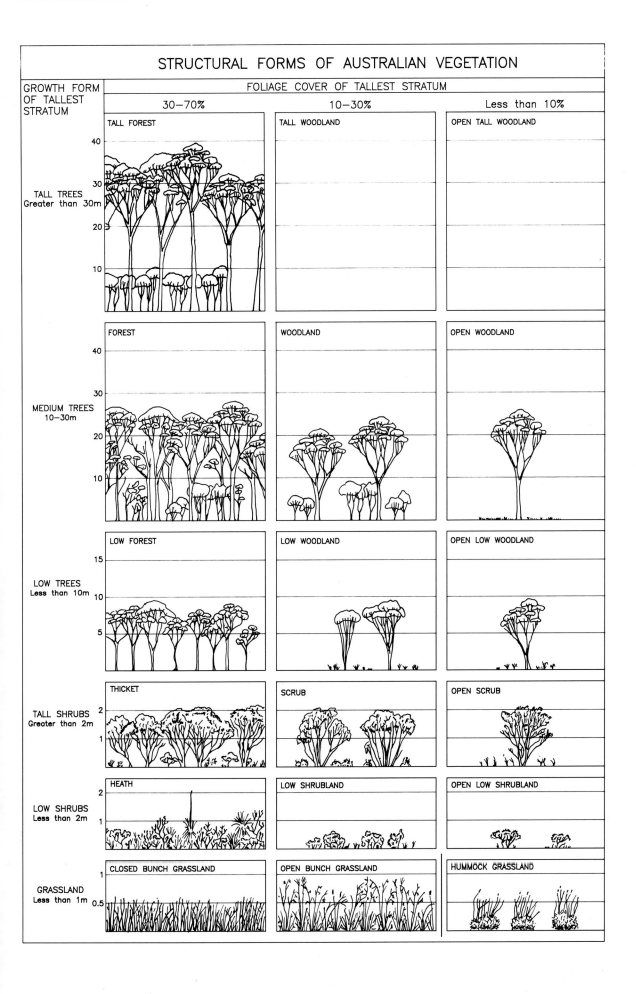

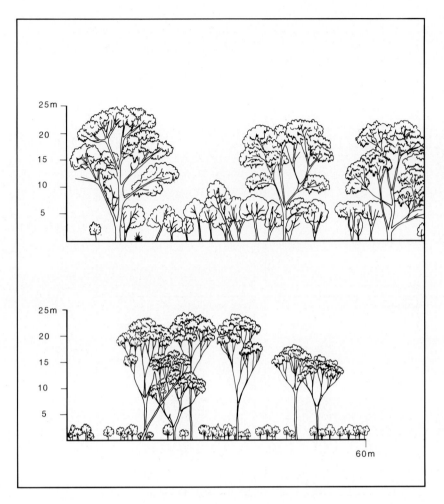

FIG 1.14
Actual measured examples showing the structure of two different types of eucalypt woodland.
Upper: Reconstructed profile of original tuart woodland in King's Park, Perth. Well-spaced, massive, branchy trees of *Eucalyptus gomphocephala* (tuart) over a low-tree layer of *Banksia* spp. and *Casuarina fraseriana*.
Lower: Profile of *Eucalyptus salmonophloia* (salmon gum) woodland in the Hyden district. Trees irregularly scattered, slender and more upright, with a shrub layer 2–3 m tall mainly of *Melaleuca pauperiflora*.
From Vegetation Survey publications[6].

There are basically two ways of looking at the plant cover. One can regard it as an assemblage of plant species, each to be identified and named, not necessarily in relation to the others present nor to the habitat. This is the *taxonomic* approach. Alternatively we can regard each example of the plant cover as a type of vegetation which has a particular structure (height, density, stratification), a particular set of plant forms (trees, shrubs, grasses, saltbush and so on) and a particular habitat. This is the ecological approach, and in this case the naming and classification of component species are interesting but not of vital importance. In the former case, if we are in an undisturbed area in the Darling Range we may see trees of *Eucalyptus marginata* and *E. calophylla*, small trees of *Banksia grandis*, *Casuarina fraseriana* and *Persoonia longifolia*, and very numerous shrubs mainly in the families *Proteaceae*, *Myrtaceae*, *Mimosaceae*, *Fabaceae* and *Epacridaceae*, with some creepers, usually leguminous and some grasses, sedges and dicotyledonous herbs. In the second case we shall say that we are in the jarrah forest dominated by jarrah and marri 25 to 30 m tall with a stem density of 125 to 150 per hectare: the forest has a subordinate tree layer at 10–15 m and a ground layer of sclerophyll* shrubs 1–2 m tall averaging 185 individuals to the hectare plus minor creepers, and herbaceous ground plants.

Diels' *Pflanzenwelt*, being a pioneer in its field, had as its objective the description and definition of vegetation types as well as taxonomic study of the flora. At the present day many well illustrated works have been produced treating the flora taxonomically, some of outstanding quality, notably by C.A. Gardner[8], M.K. Morcombe[9] and Ericson, George, Marchant and Morcombe[10], but there has been no illustrated work treating the flora as vegetation. The present book is designed to take the latter approach.

As we have outlined above, vegetation is the plant cover considered in the entirety of its structure and plant forms as well as its component species. The principal characteristics of vegetation are:

---

*Hard, brittle leaves, low in fleshy tissue.

1 *Floristic composition.* Vegetation is composed of individual plants, which can be botanically identified.

2 *Structure.* Component species will be of a certain size, and arranged in definable patterns horizontally and vertically, that is, in layers (strata) of a certain spacing (Figs 1.13 and 1.14).

3 *Growth-form.* Components will vary in their morphology, i.e. they may be grasses, succulents, mosses and so on, may be annual or perennial, evergreen or deciduous, and have leaves of a certain texture, size and shape.

The botanical identity of individual plants is of interest to plant taxonomists, plant geographers, foresters, pastoralists and many others, but plants as vegetation are components of a community which has its own characteristics as a whole. In this context plant names may not always be important, in fact it will usually suffice (for purposes of recognition of vegetation units) to identify only those plants which are most abundant or physically predominant over others, those which the ecologist calls *dominant species*. The vegetation of a given area will be found to consist of a certain list of component species with particular dominants. This is called a *plant association*. If the vegetation of another area has the same dominants, it is said to belong to the same association, but if they have significantly changed, a different association is present. The association is the basic mapping unit of vegetation. It is assumed that the environmental factors, climate, topography and soil, are sensibly uniform throughout the range of an association and that changes will be reflected in a change of association.

Vegetation structure is its second important characteristic and can be defined and described without reference, if necessary, to specific composition, since what matters is the height of the plants in different layers, if these can be recognised, their shape and their spacing. Vegetation may consist, for example, of a closed forest of tall trees of a certain height, form and range of diameter, arranged so many to the acre and with lower and more open layers of small trees and shrubs; or in another case of an open growth of short spreading trees set in a grassland.

Growth-form may be defined as the morphological characteristics of the component plants. Are they woody or herbaceous, annual or perennial, thorny or succulent, and what kind of leaves do they have?

A plant association will be found to have a characteristic physiognomy, i.e. structure and growth-form. Different associations, while differing floristically, may be of like physiognomy and are then said to belong to the same *plant formation*. A formation is thus a vegetation unit defined by its physiognomy, and is independent of specific composition. Formations were the principal mapping units of the vegetation survey, but associations were mapped if the scale permitted.

In Western Australia our studies of vegetation are greatly assisted and made more interesting by the fact that so much of the original plant cover as it was before European settlement has survived. In the wheat belt of course, most of the country has been cleared for farming but even there sufficient of the original vegetation usually survives along roadsides and as isolated trees in paddocks to enable the original types of vegetation to be identified and mapped. On the Swan Coastal Plain much clearing has taken place for urban and industrial development as well as for farming but the area affected is small relative to the size of the whole State. In some outback areas the plant cover has been modified by pastoral use but normally its essential identity is still retained. No one has precisely calculated the area of original vegetation which has been cleared in Western Australia, but an approximation can be obtained from the known area of land which has been 'alienated' (transferred from the Crown to private freehold), given in the official Yearbook for 1981 as 189 580 km². Not all this land has been cleared since farmers very often retained uncleared areas, but on the other hand the figure does not include roads, railways, power lines and so on which have been cleared on Crown land, nor the areas of mines and mining settlements. If a reasonable approximation of 200 000 km² is accepted for cleared land, this amounts to 8% of the whole State, so that in the other 92% we are fortunate in being able to travel freely through little disturbed and intact natural vegetation. This is a phenomenally high proportion for any country and is due to the arid environment of most of the State.

As we have seen, vegetation is typified into formations and associations which can be represented in mapping. The choice of units will depend to a large degree on the scale of map to be produced. The scale of a map is quoted in the form of 1:50 000, 1:250 000, 1:1 000 000 etc, meaning (in the first case) that 1 unit of measurement on the map represents 50 000 units on the ground — thus 1 millimetre on the map represents

50 000 mm (50 metres) of country. At 1:1 000 000, 1 mm represents 1 km. Maps with a small ratio such as 1:50 000 are said to be 'large scale', and with a large ratio such as 1:1 000 000 'small scale'. A larger scale map portrays a smaller area of country more enlarged in a given size map sheet, and in vegetation mapping one is able to subdivide one's units more and treat them more intensively in larger than in smaller scale mapping where one has increasingly to generalise.

When the Vegetation Survey of Western Australia had been completed in 1981 at the scale of 1:1 000 000 — 1 mm to 1 km, or in the old units 1 inch to 15.78 miles — 130 different vegetation types had been distinguished plus 5 categories of land carrying no vegetation (fresh and salt lakes, mudflats, rock outcrops and drift sand). On conclusion of the Survey the data were summarised and reduced to produce a single smaller scale map sheet for the whole State which was published by the Forests Department of W.A. in 1981. The scale of 1:3 000 000 was chosen and necessitated further generalisation. The 133 vegetation types were reduced to 50 plus 3 of bare land. The map reproduced with this book is a still smaller one reduced to the size of one page at a scale of 1:10 000 000. This has necessitated further generalisation and reduction to 25 'primary vegetation types'. This map was originally produced for the *Atlas of Human Endeavour* published by the State government for the sesquicentenary in 1979[11].

The vegetation units shown in the key in the upper left of the map are classified according to structure, growth-form and floristic composition. Strictly speaking the classification hinges upon the structure and growth form of the dominant layer. Most vegetation consists of several different layers. For example, when discussing the jarrah forest earlier in this section, we saw that it comprised layers of taller trees, low trees, shrubs and ground plants. The *dominant layer*, in this case the taller trees, is the one which most exercises a control over other layers by its competition and at the same time impresses the whole community with its essential character. The dominant layer is frequently the tallest one, but not always. Where trees and shrubs are scattered it may be a lower shrub or grass layer which is dominant. It has become fashionable in many parts of Australia to classify vegetation on the basis of the tallest layer but this can lead to unfortunate results. For example spinifex grassland is a formation which perhaps covers more of Australia than any other and is well known and only too readily recognisable. (Units nos 6, 7 and 8 of the map.) Yet it rarely occurs without scattered trees and shrubs present. In such cases classification on the tallest stratum would make it a very open woodland, and the spinifex grassland would disappear from the map!

The primary classification used in the Vegetation Survey of Western Australia is as follows:

Tall trees	>25 m tall
Medium-height trees	10-25 m tall
Low trees	<10 m tall
Tall sclerophyll shrubs	> 1 m tall
Low sclerophyll shrubs	< 1 m tall
Succulent and semi-succulent shrubs	
Bunch grasses	
Hummock grasses (spinifex)	
Forbs (herbs other than grasses and sedges)	
Lichens and mosses	

The next criterion is density. Trees forming a closed canopy constitute forest, if more open, woodland or open woodland. Tall shrubland may have scattered trees present. If not it is called thicket if closed, scrub if open, with a special category of mallee if eucalypts are dominant. Low shrubland is called heath if closed, low scrub if open, and mallee-heath or scrub-heath if taller shrubs of eucalypts or non-eucalypts are present. These classes refer to the hard-leaved or sclerophyll shrubs which are commonly encountered in most parts of Australia on nutrient-poor neutral to acid soils. A special local term, *kwongan*, is used for southwestern sclerophyll shrublands other than mallee[11] to match the special terms used in other mediterranean regions, e.g. maquis, macchia (Europe), chaparral (California), fynbos (South Africa) and matorral (Chile). Another local term, *pindan*, is used for the equivalent vegetation of sandplains in the Northern Province. A separate category of succulent and semi-succulent shrubs is required for the soft, fleshy-leaved saltbush and bluebush shrubs and the succulent samphires which characterise vegetation of alkaline and saline soils and constitute a radically different vegetation type. For want of a better name these saltbush and bluebush communities are labelled 'succulent steppe' from their resemblance to some overseas models. Scattered small trees and shrubs are frequently present. In Western Australia bluebush communities are principally developed on the limestone of the Nullarbor Plain, saltbush and samphire communities near to and bordering salt lake systems.

Figures 1.15 to 1.38 illustrate photographically the major structural types of vegetation in Western Australia. Please compare with the frontispiece map. Figs 15–23, Southwest Province, 24–32 Eremaean Province, 33–38 Northern Province.

Fig 1.15

Tall forest of karri. Trees may reach 80 m tall and require highest rainfall and deep soils, only found in the lower southwest. This equates with the wet sclerophyll forest of eastern Australia.

Fig 1.16

Forest (medium height) of jarrah. Trees less than 30 m tall, medium rainfall, poorer soils. Makes up the greater part of the forest area of W.A. Equates with dry sclerophyll forest.

Fig 1.17

Low forest of *Casuarina fraseriana* on poor sandy soil, between Shannon and Walpole. The trees are about 10 m tall, but still growing in relatively dense stands. If more widely spaced, they are classed as woodland.

FIG 1.18
Woodland; trees 10–20 m tall, irregularly and widely spaced, understorey of scattered shrubs, smaller herbs and annuals. This example is of *Eucalyptus salmonophloia* (salmon gum) at Zanthus east of Kalgoorlie.

FIG 1.19
Low woodland; trees less than 10 m tall, scattered. Dominants here are *Banksia attenuata* and *B. menziesii*, the understorey is of dense sclerophyll low shrubs of which *Verticordia nitens* is in flower. Moore River National Park.

FIG 1.20
A shrub is defined as a woody plant branching from near the base, without a definite trunk. Shrub formations are classed as thicket if closed, scrub if open. Most thickets are dominated by species of *Acacia*, *Casuarina* and/or *Melaleuca*. This example is of *Melaleuca uncinata*, from Beacon in the Wheatbelt.

Fig 1.21

Mallee is classed separately, being eucalypt-dominated, and may form either thickets or scrub. The many small stems are formed by resprouting from a massive underground rootstock after a bushfire. Mallee species greatly outnumber tree species of eucalypts, and are often hard to distinguish. Ravensthorpe District.

Fig 1.22

Closed low shrubland is classed as heath, but scattered taller shrubs may be present. If these are small mallee species, they constitute mallee-heath, and if other mixed shrubs, scrub-heath. Mallee-heath with *Eucalyptus tetragona* shown here is characteristic of south coast sandplains. Fitzgerald National Park.

Such communities with a dominant heath layer are also known as *kwongan*, an Aboriginal word from the southwestern Nyungar language.

Fig 1.23

Scrub-heath has a fairly dense heath layer, here with *Thryptomene* sp. in flower, and scattered taller shrubs which may be of *Acacia, Casuarina* and *Melaleuca* and even occasional mallee in the inland, or species of Proteaceae such as *Banksia* and *Grevillea* in the northern sandplains. This example from a sandplain near Southern Cross.

Eremaean Province

Fig 1.24
The southern part of the Eremaean Province is characterised by low woodland of *Acacia aneura*, known as mulga. The scattered shrubby trees are mixed with other species of *Acacia*, with a lower layer of occasional *Eremophila* and *Cassia* spp., and herbaceous annuals in season, such as the pink *Velleia rosea* shown here. Near Leonora.

Fig 1.25
The laterite plains of the Gibson Desert are mainly covered by a more open form of mulga growing as scattered clumps in spinifex grassland, which appears on the map as mulga parkland. Gibson Desert, on the Gary Highway.

Fig 1.26
Calcareous plains in the south have a vegetation of semi-succulent, soft leaved shrubs belonging to the family Chenopodiaceae, which are a radically different form from the sclerophyll shrublands and are treated separately. Note the pieces of limestone and the whitish tinge to the soil in this picture. The shrubs here are bluebush *(Maireana sedifolia)* replaced by saltbush *(Atriplex)* if the soil becomes more saline. East of Kalgoorlie.

Fig 1.27
On the Nullarbor Plain there are expanses of treeless bluebush in the centre, but on most of the plain there are scattered small trees of myall *(Acacia papyrocarpa)* and in a good season abundant annual grass and herbs. Nullarbor near Rawlinna.

Fig 1.28
Where the soil is strongly saline we find a fully succulent vegetation of small shrubs in the family Chenopodiaceae which are leafless and have jointed succulent stems. These are called samphires, using the English word for these marsh plants, a corruption of the French *herbe de Saint Pierre* (St Peter's herb).

Fig 1.29
The vegetation of salt flats consists of scattered samphires with other salt-tolerant plants such as *Senecio lautus* shown flowering, and annual grasses. On slightly higher ground larger shrubs come in. Nambi station.

FIG 1.30

The northern half of the Eremaean Province is almost entirely covered with spinifex forming hummock grasslands. Spinifex, belonging to the genera *Triodia* and *Plectrachne*, is a stiff, spiny grass interlacing into domed hummocks. It is a growth-form unique to Australia due, it is believed, to the poverty of Australian soils.

FIG 1.31

The spinifex hummocks grow outward and begin to die out in the centre. Eventually crescentic patterns are formed.

FIG 1.32

Spinifex country normally grows with scattered trees, mallees and shrubs between the hummocks, as in this case with *Eucalyptus youngiana* and *Acacia* spp. in the Great Victoria Desert.

Grasslands are classified basically into bunch grassland and spinifex or hummock grassland but there is an intermediate type known as 'curly spinifex'. Bunch grasses occur widely in the tropics where they form the tropical savannas. The grasses grow from perennial tussocks and are drought-evading, drying off during the dry season and sprouting again from the tussock during the following wet season. Hummock grasses grow as intertwined stems forming a domed mass and are drought-resisting, being evergreen and perennial. Their leaves are rigid and sharp-pointed. Hummock grasslands are in the main typical of desert areas, as the map shows. The intermediate curly spinifex is perennial, drought-resisting, but not a hummock grass.

A savanna, incidentally, is any tropical bunch grassland, and it is incorrect to suppose that scattered trees must be present. Treeless savannas are not uncommon on cracking clay and other soils. Elsewhere the tree component varies in height and density, as indicated by the small tree symbols in the map key. If density increases to the point where the tree layer becomes dominant rather than the grass layer, the vegetation is classified as woodland.

The map shows that the vegetation of Western Australia varies rather widely throughout the State. It does so, of course, as anyone will appreciate, because growing conditions vary equally widely. In warm countries where there are no severe winters with frost to place a limitation on plant growth, vegetation is controlled by and essentially responds to the supply of available moisture. Climate is therefore of prime importance for rainfall declines from the wet, humid

Northern Province

FIG 1.33

The Northern Province is essentially the land of tropical bunch-grass savannas. The bunch-grass, annual or perennial, forms a tussock of upward-growing stems with soft leaves. Typically, it is drought evading by drying off during the dry season, unlike the evergreen, drought-resisting spinifex. Spinifex continues to occur on rocky and inhospitable sites.

FIG 1.34

Savannas are treeless on certain soil types, especially on 'black soil' plains, but are wooded for the most part forming savanna woodland or tree savanna according to tree density. Tree height varies also. Communities are distinguished by tree height and density, and species composition, but the growth form of the grasses is the primary basis for classification.

Fig 1.35

High-grass savanna, in which the species of *Sorghum, S. plumosum* and *S. stipoideum*, are dominant, may reach 3 m tall in the wet season, collapsing into a tangled mass in the dry season. It is typical of the North Kimberley, where rainfall is highest.

Fig 1.36

In tall bunch-grass savanna, the grasses are 50–100 cm tall, perennial, drought evading. This occupies the better soils south of the North Kimberley Region and is of the most value for pastoral purposes. *Eucalyptus tectifica* (grey box) savanna woodland, Mount House station.

Fig 1.37

Short bunch-grass savanna, grasses less than 50 cm tall. Found on calcareous soils of drier areas, mainly in the East Kimberley. Trees here are *Eucalyptus argillacea* and *Terminalia* spp. Rosewood station.

FIG 1.38

Curly spinifex country, typical of the Central Kimberley Region, shown looking across to Bold Bluff in the King Leopold Range. *Eucalyptus dichromophloia* at left, ground layer of tussocks of *Plectrachne pungens*, tough, wiry, perennial, drought resisting, but not adopting a hummock grass form.

FIG 1.39

The sandy plains of Dampierland bear a distinctive formation called *pindan* in which the grass layer is only dominant for a short period after fire. A thick layer of *Acacia* quickly springs up and suppresses the grasses until fire again sweeps through. Scattered low trees of various species (*Gyrocarpus americanus* at left) are fire resistant.

southwest into the aridity of the central desert, and then increases again northwards into the better watered Kimberley District. Not only the amount of rainfall is significant but its distribution. In the south rain falls mainly in winter, in the north in summer, while in the desert it is erratic and unpredictable. Temperature and relative humidity are other significant climatic factors. The moisture supply is also affected by the factors of topography, geology and soil. The geological structure and hardness of the rocks usually determine the topography, the hilliness of an area. The *lithology*, the nature of the rocks, influences soil forming processes and determines the physical characteristics and fertility of soils. Geological history affects landforms, weathering and soil fertility. Topography influences soil formation. The interaction of these factors, called the factors of the environment, determines the moisture-supplying ability of the soil at any given site, which in turn determines the nature of the plant cover. To these physical factors we must add the effects of fire, which occurs frequently in most Australian types of vegetation and has done much to shape their character.

In the chapters which follow, the factors of the environment will be outlined for different areas of the State, and it is hoped to relate these meaningfully to the characteristics of the plant cover. The aim of this book is to promote understanding of the different types of vegetation encountered, and of their ecological relationships, in the hope that this may add to the enjoyment of those who travel about the State and look around them.

A word must be added about plant names. Every effort has been made to ensure correctness of nomenclature and all botanical names used in this book have been checked against J.W. Green's *Census*[13], second edition 1985 with supplements, which now serves as the standard authority on such matters. In the course of earlier field work the writer listed plant names on the basis either of voucher specimens subsequently identified in the Herbarium or of personal recognition where the plant was considered to be already well known. In both cases the possibility exists that the plant was not correctly named in the first place. Uncertainty may be introduced later by taxonomic research if this shows that several different names should now be used where a single wider, more embracing name was used before; it may now be impossible to be certain which successor name is the correct one to apply. Generally, taxonomic research is a necessary ongoing procedure which gradually refines the accuracy of plant names, however much name changes may infuriate the layman.

Some name changes have to be accepted under the International Rules of Plant Nomenclature, others are clearly justifiable and acceptable though not obligatory, while others seem due more to pedantry than common sense. The public should realise that none of us is compelled to accept name changes of the kind which are not obligatory under the International Rules but only represent a taxonomist's opinion. Under this heading I personally refuse to recognise the genus *Allocasuarina*. In this respect I differ from Green's Census and all sheoaks in this book are treated as *Casuarina*. A similar difficult situation has arisen from the revision of the bloodwood group of eucalypts by Drs Denis and Maisie Carr. *E. dichromophloia* was a name widely applied to tropical bloodwoods throughout the north of Western Australia and occurs in this book about 30 times; yet now this species is supposed to be confined to the Northern Territory and replaced in Western Australia by 10 or more distinct species. It has become in practice virtually impossible to tell which of these names should now be used, so that throughout this book I have retained *E. dichromophloia* which should be understood technically as *sens. lat.* (in the broad sense).

BIBLIOGRAPHY

1 Mueller, F. von, 1867. Australian Vegetation. *Intercolonial Exhibition Essays 1866-7*, No 5, Blundell & Co, Melbourne.
2 Diels, L., 1906. Die Pflanzenwelt von West-Australien südlich des Wendekreises. *Vegn. Erde* 7, Leipzig.
3 Jutson, J.T., 1914. An outline of the physiographical geology of Western Australia. *Geol. Surv. W.A. Bull.* 61. Reissued as Bulletin 95 in 1934 as 'The physiography (geomorphology) of W.A.'
4 Gardner, C.A. 1944. The vegetation of Western Australia with special reference to the climate and soils. *J. Roy. Soc. West. Aust.* 28: 11-87.
5 Beard, J.S., 1965. Descriptive Catalogue of West Australian Plants. Society for Growing Aust. Plants, Sydney.
6 Beard, J.S., 1974-80. Vegetation Survey of Western Australia, 1:1 000 000 Series. Univ. W. Aust. Press, Nedlands.
1974a Sheet 2, Great Sandy Desert
1974b Sheet 3, Great Victoria Desert
1975a Sheet 4, Nullarbor
1975b Sheet 5, Pilbara
1976   Sheet 6, Murchison
1979   Sheet 1, Kimberley
1981   Sheet 7, Swan

7a Beard, J.S., 1972-80. Vegetation Survey of Western Australia, 1:250 000 Series. 21 titles published by Vegmap Publications, Applecross, W.A. 6153.
7b Smith, F.G., 1972-4. Vegetation Survey of Western Australia, 1:250 000 Series. Pemberton & Irwin Inlet 1972: Busselton/Augusta 1973: Collie 1974. Dept of Agriculture, South Perth.
8 Gardner, C.A., 1968. *Wildflowers of Western Australia*. W.A. Newspapers Ltd, Perth.
9 Morcombe, M.K., 1968. *Australia's Western Wildflowers*. Landfall, Perth.
10 Erickson, R., George, A.S., Marchant, N.G., & Morcombe, M.K., 1973. *Flowers and Plants of Western Australia*. Reed, Sydney.
11 Jarvis, N.T. (editor), 1979. *Western Australia, an Atlas of Human Endeavour 1829-1879*. Govt Printer, Perth.
12 Beard, J.S., 1976. An indigenous term for the Western Australian sandplain and its vegetation. *J. Roy. Soc. West. Aust.* 59: 55-7.
13 Green, J.W., 1985. *Census of Vascular Plants of Western Australia*, 2nd. Ed. Dept. of Agriculture, Perth.

2
THE ENVIRONMENT

1 *The Natural Regions*

The vegetation of Western Australia varies widely from one part to another, and if it is to be described clearly, avoiding confusion which may arise from the extensive area concerned, some meaningful system of regionalisation will be necessary. Diels, in fact, already took this approach (Fig. 2.1). Ferdinand von Mueller in the previous century had remarked in several publications on the distinctive character of the flora and vegetation of the southwest, south of an undefined line running from Shark Bay to Israelite Bay. Diels named this portion the Southwest Province, distinguishing it from the Eremaea or Eremaean Province of the dry interior. The latter term is from the Greek, meaning solitude or wilderness. R. Tate in 1890[1] had used the word in the form Eremia for a portion of arid northern South Australia. The reason for the change of spelling by Diels is not known: he was aware of Tate's earlier usage (Diels 1906, p.372). The boundary between the two provinces, said Diels, coincided approximately with the 300 mm rainfall line, separated the internal drainage area of the country from the portions draining to the sea (not strictly true), was of importance to human settlement in being the boundary of cereal cultivation and had also recently (at that time) been picked up in zoogeography, by B.H. Woodward in 1900[2]. 'It follows,' said Diels again, 'that the biological boundary between the Eremaea and the Southwest Province is essentially climatically determined.'

Diels proceeded to divide the Southwest Province into 6 Botanical Districts and the southern Eremaea so far as known to him into 2. A sketch map was provided (Fig. 2.1). Each district was characterised by a range of rainfall, by particular types of vegetation and by species distribution.

In 1926 a geologist, E. de C. Clarke, proposed a regional subdivision of the State by a synthesis of all ecological (environmental) factors to determine 'natural regions'. There were 15 of them and he gave a general description of the geology, topography, rainfall, soils and vegetation of each[3]. This was an important step forward in geographical knowledge of the State which had become greatly opened up since Diels' time (compare Fig. 1.4).

In 1944 the Government Botanist C.A. Gardner added the concept of a Northern Province comprising the Kimberley and Pilbara districts, and later in 1956[4] he expanded Diels' Botanical Districts on a State-wide basis, so that there were now 5 districts in the Northern Province, 5 in the Eremaea (which he wrote as Eremea, introducing a third spelling) and 6 in the Southwest. Gardner's map is reproduced here as Fig. 2.2. Comparison with Fig. 2.1 shows that Gardner had

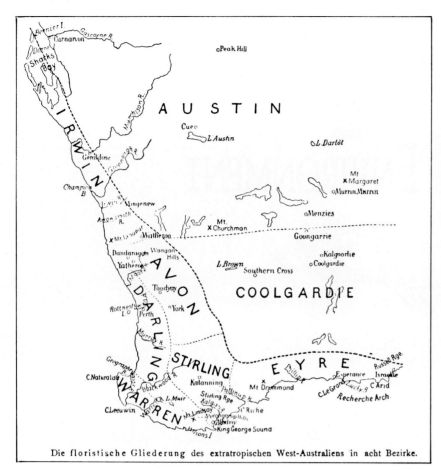

FIG 2.1

It is convenient for many purposes to divide a large area like the State of Western Australia into regions of varying size and status. Ferdinand von Mueller in the previous century had remarked on the distinctive character of the flora and vegetation of the Southwest, south of an undefined line running from Shark Bay to Israelite Bay. L. Diels in his book of 1906 named this portion the South-West Province, distinguishing it from the Eremaean Province of the dry interior. Diels further divided the South-West Province into six botanical districts, and the southern Eremaea as far as known to him into two, as shown in the figure above. Each district was characterised by a particular range of rainfall, by types of vegetation and by species distribution.

moved the boundary of the Southwest Province somewhat further east. In view of Gardner's ecological approach (discussing plants in relation to environment) it is not surprising that outside the Southwest Province in which the Botanical Districts had been taken up from Diels' work, they frequently agreed rather closely with Clarke's Natural Regions.

In 1960 Nancy T. Burbidge published a classic work on Australian plant geography[5] in which the Western Australian regionalisation was related for the first time to a continental model. Burbidge proposed as shown in Fig. 2.3 three 'Principal Floristic Zones'. The Northern Province of W.A. is seen to form part of the Tropical Zone extending round through Queensland; the Eremaean Province is part of the Eremaean Zone of Central Australia, while the Southwest Province belongs to the Temperate Zone with its special character recognised as the S.W. Focal Area. Burbidge also recognised three Interzone Areas where the flora and vegetation are of an ecotonal or intermediate character. One of these is situated in the west. Burbidge wrote: 'In Western Australia there is a triangular area lying between the Southwest Province and the Eremaea proper ... within it there is some mingling of the genera of the sand-heath associations of the Province with the aridity-tolerant genera of the Eremaea.' Reference to Figs 2.1 and 2.2 will show that this triangular area is the Coolgardie Botanical District which now does double duty as the *Southwestern Interzone*, distinct from both adjacent Provinces.

When the present writer began field work for the King's Park Botanic Garden in the 1960s the reality of the ecological regionalisations of Clarke and Gardner was quickly appreciated. Later when vegetation maps were prepared it was seen that in most cases the regionalisations could be quite clearly seen on the maps, so much so that actual boundaries could be drawn. In 1969 attention was drawn to this in a paper on the desert areas of the eastern half of the State[6], to quote:

The desert area is divisible naturally into a number of quite distinct regions, each having a characteristic landscape and vegetation. The boundaries of these regions, as shown in Fig. 5, are accurately derived from vegetation mapping. Each of them can be said to be a Natural Region in the sense of

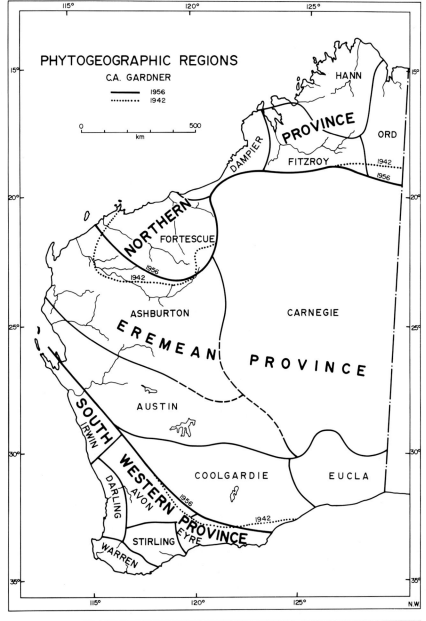

FIG 2.2

In 1944 the Government Botanist C.A. Gardner added the concept of a Northern Province comprising the Kimberley and Pilbara Districts and later, in 1956, in a book on *Toxic Plants* with H.W. Bennetts, he expanded Diels' Botanical Districts on a State-wide basis, so that there were now 5 districts in the Northern Province, 5 in the Eremaea and 6 in the South-West. These were shown in the small-scale sketch map above. Comparison with Fig. 2.1 shows that Gardner had moved the boundary of the South-West Province somewhat further east.

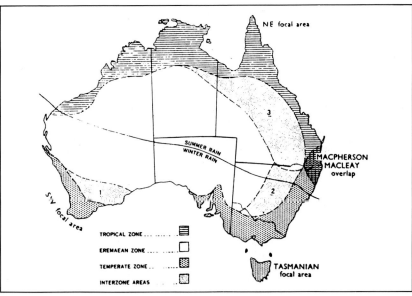

FIG 2.3

The principal floristic zones of the Australian region, according to Burbidge (1960). In 1960 Nancy T. Burbidge published a classic work on Australian plant geography in which the Western Australian regionalisations were related for the first time to a continental model. She proposed three 'Principal Floristic Zones'. The Northern Province of Western Australia forms part of the Tropical Zone extending round through Queensland. The Eremaean Province is part of the Eremaean Zone of Central Australia, while the South-West Province belongs to the Temperate Zone with its special character recognised as the South-West Focal Area. Burbidge also recognised three Interzone Areas where the flora and vegetation are of an ecotonal or intermediate character. One of these is situated in the west, and is referred to as the Southwestern Interzone. It equates with the Coolgardie Botanical District of Diels and Gardner.

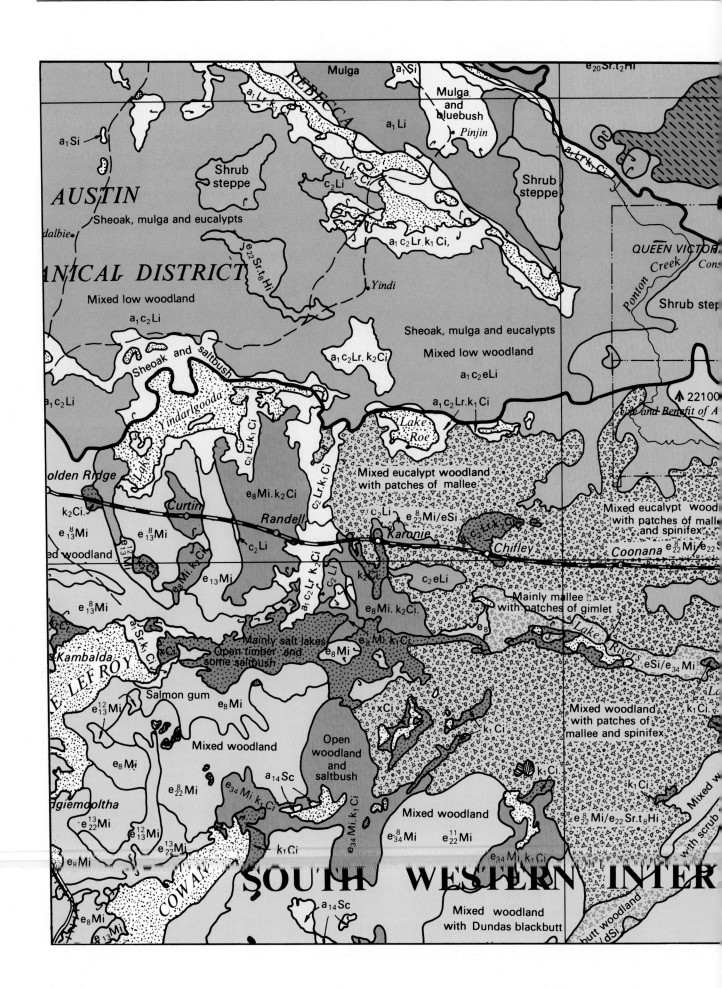

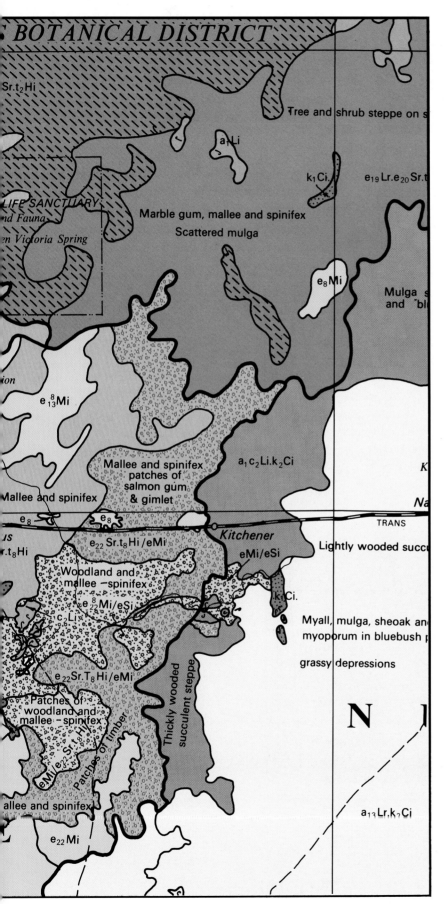

FIG 2.4

Vegetation maps usually show quite clearly the distinctive vegetation of adjacent regions, and the author made a practice of marking the boundaries between them on the maps. This figure is an excerpt from the 1:1 000 000 map sheet *Nullarbor*. Along the eastern side there is a portion of the Nullarbor Plain, a limestone plateau with a cover of bluebush steppe partly with and partly without scattered trees. This vegetation appears in brown colours and shows the extent of the Plain as a natural region, and its precise boundaries. North of it lies the Great Victoria Desert (= Helms Botanical District), a sandy region in which spinifex steppe coloured in red shades is predominant. West of the Plain there is a large area in which eucalypt woodlands coloured light green are the principal cover, and constitute Burbidge's Southwestern Interzone, otherwise called the Coolgardie Botanical District by Diels and Gardner. In the northwest corner of the map there is a transition from these woodlands to low woodlands coloured orange in which mulga *(Acacia aneura)* is dominant, making up the Austin Botanical District.

Clarke (1926) as well as a Botanical District in the sense of Diels (1906) and of Gardner & Bennetts (1956), and to represent a refinement of the earlier work in both cases with the aid of more advanced geographical and botanical knowledge. Each of the regions recognised here is being named as a natural region and a botanical district.

The mapping thus established exact boundaries for the regions and later when the vegetation maps were published these boundaries were marked on them. For the most part they coincide with the boundaries of vegetation units and only rarely are drawn to cross them. This method greatly reduces the element of subjectivity in the definition of regions which are normally quite clearly apparent from the map itself. The map 'Nullarbor' gives particularly clear cases, and a portion is reproduced here as Fig. 2.4.[7] A large area is occupied by the Nullarbor Plain, a limestone plateau which has a vegetation of chenopodiaceous steppe partly with and partly without scattered low trees. This vegetation appears in brown colours and shows the extent of the Plain as a natural region and its precise boundaries. North of it lies the Great Victoria Desert, a sandy region in which spinifex steppe coloured in red shades is predominant. West of the Plain there is a large area in which eucalypt woodlands coloured light green are the principal cover, and constitute Burbidge's Southwestern

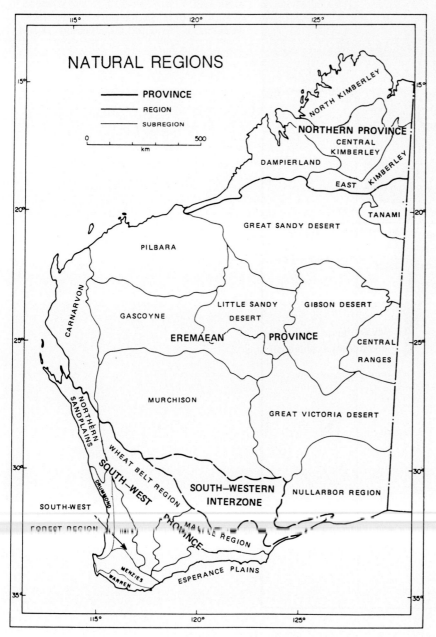

FIG 2.5 THE NATURAL REGIONS OF WESTERN AUSTRALIA

The end result is a map of natural regions of the State, each having a characteristic landscape and vegetation. The boundaries have been accurately derived from vegetation mapping and are therefore tortuous, unlike the sweeping curves of the sketch maps of earlier authors. Some revision of the regions was necessary in the light of more recent information.

Interzone, otherwise called the Coolgardie Botanical District by Diels and Gardner. In the northwest corner of the map there is a transition from these woodlands to low woodlands coloured orange in which mulga (*Acacia aneura*) is dominant, along a line which is regarded as the boundary of the Eremaean Botanical Province. Towards the south coast (not shown in Fig. 2.4) there is a transition from the eucalypt woodlands to mallee (yellow-green) along a line which is regarded as the boundary of the Southwest Botanical Province. Still further south the mallee changes to heath formations (yellow shades) and this transition defines the boundary between the Roe and Eyre Botanical Districts.

In 1980 when the writer published a detailed discussion of this phytogeographic regionalisation[8] a diagram was included showing the treatment of Botanical Provinces and Districts derived from the vegetation mapping. This diagram was reproduced again by Beard and Sprenger in 1984[9] with the Botanical Districts given alternative geographical names and the status of Natural Regions, and it appears here as Fig. 2.5. The boundaries of the regions are tortuous as they have been derived from detailed mapping at a larger scale. They differ from the sweeping curves of the earlier authors' sketch maps. A natural ecological region possesses its own characteristic landscape due to its particular features of climate, geology, landforms, soils and vegetation. In later chapters where the plant life is being discussed region by region, a summary of the characteristics of each region will be given. Table I lists the 21 regions and 4 sub-regions with their areas in square kilometres. It may be of interest that the Eremaean Province, the Arid Zone, occupies 70% of the State, the Northern and Southwest Provinces 12% each and the Southwestern Interzone 5%.

The Northern Province has had to undergo substantial revision since the vegetation map of the Kimberley District on completion proved not to be in accord with Gardner's earlier treatment (Fig. 2.2). With the scanty information of earlier days it is evident that Gardner was too greatly influenced by Clarke[3] whose treatment mainly followed the geological structure. The distribution of vegetation is more strongly influenced by rainfall. After discussion among the Western Australian botanists it was decided to conserve the name of the Dampier district from Gardner but to rename the other three districts which are entirely new in concept. One is named after C.A. Gardner himself in honour of his pioneer botanical work in that area in 1921.

In the Eremaean Province there are now 11 districts. In the remote interior there has been substantial revision based on information not previously available[6]. The Fortescue district has been transferred from the

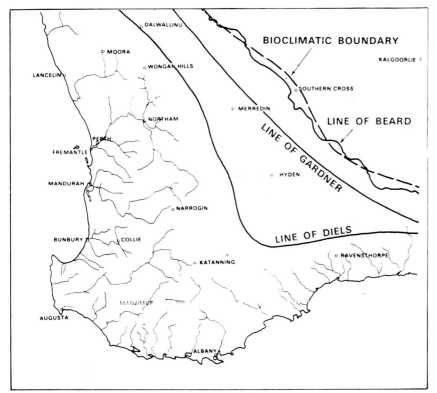

FIG 2.6

Comparison of the positions of the boundary of the Southwest Botanical Province according to Diels (from Fig 2.1), Gardner and Bennetts (Fig 2.2) and Beard (Fig 2.5), with the bioclimatic boundary (broken line) from Fig 2.8. These successive movements represent improving accuracy as the botanical and geographical knowledge of the State has increased.

TABLE 2.1

Vegetation survey of Western Australia, area calculations
Regional summary

			km²	km²
Northern Province	12.57% of State			
North Kimberley Region	(Gardner Botanical District)		99 100	
Central Kimberley Region	(Fitzgerald Botanical District)		83 330	
Dampierland	(Dampier Botanical District)		84 400	
East Kimberley Region	(Hall Botanical District)		50 510	317 340
Eremaean Province	70.16% of State			
Murchison Region	(Austin Botanical District)		316 239	
Gascoyne Region	(Ashburton Botanical District)		181 453	
Pilbara Region	(Fortescue Botanical District)		178 017	
Carnarvon Region	(Carnarvon Botanical District)		91 046	
Great Sandy Desert	(Canning Botanical District)		291 017	
Tanami Desert	(Mueller Botanical District)		35 192	
Little Sandy Desert	(Keartland Botanical District)		110 314	
Gibson Desert	(Carnegie Botanical District)		149 784	
Central Ranges	(Giles Botanical District)		60 788	
Great Victoria Desert	(Helms Botanical District)		209 206	
Nullarbor Region	(Eucla Botanical District)		148 764	1 771 820
Southwest Province	12.27% of State			
Southwest Forest Region	(Darling Botanical District)			
Swan Coastal Plain	(Drummond Subdistrict)	14 637		
Northern Jarrah Forest	(Dale Subdistrict)	19 473		
Southern Jarrah Forest	(Menzies Subdistrict)	26 572		
Karri Forest Subregion	(Warren Subdistrict)	8 323	69 005	
Northern Sandplains	(Irwin Botanical District)		39 656	
Wheat Belt Region	(Avon Botanical District)		93 520	
Mallee Region	(Roe Botanical District)		78 957	
Esperance Plains	(Eyre Botanical District)		28 702	309 840
Southwestern Interzone	(Coolgardie Botanical District)			126 500
	5.00% of State			
		Total area km²		2 525 500

*Areas revised after consultation with W.A. Dept of Land Administration to conform to regionalisation in the Northern Territory.

Northern Province according to arguments advanced by the writer and a new Carnarvon district created. The Ashburton, Austin and Eucla districts are essentially as Gardner conceived them.

The Coolgardie district is equated with Burbidge's Southwestern Interzone and is essentially as conceived by both Gardner and Burbidge.

The Southwest Botanical Province is bounded essentially as conceived by von Mueller, Gardner and Burbidge. Vegetation mapping has closely confirmed the views of these authors. Diels (see Fig. 2.1) took a boundary running much closer to the coast. On the continued criterion of mapping according to vegetation it was found necessary to revise Diels' districts very considerably and again, after discussion at the Western Australian Herbarium, it was decided to adopt five Botanical Districts of which one (Darling District or Southwest Forest Region) would be divided into four sub-districts. Diels' names have been conserved in four cases (Irwin, Darling, Avon, Eyre) but his Stirling District has disappeared and his Warren District has been reduced to a sub-district. One new district was proposed, the Roe district named after the famous early surveyor-general John Septimus Roe who was also a botanical explorer: it comprises the mallee country.

Another point of explanation concerns the progressive eastward movement of the boundary of the Southwest Province shown in Fig. 2.6. Diels' concept of it followed approximately the 300 mm rainfall line, as it was established in his day, but evidently a relatively dry cycle was then being experienced, and the rainfall figures quoted by Diels in his page 81 are lower than subsequent averages. Diels was also influenced by other considerations as he tells us, such as the inland limit of cereal cultivation which had not spread very far at that time. There may have been a popular impression that it would be too dry for crops further inland, so that the Eremaea was thought to begin where Diels showed it. In the light of subsequent knowledge both of vegetation and the environment later authors moved the boundary further east. Beard's line in Fig. 2.6 established by vegetation mapping is shown to coincide closely with a *bioclimatic boundary* between two major rainfall zones as will be described below.

We must now consider the major outlines of those factors of climate, geology, landforms and soils which characterise the various regions and determine the nature of the plant cover.

2 Climate

If we wish to assess the effects of climate we must consider not only the average annual expectation of rainfall and temperature but the variation of these throughout the year, in particular the season or seasons when rainfall is received, the length of such seasons and their reliability. Attempts have been made at various times to produce climatic indices embodying the characteristics of climate which are the most important particularly for vegetation. Those which attracted most attention and still command much respect were the systems of Köppen (finalised 1936) and Thornthwaite (1931, 1948). It has been found however that for Australia at least these indices are not accurate enough. They show the arrangement of various intergrading climatic zones reasonably correctly, but do not delineate them to correspond with observed vegetation zones. An important improvement was effected by a school of biogeographers at the University of Toulouse in France with which the name of Professor H. Gaussen is associated[10]. They argued that as the supply of moisture is of overriding importance for vegetation, the critical factor will be the season and length of the dry period. This they calculated from records of average monthly rainfall and temperature, using for each weather station a chart known as an *ombrothermic diagram*. Some examples of these are shown in Fig. 2.7. In the diagram mean monthly rainfall in mm and temperature in °C are both plotted, using double the scale for rainfall as for temperature ($r = 2t$), as shown in the right and left-hand margins of the diagrams in Fig. 2.7. Months in which the rainfall line falls below the temperature line are assumed to be 'dry', i.e. that precipitation is inadequate to sustain plant growth. The validity of this assumption has been well tested in practice. The number of dry months is read from the diagram for each recording station and used to classify it for *bioclimate*. Classification is based on the type of rainfall distribution, whether with a summer maximum, a winter maximum, or bimodal and so on, divided into types according to the number of dry months, and further divided as required according to other criteria.

Fig. 2.8 shows a bioclimatic map of Western Australia prepared following Gaussen's methods, which is found to accord very well with vegetation boundaries. In the Northern Province, rainfall occurs almost entirely in summer, with a very long dry season of eight months, leading to a classification of Dry Hot Tropical. Rain

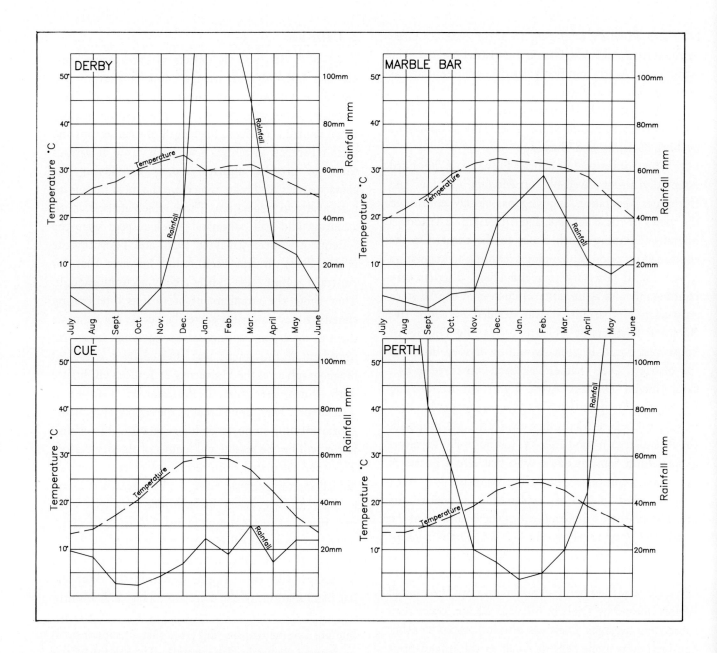

FIG 2.7

Climate diagrams for representative stations. Temperature, mean monthly, is shown by the broken line and the scale at left. Rainfall, mean monthly, is shown by the continuous line and the scale at right.

Derby: Dry hot tropical regime with short summer rains
Marble Bar: Desert with summer rain
Cue: Desert with summer and winter rain
Perth: Dry mediterranean, winter rain

Climate is the primary determinant of vegetation, the most important factor being the season and length of the dry period. Types of climate can be determined from diagrams such as those above, and used to construct the climatic map in Fig 2.8.

results from monsoonal southward movements of moist tropical air, and is normally received from thunderstorms and cyclones. There is a narrow intermediate band of semi-desert as the Eremaean Province to the south is entered. Rainfall stations in the main body of the Eremaea (except for the northern and southern transitions) invariably attract a classification of Desert where all months in the year are 'dry'. This does not mean of course that there is no rainfall, merely that on the average it is inadequate for plant growth. This also implies that there is no assured growing season. Erratic rainfall does occur and may at times be heavy enough to bring plants into growth. In practice it is found that rainfall of sufficient amount for this may be expected on the average three times a year, but droughts of three

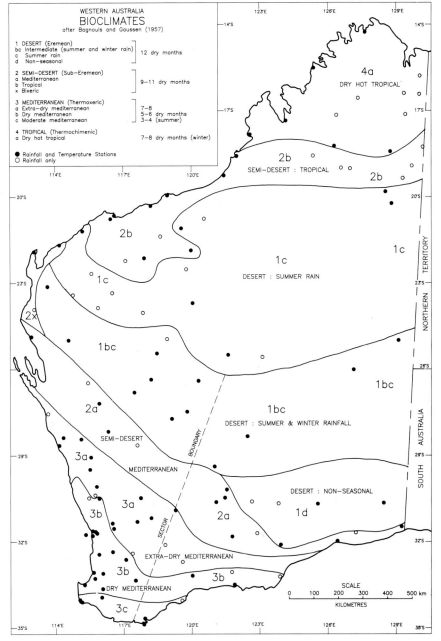

FIG 2.8

Map of Bioclimates, that is, types of climate as affecting living things. The classification adopted here is based on the season and length of the dry period and is found to accord very well with vegetation boundaries, rather better in fact than other well known forms of climatic indices.

years' duration can also be experienced. In the northern half of the desert any heavy rainfalls are normally received from tropical cyclones in the summer months. After forming over the ocean north of Australia and travelling initially southwest, the cyclones may change direction to southeast, cross the coast and become tropical rain-bearing depressions. The southern half of the Eremaea on the other hand receives most of its beneficial rainfall in late summer to early winter from occasional irruptions of moist tropical air streaming down from the northwest across just that belt of country, resulting in a skewed rainfall distribution (see diagram for Cue). The spring months September and October are the driest. Lesser falls of rain may be received from summer thunderstorms and winter cold fronts. In the extreme south of the Eremaea rainfall becomes non seasonal in distribution, that is, there is an equal chance of rain in any month of the year. None of the Australian desert has a winter rainfall maximum.

The latter is characteristic of the southwest portion of the State where the wet winter and dry summer create a regime known as mediterranean from its prevalence around the Mediterranean Sea. Winter rain, summer drought, is an uncommon regime in the world as a whole and in addition to Western Australia, part of South Australia and the Mediterranean it is only found in California, Central Chile and the tip of Southern Africa, 'The Cape'. It occurs on the west coasts of continents in middle latitudes and results from anticyclonic circulation over the continent during summer. Four major mediterranean climatic zones have been recognised in Western Australia as shown in Fig. 2.8, classified by the average number of dry months in the year as follows:

Semi-desert mediterranean   9–11 dry months
Extra-dry mediterranean   7–8 dry months
Dry mediterranean   5–6 dry months
Moderate mediterranean   3–4 dry months

The boundaries are drawn at 4.5, 6.5 and 8.5 dry months. There is some practical value in this classification. It expresses a natural ecological division showing the effect of climate on vegetation and human activity, as the following details show:

Semi-desert mediterranean zone: the southwestern edge of this zone coincides with the boundary of the Southwest Province. In its northern, warmer section the semi-desert zone is Eremaean and its vegetation comprises mainly acacia thickets and scrub transitional to the mulga of the interior. The southern, cooler, section comprises the Southwestern Interzone whose vegetation is predominantly of dry eucalypt woodlands. The zone is too dry for farming and it has only limited pastoral use. Except for goldmining settlements it is largely unoccupied.

Extra-dry mediterranean zone: includes the wheat belt, the prime agricultural region of the State. Original vegetation was eucalypt woodland, mallee, thickets and heath according to soil. The inland boundary coincides with that of the Southwest Province.

Dry mediterranean zone: relatively humid zone of the southwest. Forests of jarrah, woodlands on poorer and drier soils. It is used for forestry, dairying and fruit-growing.

Moderate mediterranean zone: high-rainfall zone of the extreme southwest. Forests of karri, tingle and jarrah mingled with low forests on poor soils, and swamps. Forestry is the chief activity, with little agriculture.

All the mediterranean zones are shown on the map in Fig. 2.8 divided by a 'sector boundary' into eastern and western sectors. This is because rain essentially comes in from the west, so that the western sector receives more than the eastern. Over an equivalent number of rainy months the eastern sector receives a lower total. This results in less groundwater recharge and a lower ability to support tree growth. The vegetation of the eastern sectors is in consequence typically mallee and mallee-heath instead of forest and woodland.

3 Geology

The underlying geology is important for vegetation in many ways. If there have been past periods of mountain-building activity, these may have left a legacy of mountains and hills to this day, and if there have not the country may be correspondingly flat and monotonous. Hard rocks resistant to erosion tend to create hilly topography, and soft rocks the reverse. Hard rocks may not weather readily to form soil, and may be an inhospitable environment; and again, soft rocks the reverse. Some rocks are chemically rich and disposed to form fertile soils, while others are chemically poor and form soils low in nutrients. Rocks of volcanic origin such as basalt and dolerite are typically rich while sandstones and quartzites are typically poor. Limestones in general tend to produce fertile soils but as these are normally somewhat alkaline they may favour only specialised plants.

Fig. 2.9 shows the geological structure of the State in simplified form[11]. *Precambrian* rocks, formed before the beginning of the Cambrian period 570 million years ago underlie the whole State as the *Precambrian Basement* but have subsided in certain areas shown in Fig. 2.9 to form *basins* which have filled up in the past with sediments transported from adjacent highlands by rivers. These sediments have formed rocks in turn which are post-Cambrian or *Phanerozoic*. The Cambrian is considered an important threshold in geology as it marks the beginning of readily recognisable fossils in rocks. Phanerozoic means the period of visible life. Earlier rocks were once thought to be devoid of signs of life but we have since learned to recognise traces of simple forms of life stretching back for a very long period into the past. Such rocks are called *Proterozoic* meaning 'first life'. Colonies of simple unicellular organisms called

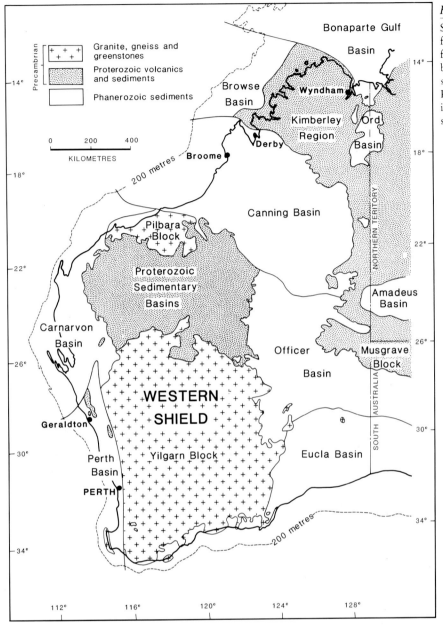

FIG 2.9

Simplified geological structure of Western Australia, from (12). Very hard, ancient Precambrian rocks form a basement which has subsided locally into basins which have filled with Phanerozoic sediments—sandstone, siltstone and limestone. The Precambrian is divided into basal granite (with included gneiss and greenstone) and Proterozoic sediments and volcanics deposited in older basins.

Cyanobacteria trap sediment to build up domed rocky structures for example on the shores of Hamelin Pool, part of Shark Bay, at the present time, yet similar domed structures can be identified in fossil form in rocks in the Pilbara dated as 3 600 000 years old. The earth itself is believed to have formed about 4 600 000 years ago. Since life first began on earth it has evolved at an accelerating rate, a fascinating story which is ably told by Mary White in her book *Greening of Gondwana*[12].

As Fig. 2.9 shows, the western part of the State is underlain largely by Precambrian rocks forming the *Western Shield*, so called because it is a large rigid and inert section of the earth's crust. In the Yilgarn Block it consists mainly of granite and gneiss with infolded belts of altered volcanics and sedimentary rocks which are mineralised and mined for gold, nickel, iron and other metals. The altered volcanics (amphibolites) are known as *greenstones* and give the name greenstone belts to the mineralised zones. Extremely hard rocks very rich in iron known as banded ironstones may occur within the greenstone belts and tend to form abrupt ridges standing up from the plain. Rocks of the Yilgarn Block date within the range of 2900 to 2500 million years old. In the far north, rocks of the Pilbara Block are even older. Between the two blocks there is a substantial area of Proterozoic volcanics and sediments laid down between 1800 and 1300 million years ago and again around 1000 million. These rocks range from crystalline

FIG 2.10

Geological structure has a strong influence upon the landscape. Granite country of the Yilgarn Block is gently undulating with rock outcrops known as 'tors' but in general has well developed soils over deep profiles.

FIG 2.11

Proterozoic rocks, seen here on the Kimberley Plateau, tend to be hard and not as readily weathered as the granite, forming rocky plateaux and scarps with little soil cover.

Fig 2.12

Phanerozoic rocks, as in the Moresby Range near Geraldton, are softer and give country of more rounded outline with deeper soils though harder bands of sandstone may still form tabular hills and scarps.

to both fine and coarse-grained sedimentary, and are the source of the iron ore now mined. The very hard rocks form hilly to semi-mountainous terrain. The Kimberley Region in the northeast is formed of Proterozoic rocks typically in the age range 1900 to 1650 million years. The Musgrave Block is a major Proterozoic structural unit of Central Australia extending much further to the east. Within Western Australia the structure and composition are complex, and include both volcanics and altered granite-like rocks. The harder members form ranges of hills rising from sandy plains.

The Canning, Officer, Carnarvon and Perth Basins were intermittently beneath the sea from early Phanerozoic times until the end of the Cretaceous period 65 million years ago, since when they have been dry land. The rocks are mainly coarser sediments, i.e. sandstones. The topography formed is usually of rather little relief but occasional tabular hills and scarps afford some diversity. The Eucla Basin is different. It did not subside till Cretaceous time, and received sediments until Miocene time (about 15 million years ago). The uppermost beds are limestones and form today the very extensive Nullarbor Plain, an almost featureless expanse which is one of the largest limestone plains in the world.

4 Landforms

Since the Western Shield is an ancient and rigid portion of the earth's crust, it has been undergoing erosion for a very long period. The earliest event which is still relevant for the modern landscape is the ice age which occurred in the early Permian Period about 270 million years ago. Australia lay in high latitudes at that time, with the South Pole in Tasmania, and it seems that a continental ice sheet similar to that of Antarctica today covered the Western Shield for a time. Sediments of Permian age containing material of glacial origin were laid down in the sedimentary basins, sometimes in great thickness, so that we can assume that the ice extensively wore down and levelled the landscape. After the retreat of the ice, the country probably resembled what we see today in northeast Canada and in Finland where similar crystalline rocks have been subjected to the ice ages of

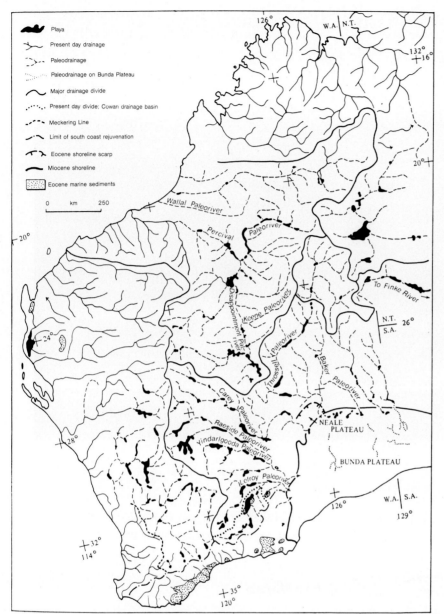

FIG 2.13

Drainage basins of Western Australia, after van der Graaf et al. 1977[13]. Existing active rivers are shown by continuous lines, paleorivers by broken lines. Paleorivers were active in the distant past when rainfall was higher but are now represented only by dry valleys, usually with chains of salt lakes. Although there is no coordinated drainage today in the dry interior, none the less the landscape has been shaped by river action in the past.

the Pleistocene and Recent periods. The retreating ice there has left a landscape of little relief, excavated into bosses and hollows, the latter filled with long narrow lakes. Such country has at first no organised drainage, but gradually river systems can be expected to take shape, and drainage basins to develop.

It is thought that during the geological periods subsequent to the Permian, down to the end of the Cretaceous 65 million years ago, the river systems that we see today gradually developed. The Western Shield today has a central watershed running north-south roughly down the middle (Fig. 2.13), standing at about 600 m above sea level in the north, falling gradually to 300 m in the south. The surface of the Shield, dissected by rivers, slopes down gradually both to east and west from the watershed. On the western slope, towards the Indian Ocean, the rivers are still active, principally the Swan, Murchison, Gascoyne, Ashburton and Fortescue. On the eastern slope rivers which formerly drained into the Canning, Officer and Eucla Basins have dried up as the interior of the continent became more arid. Today their valleys contain chains of salt lakes, more often dry than full, which only rarely produce an organised flow from one to the other. There are no rivers of any magnitude flowing to the south coast of the Western Shield since the drainage there is much younger. The Australian and Antarctic continents were formerly joined together along the south coast of Australia so that southerly drainage only began to develop when they broke apart, beginning perhaps in late Cretaceous time.

The Canning and Officer Basins were uplifted at the end of the Cretaceous (say 65 m.y.a.) and then developed river systems of their own, the Canning draining northwest and the Officer southward. These rivers have dried up but their courses can still be traced[13]. The Eucla Basin was uplifted very much more recently to form the Nullarbor Plain which bears extremely little trace of any former river action on its surface. It is assumed therefore that by the time the Plain was established desiccation had greatly reduced run-off from the interior.

By the end of the Cretaceous the elements of the Western Australian landscape as we know them today seem already to have come into being. The principal feature is a gently undulating plateau, its monotony broken only here and there by hills of harder and more resistant rock — bosses of granite, ridges of banded ironstone. Different country occurs only where the granite basement is overlain by surviving Proterozoic rocks, in the Stirling Range in the south, in the Hamersley and other ranges of the Pilbara in the north, and in the Kimberley region and the Musgrave Block of the far east (Fig. 2.9). Since the Cretaceous the landscape seems to have changed relatively little due to its pre-existing flatness undisturbed by any mountain building activity or major uplifts and subsidences of the earth's crust. It is this *tectonic inertness* — lack of movement — which creates the well known impression of the immense age of the Australian landscape, with its little mountains worn down almost to the very root and surrounded by wide plains carrying the erosional debris of millions of years which has not been transported into the sea.

The evidence indicates that the Cretaceous was mostly a period of high temperature and humidity all over the world and that subsequently the world's climate has become episodically (by fits and starts) cooler and drier, so that during the last 2 million years the earth has been subject to intermittent ice ages. During the late Cretaceous and subsequent early Tertiary Australia was for the most part covered with a rainforest similar to that found today in the North Island of New Zealand, and the highlands of New Caledonia and New Guinea. The earlier flora has taken refuge in such parts, we suppose, because the climate has continued much the same as in early Tertiary time whereas it has died out in Australia. The southern beech *(Nothofagus)* was common in our rainforest as were southern conifers such as *Podocarpus* and *Dacrydium*, many broadleaved trees of genera considered subtropical today, also cycads and ferns. Ecological diversity must have existed then as now, some habitats were probably too adverse for rainforest and the progenitors of our modern sclerophyll woodland and kwongan* floras are likely to have been locally present. Fossil banksia cones have been found in the Kennedy Range (inland of Carnarvon) which are dated as late Eocene, about 40-45 m.y.a.[14]

Something of great importance for our landscape took place in that early time of the rainforest period under the warm, wet climate — the process of *deep weathering*. The surface rock beneath the soil softened and decayed to depths as great as 50 metres by chemical weathering and alteration of the original minerals, and leaching out of the more soluble constituents. Unfortunately the latter included the principal plant nutrients, so that a weathered profile may be very infertile indeed. What remains consists partly of quartz, because it is inert. Quartz crystals in the original rock survive to become sand grains. Other principal constituents are oxides of iron and aluminium, and kaolin (complex aluminium silicates) which is white and colours the weathered profile so strongly that it is known as the *pallid zone* (Fig. 2.18). Deep weathering has affected not only the granite and other rocks of the Western Shield but the younger sedimentary rocks of the basins also, except for the much younger limestones of the Eucla Basin which emerged long after the rainforest period.

A second process, that of *surface lateritisation*, took place later, it is thought mainly during the Oligocene (38 to 26 m.y.a.), when the climate was becoming drier and more seasonal. This process resulted widely in the formation of a hardened *laterite* (ironstone) layer on the surface, called a *duricrust*. It is conspicuous in Fig. 2.18. It is formed of quartz, iron oxide, alumina and clay pellets strongly cemented together. If examined it can be seen that it has been formed by the growth of nodules which have grown larger and coalesced. Sometimes the duricrust is formed only of nodules *(pea ironstone)* while at the other extreme these have set into a solid mass *(massive laterite)*. There is no unanimity among experts as to when and how surface duricrusts have formed. We have to accept that they are there and extremely

---

*Kwongan: See Chapter 1, p.18.

FIG 2.14
Fossil evidence shows that during earlier Tertiary time, 40–60 million years ago, Australia mainly enjoyed a rainforest climate and possessed a flora whose descendants are found today in Tasmania, New Zealand, New Caledonia and New Guinea. Tall rainforest dominated by the southern conifers *Podocarpus* and *Dacrydium* must have been common, as we see here in the south island of New Zealand.

FIG 2.15
Forests dominated by southern beech *(Nothofagus)* are also indicated by the fossil record, similar to this example from Tasmania.

K.J. McNamara

Fig 2.16 and 2.17

It is likely that ecological diversity existed in the early Tertiary as today, and the general rainforest cover was broken by lesser types of vegetation adapted to poor sites of various kinds. The ancestors of our modern sclerophyll species seem to have evolved early as evidenced by the fossil banksia cone (Fig 2.16) found in the Eocene beds of the Kennedy Range near Carnarvon. Named *Banksia archaeocarpa* (ancient fruit) it is unmistakably similar to the modern *B. attenuata* in Fig 2.17, although some 40 million years old.

Fig 2.18

A major effect of the prolonged period of high rainfall combined with tectonic inertness was the deep-weathering of the surface mantle into a white, kaolinised 'pallid zone' which may be 50 m thick and frequently capped with a ferruginous duricrust or laterite. The weathered layer has been leached of most soluble constituents including plant nutrients, leaving mainly quartz and the oxides of iron and aluminium. Most of Western Australia is mantled with this exhausted material.

widespread except where they have been subsequently eroded away as has happened in the hilly regions of the Pilbara and Kimberley, or did not originally form e.g. on the Nullarbor Plain. Lateritic duricrust is normally present in most areas of Western Australia capping the higher ground. In the wheatbelt for example it is absent from valleys but caps all the watersheds, its edge often marked by a small scarp or breakaway (Fig. 2.19).

A third process, if we can separate it from the others, has been the formation of a surface layer of sand above the duricrust, and this is normally present to varying depth except close to breakaways. Again experts disagree as to its formation and origin, but ultimately it must be derived from weathering of the country rock. It is the end-product of disintegration when almost all minerals except quartz grains have been leached away. There is evidence for much reworking of the surface sand, that is, for movement by wind during previous arid periods. In the southwest, any deep yellow sands appear to have been wind-transported, and in the deserts of the interior systems of linear dunes have been built up from the surface material. This is thought to have happened during the peak of the last ice age 15-18 000 years ago: not that Western Australia was glaciated, but climate all over the world was cool and dry at that time and in the Australian interior evidently very windy.

FIG 2.19

On the Yilgarn Block, the higher ground is normally capped with sheets of laterite. Erosion of these in progress today at the edges may create a small scarp or 'breakaway'. In this example, a sheet of bare laterite is seen at upper right, and a white leached sandy soil below the breakaway at left. Trees of the eastern Wandoo, *Eucalyptus capillosa*, find a congenial habitat immediately below the little scarp.

FIG 2.20

Decay of the laterite in later geological time has created sheets of leached sand called 'sandplains' which cap the higher ground, may extend for many kilometres and have a low shrubby vegetation called 'kwongan'. Sandplains are a major feature of Western Australia, covering 39% of the State.

The presence of extensive sheets of sand overlying duricrust in the higher ground gives rise to a peculiarly Western Australian phenomenon, the *sandplain*. (Fig. 2.20) Except for some cases where sand has been transported into the valleys, sandplains are high-lying, flat or almost so, extending often for many kilometres and with a low, open vegetation of kwongan in the southwest, acacia scrub or spinifex grassland in the Eremaea, pindan in the Kimberley. An entire book has been devoted to the subject[13], and a map of sandplains accompanies a paper by the writer[15] which also shows that they cover 39% of the whole State.

The other 61%, comprising sand-free country, consists of the hilly tracts already mentioned where the earlier weathered mantle has been eroded away, and the valleys of the granitic plateau where duricrust and its derived surface sand are normally absent. Here too it may have been eroded away, but for the most part the pallid zone is still intact in the valley beneath a top soil of red loam, suggesting that the duricrust never formed. It is difficult to see how erosion could have so completely removed the duricrust from wide areas without also destroying the pallid zone beneath it. In some areas, particularly in the Murchison and Gascoyne catchments, complete erosion to bedrock can be seen in process, but this is rarely the case further south where pallid zone soils are normally present beneath the valleys.

5 Soils

The widespread arrangement of sandplain on the higher ground, red loam soils in the valleys, gives rise to systems of soils called *catenas*. Catena is Latin for chain, and if we imagine a chain strung from one hilltop to another across a valley, we shall find a sequence of different soil types arranged like links in the chain down the slope, and up again the other side in reverse order. A particular soil type is always linked to its position on the slope. Catenary systems were first identified in Africa and appear to be characteristic of old landscapes. In Western Australia they occur widely and are most conspicuous in the wheatbelt region. Here, as the landscape undulates gently we may begin at the margin of a salt lake in the valley bottom with a saline loam, and pass gradually up slope through red loam soils, very heavy at first with a high proportion of clay, becoming more sandy until, quite high up a *duplex soil* with a sandy surface layer is encountered. Shortly above this point it is usual for the breakaway marking the edge of the duricrusted upland to be encountered. After a rise of 1–2 m there will be exposed laterite or thinly covered laterite for 50–100 m, after which the sandplain is entered with its sand soils of varying depth overlying laterite or clay. This basic sequence is repeated up and down, over and over again, and may be readily observed when traversing roads in the wheatbelt. A diagram showing a typical catena is given in Fig. 2.21.

Climate exerts the primary determining influence on vegetation. Within each climatic zone and sector, soil type exerts the second degree of influence. Therefore, for each soil type there is a corresponding vegetation type, and as we have catenary sequences of soils, so also we can distinguish catenary sequences of plant communities. In Fig. 2.21 the catena embraces two types of woodland, one of mallee and two types of kwongan. Elsewhere under a different rainfall but similar soils there will be a different catena of vegetation.

The soils of the sandplain are open-textured, freely draining and extremely deficient in nutrients. The red loam soils of the valleys are more heavily textured, hard-setting when dry, have a higher water-holding capacity and are less impoverished. The intermediate duplex sand-over-clay soil has intermediate characteristics. It is the typical soil type for mallee vegetation throughout the southwest. Valley soils in the present climatic epoch tend to accumulate material instead of losing it by leaching. Soil beneath salmon gum is typically accumulating calcium carbonate in round nodules called *kunkar*. In the east on the Goldfields this may be massive enough to coalesce to form a calcareous hardpan. In the Murchison Region of the Eremaea red loam valley and plains soils develop a siliceous hardpan *(silcrete)* which builds up in dense layers in the soil. Pastoralists who have to dig post holes into it call it 'the cement'. Later erosion of the topsoil may leave the silcrete exposed in platforms.

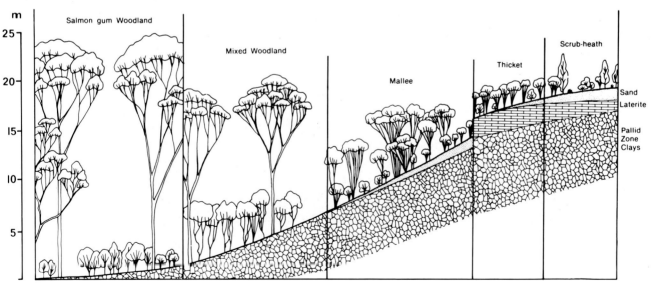

FIG 2.21

Catenary sequences of soil and vegetation, illustrated for the Wheatbelt Region[15]. The presence of laterite and sandplains on the higher ground, and their general absence from the valleys, is expressed in topographic sequences of soils called *catenas*, where soil types are repeated over and over at the same position on the slope. Every soil type has a corresponding vegetation type, so that there are also catenary sequences of vegetation.

A dominant feature of Australian soils and of western soils in particular is their poverty in the agricultural sense. Before founding the colony in 1829 Captain Stirling had made a reconnaissance of the Swan River and its vicinity, but looking at it alas with determinedly rosy-tinted spectacles. On arrival of the settlers it was found that the only suitable land within reach among the useless sandy wastes of the coastal plain was on alluvial terraces of the upper Swan above Guildford, and that of limited extent. The situation was saved by Ensign Dale who went exploring over the Darling Range and discovered the Avon Valley (actually a higher reach of the Swan) which from Beverley to Toodyay has relatively young and unweathered soils developed from a mixed suite of rocks of greater fertility than the common granite. Other odd pockets of fertile soil were found such as the Greenough Flats south of Geraldton and for many years actual farming had to be confined to them. Otherwise the settlers spread out into huge sheep runs and agriculture was limited to small food-growing plots near each homestead which could be fertilised with sheep manure. Only the introduction of superphosphate in the 1890s made the extension of farming feasible. It then became possible to cultivate the red loam soils of the valleys of the wheatbelt and a rapid expansion took place aided by development of the railway system. Even then the settlers could still do nothing with the sandplains, so-called 'light land'. They were puzzled. Addition of superphosphate plus the conventional nitrogen and potash still produced no crops. It was only in the 1930s when the role of trace elements became understood that this problem was solved. Light land must be given in addition small quantities of zinc, copper, molybdenum and cobalt. This discovery contributed after the second world war to a further wave of agricultural expansion when light land was much in demand.

The soil poverty is due in the first place to a general deficiency in Australian rocks of the essential plant nutrient phosphorus, secondly to the deep weathering and leaching of the soil mantle which has removed most of the readily available nutrients. The inertness of the landscape has left most of this exhausted material in place as a thick overburden preventing access to fresh rock beneath. However, fertility means one thing to agricultural crops and another to native vegetation. The latter has evolved over a long period to be adapted to these conditions so that native plants are capable of growing successfully and flowering brilliantly at astonishingly low levels of nutrition. The adaptation is shown in many different ways. There is an unusual number of insectivorous plants exploiting a source of nutrients independent of the soil, and of parasitic plants robbing nutrients from each other. Leguminous plants and *Casuarina* fix nitrogen in root nodules, adding to the nutrient pool. Members of the Protea family (*Banksia, Dryandra,* etc) possess specialised roots which apparently aid maximum extraction of nutrients. *Mycorrhizas* (fungi associated with roots which promote nutrient uptake) are common. The most obvious and widespread feature is the sclerophylly or hard brittle leaved character which was mentioned in Chapter 1. At one time this was thought to be an adaptation to withstand drought but its accentuated form throughout Australia has led to further investigation concluding that it is more likely a response to nutrient deficiency. It is to be seen in the harsh, often spiny leaves of a majority of shrubs of all sizes, in the hard brittle leaves of *Eucalyptus*, and in the rigid, sharp-pointed leaves of the desert spinifex grass. Sclerophylly is so dominant in Australia that this special form of sclerophyll grass has evolved, quite unlike any other desert vegetation in the world.

Nutrient deficiency has shaped plant forms in our vegetation but its effect on vegetation structure appears to be much less. At first sight, the contrast of low, open kwongan vegetation on a sandplain with eucalypt woodland in an adjacent valley under the same rainfall might suggest that the more extreme deficiency of the sandplain soil would be the causal factor for its poorer vegetation. The sandplain however has a much lower water-holding capacity than the valley soil, and it has been concluded that the moisture supply through the dry season is actually the factor responsible[15].

The effects of deep weathering, too, are not all adverse. The soil profile is never actually sealed off from penetration of moisture or tree roots by its overlying duricrust, and the pallid zone beneath, by its depth, has a great capacity for moisture storage. It has been demonstrated that jarrah trees in the forest put down their roots deeply into the pallid zone and that this provides them with a moisture supply sufficient to carry them through the long dry season. It is evident that it is this factor which is responsible for southwestern Australia being so much better wooded than any of the other mediterranean regions of the world. Trees 20 m tall grow right out to the edge of the Nullarbor Plain where the average rainfall has declined to 175 mm per

Australian native plants are adapted in many different ways in response to the prevailing deficiency of nutrients. These adaptations give the Australian flora its special character.

FIG 2.22

There is an unusually large number of insectivorous plants in the Western Australian flora. This adaptation enables them to tap a source of nutrients independent of the soil. The two species of *Byblis*, southern and northern, are herbs with leaves covered with sticky glands to trap and digest insects.

FIG 2.23

The sundews, genus *Drosera*, also have their leaves modified in various ways to trap insects. They are only small herbaceous plants but with showy white or pink flowers. About a third of all the world's species of *Drosera* occur here.

FIG 2.24

Another effective strategy is to rob other plants. In the family Loranthaceae the numerous species of mistletoe attach themselves to the aerial portions of trees or shrubs, except for this one, the W.A. Christmas tree, *Nuytsia floribunda*. As well as being one of the finest flowering trees in the world, *Nuytsia* is botanically a mistletoe and operates by attaching its roots to the roots of other plants. This adaptation is used also in the sandalwood family, Santalaceae, where many such as *Exocarpos* and *Leptomeria* are root parasites.

FIG 2.25

When the roots of *Nuytsia* ramifying through the soil encounter the roots of other plants, they attach themselves by developing a fleshy collar from which the tissue of the host is penetrated to draw off water and nutrients. This photograph was taken in the wild at Mt Merivale, Esperance, and shows roots of *Pelargonium* with two *Nuytsia* sucking organs attached, a complete collar at the upper left and a partial one at lower right, the latter with a thinner white *Nuytsia* rootlet leading away from it. *Nuytsia* has been known to attach itself in this way to plastic-covered underground electric cables, dissolving the insulation and causing shorting!

J.S. Pate   J.S. Pate

Fig 2.26 and 2.27

Symbiosis is a gentler form of robbery than parasitism, and some groups of plants have developed symbiotic relationships with fungi or bacteria which are harboured in the roots, often in special tissues. The symbionts are able to fix atmospheric nitrogen which becomes available to the host plant, while the symbiont receives water and other nutrients in return. Fig 2.26 shows bacterial nodules on the roots of a native legume, *Viminaria juncea*. Such nodules are normally produced on the roots of leguminous plants in the families Mimosaceae and Papilionaceae, and have been recorded in at least 300 Australian native species. Fig 2.27 shows so-called coralloid roots of the zamia palm *Macrozamia riedlei* which harbour a blue-green alga in this case and also have a nitrogen-fixing role. Nodulation similar to that of legumes but involving an actinomycete fungus occurs in species of *Casuarina*.

Byron Lamont

Fig 2.28

Proteoid roots are special organs produced by members of the family Proteaceae which includes *Banksia, Hakea, Grevillea, Dryandra, Isopogon* and *Petrophile*. They consist of dense clusters of hairy rootlets, which form a 2–5 cm thick mat at the soil surface and are most prolific where there is decomposing litter. Their prime function is to enhance nutrient uptake but they probably have a major role in water uptake as well. They are therefore an adaptation designed to confer advantage in poor soils.

Fig 2.29

Sclerophylly, the character of hard leaves, is a very marked feature of Australian plants, and has come to be interpreted as a response to soil poverty and lack of nutrients. Soft leaf tissue is relatively demanding in nutrients, and is replaced as soon as possible by hard lignified tissue. In many shrubs such as this *Isopogon dubius*, the leaves are not only hard, tough and brittle but are deeply divided with narrow lobes ending in sharp points.

Fig 2.30

Another strategy is to get rid of leaves altogether, with the stem taking over the photosynthetic function. The needle-like foliage of sheoaks *(Casuarina)* consists of thin jointed stems. The actual leaves are represented by minute scales at each joint.

Fig 2.31

Leafless plants are particularly common in the pea family, the Fabaceae. In the case of *Brachysema aphyllum* the organs which appear to be leaves are flattened green stems, shown by the fact that they branch and bear the flowers at nodes. These are known as *cladodes*. They remain sclerophyllous, without developing the succulent character found in leafless plants in other parts of the world, for example in the cacti and euphorbias. Succulence is very little known in Australia; this is attributed to nutrient deficiency.

Western Australian Herbarium

Fig 2.32

In the genus *Acacia* sclerophylly has been attained, in most Australian species, by a strategy of suppressing the normal leaf blade in favour of an organ called a phyllode formed by lengthening and broadening the petiole or leaf stalk. The strap-shaped 'leaves' with which we are familiar in Australian acacias (*A. rostellifera* above) are in fact phyllodes. This can be proved by reference to *Acacia* seedlings.

Fig 2.33

Acacias occur world wide and outside Australia normally have bipinnate leaves with soft, narrow pinnules, each leaf subtended by a pair of spines. Only a few Australian species exhibit these forms in the adult plant, though all Australian species have bipinnate leaves in the seedling. The illustration above is of a seedling of *Acacia farnesiana*, a pantropical species occurring in Australia which has bipinnate leaves and spines in adult life. There will therefore be no change from the seedling to the adult form.

Fig 2.34

In *Acacia subcaerulea* the first few seedling leaves (usually 4) are bipinnate. Then commences a progressive lengthening and broadening of the leaf-stalk until finally the pinnae appear only as a remnant at the tip, and in the next succeeding leaf will be discarded altogether leaving the phyllode. All subsequent foliage will consist of phyllodes.

FIG 2.35

In *Acacia holosericea* a sudden jump to phyllode is made after the first four bipinnate seedling leaves, without the intermediate stage. All species of *Acacia* grown in the King's Park nursery started life with bipinnate leaves, changing later to the adult phyllodal form. This repeats in each generation the evolution of the genus.

FIG 2.36

Spinifex is a sclerophyll grass, a growth form uniquely evolved in Australia. This photograph shows a clump of *Triodia scariosa* cut away to show the structure of the hummock. Numerous stolons arise from the root and intertwine. The leaves are tightly rolled into rigid cylinders terminating in sharp points which readily break off on contact and become embedded in human or animal flesh, causing painful festering. Normal gramineous inflorescences are produced in summer following adequate rainfall.

FIG 2.37

Since land clearing for agricultural purposes, remaining native vegetation has been increasingly affected by salinisation. Moisture derived from rainfall is no longer fully utilised by the plant cover and seeps into the bottom lands carrying salt held in the subsoil. This results in death of existing vegetation and its replacement by more salt-tolerant forms.

FIG 2.38

Stands of large trees may be affected and die off, even without obvious signs of surface salt. The problem in its acute form is confined to bottom lands where seepage takes place.

annum. Nowhere else could this happen under comparable climatic conditions.

On the other hand the deep pallid zone creates problems because it stores salt. Salt particles are carried inland on the winds from the ocean, come down with the rain and are washed into the soil. In the drier interior, before land clearing, native vegetation more or less used up the whole season's rainfall each year by transpiration. Except for occasional flash floods there was no run-off, no seepage, no streamflow, so that there was nothing to wash the salt out of the soil, and it accumulated. Calculations show that at the present rate of accretion, salt has been accumulating for 10 000 years — a significant date as it takes us back to the end of the last ice age. This does not apply to the high rainfall areas in the western Darling Range and the karri forest where there is an excess of rainfall causing regular stream flow and leaching of the soil's salt content. In the drier areas land clearing has substituted crops and pasture for native vegetation. The year's moisture supply is no longer fully used, and the excess seeps to the bottomlands carrying some of the accumulated salt with it. The result is creeping salinisation of bottomlands, death of any remaining native vegetation and inability to grow planted crops. The salt lakes of the interior have originated from the same source of salt and accumulate it after it is carried into them by flash floods.

However salinity like nutrient deficiency is an agricultural problem. Native vegetation growing upon pallid zone with accumulated salt storage shows no evidence of being adversely affected. Specialised and highly adapted native plants colonise the saline flats around salt lakes and pans, and only the hypersalinity of the lake bed defeats them.

6 Fire

We are all aware that most Australian vegetation becomes very inflammable during dry seasons, so that bushfires are a well-known occurrence. Repeated fires are likely to have a marked influence on vegetation, eliminating sensitive species, encouraging the spread of resistant species and perhaps modifying vegetation structure. It is therefore one of the natural ecological factors.

Inflammability and exposure to fire vary widely. A few plant communities such as mangrove swamps for example are never burnt, others only rarely. In a previous book[16] the writer suggested the following classification.

a *Subject to frequent fires* — Sites receiving fires set by man or by lightning at intervals from 1 to 20 years.

b *Subject to occasional fire* — The mean interval between fires is in excess of 20 years. Fires spread into such sites from adjacent more fire-prone areas, or are set by lightning and burn a small area before going out, or thirdly can be extensive following the rare seasons when there has been a lush growth of understorey plants.

c *Not normally subject to fire* — As mentioned above, mangroves; some coastal salt marshes and reed swamps; samphire communities round salt lakes; and the islands off the southwest coast of the State which could not be reached by Aborigines and remained undisturbed until recent times.

Fire in forest or woodland is always dependant on the inflammability of the ground layer. Fire will not travel through tree crowns unless there is a blazing understorey.

In the Northern Province the prevalence of dense grassy ground layers makes almost all vegetation fire prone, and under Aboriginal occupation the country was regularly burnt off in patches. Aboriginal practices have been documented in the Northern Territory[17]. Dry season burning has been continued by pastoralists. In the Eremaean Province spinifex country was also regularly burnt by Aborigines who would set fire to the grass hummocks one by one to drive out and catch any small creatures harboured inside. Although desert Aborigines have now left the area, satellite imagery shows a high proportion of such country to be fired by lightning each year. In the southern half of the province, the opposite obtains; mulga low woodland is only rarely burnt, after an unusually favourable season has produced an abnormally dense growth of the annuals which form the ground layer. The mulga is fire tender and easily killed by such fires. In the Southwest Province almost all vegetation must be considered fire prone except for the succulent halophytic communities of saline areas, and some islands off the coast. All forests are fire prone owing to their dense shrubby understoreys. Kwongan is all very inflammable, but eucalypt woodlands become less subject to fire in an easterly direction as their understoreys become more sparse. Woodland in the Interzone, like the mulga, is only

FIG 2.39
Dept of Conservation and Land Management

During the dry season, Australian vegetation becomes extremely inflammable, not only because of the dry weather but also because of the sclerophyllous leaf tissue. Far into the remote past fires have been set by lightning. On the arrival of Aboriginal man in Australia the fire pattern was intensified by deliberate burning off to drive out game and to keep the country open. In European times fires still occur either accidentally or caused by firebugs. In forest areas a policy of 'fuel reduction burning' is practised, to burn when fires will be light and cause minimum damage. Most Australian vegetation is highly adapted to live with fire as an ecological factor.

FIG 2.40
A bush fire normally scorches or burns all foliage and kills smaller stems of shrubs and young trees. Following the fire *Eucalyptus* trees resprout from dormant buds on the trunk and branches; this is called epicormic growth. Blackboys in the foreground are shown resprouting from the leading bud. Shrub species may either resprout from the rootstock or regenerate from seed.

FIG 2.41
The kwongan is frequently swept by fire, which it is well adapted to withstand in spite of the desolate aspect immediately afterwards. In this case at Eneabba, the blackboys *(Xanthorrhoea drummondii)* are already sprouting from their apical buds. The small trees of *Eucalyptus todtiana* will resprout from the stem and branches, many of the other woody plants will resprout from the rootstock, the remainder regenerate from seed.

FIG 2.42

Liberation of nutrients by the fire and temporary removal of competition usually lead to a flush of growth and flowering. In this kwongan north of the Moore River, banksias and eucalypts have resprouted, while *Beaufortia squarrosa* has regenerated copiously from seed into a spectacular flowering.

FIG 2.43

In drier inland areas, burning of kwongan may be followed by a regeneration of short-lived fire weeds along with the climax species. Burning of an *Acacia-Casuarina* thicket here has caused a proliferation of native poplar *Codonocarpus cotinifolius* which will outgrow the thicket and flourish for a few years.

FIG 2.44

Because of the danger to human life and property, bush fires are a matter of anxiety. This sign erected by the Bush Fires Board however appears to offer a strange recipe for their suppression, which in this case seems to have been all too successful!

exposed to fire during rare good seasons, and the same applies to the vegetation of the Nullarbor Plain further east.

Professor Sylvia Hallam has published collected accounts by early explorers and settlers of Aboriginal burning practices in the southwest, showing that these amounted to regular burning off much as in the Northern Province[18]. The burning was done in controlled sections as an aid to hunting, and the young plant growth which followed both attracted and nutritionally favoured kangaroos and their reproduction. This could be regarded as a form of farming the country. Recorded accounts apply to the coastal belt from Perth to Albany, but there is no information on Aboriginal fire practices further inland.

As may be expected, plants in fire-prone vegetation are highly adapted to withstand or to recover from fire, whereas they may not be where fires are a rare occurrence. The karri tree *(Eucalyptus diversicolor)* which is found in the south coastal area of high rainfall is resistant to light ground fires but may be killed or badly damaged by crown fires, whereas the jarrah *(E. marginata)* shows a high degree of ability to recover from the most severe fires. The salmon gum *(E. salmonophloia)* which is one of those in the 'subject to occasional fire' class is markedly fire tender as is the mulga *(Acacia aneura)* as remarked above.

The sclerophyll shrub communities, which occur either as kwongan or as understoreys in forest and woodland, regenerate rapidly after fire. Accumulated vegetable matter both in the standing crop and in surface litter is reduced to ash which when washed into the soil has a temporary fertiliser effect and stimulates a flush of growth. Studies have shown[15] that 60% of kwongan plants resprout from the rootstock after fire. The remainder of these are *obligate seeders* and must depend upon regeneration from seed for their survival. An adaptation to assist this process is the habit of *bradyspory*, meaning that seed formed each year is not shed but is retained until the plant is killed by fire or dies for some other reason, after which the seed vessels open and the seed is scattered. Banksia species growing in kwongan are all bradysporous whereas species in the humid southwest are not and shed seed annually. The storing of the seed by bradysporous species greatly increases the quantity available for regeneration after fire.

BIBLIOGRAPHY

1 Tate, R., 1890. *A Handbook of the Flora of extratropical South Australia*, Adelaide.
2 Woodward, B.H., 1900. Guide to the contents of the Western Australian Museum (with zoogeographical provisional sketch map of Western Australia). W.A. Museum, Perth.
3 Clarke, E. de C., 1926. Natural Regions in Western Australia. *Journ. Roy. Soc. West. Aust.* 12: 117–32.
4 Gardner, C.A. and Bennetts, H.W., 1956. *The Toxic Plants of Western Australia*. W.A. Newspapers, Perth.
5 Burbidge, N.T., 1960. The phytogeography of the Australian region. *Aust. J. Bot.* 8: 75–211.
6 Beard, J.S., 1969. The natural regions of deserts in Western Australia. *J. Ecol.* 57: 677–712.
7 Beard, J.S., 1975. The Vegetation of the Nullarbor Area. Vegetation Survey of W.A. 1:1 000 000 series, Sheet 4. Univ. W.A. Press, Nedlands.
8 Beard, J.S., 1980. A new phytogeographic map of Western Australia. *W.A. Herb. Res. Notes* 3: 37–58.
9 Beard, J.S. and Sprenger, B.S., 1984. Geographical Data from the Vegetation Survey of Western Australia. *Veg. Surv. W.A. Occ. Paper No 2*, Vegmap Publications, 6 Fraser Road, Applecross.
10 Bagnouls, F. and Gaussen, H., 1957. Les climats écologiques et leur classification. *Annls. Geogr.* 66: 193–220.

11 Geological Survey of Western Australia, 1975 Geology of Western Australia. *W.A. Geol. Surv. Mem. 2.*
12 White, M.E., 1986. *The Greening of Gondwana.* Reed Books, Sydney.
13 Van de Graaff, W.J.E., Crowe, R.W.A., Bunting, J.A. and Jackson, M.J., 1977. Relict early Cainozoic drainages in arid Western Australia. *Z. Geomorph. N.F.* 21 (4): 379–400.
14 McNamara K.J. and Scott, J.K., 1983. A new species of *Banksia* (Proteaceae) from the Eocene Merlinleigh Sandstone of the Kennedy Range, Western Australia. *Alcheringa* 7: 185–193.
15 Pate, J.S. and Beard, J.S., 1984. *Kwongan, the Plant Life of the Sandplain.* Univ. W.A. Press, Nedlands.
16 Beard, J.S., 1984. Aeolian Landforms. part 2 of Geographical Data from the Vegetation Survey of W.A. by J.S. Beard and B.S. Sprenger, *Veg. Surv. W.A. Occ. Paper 2*, with map 1:3 000 000. Vegmap Publications, 6 Fraser Road, Applecross.
17 Haynes, C.D., 1985. The pattern and ecology of *munwag*: traditional Aboriginal fire regimes in north central Arnhemland. In *Ecology of the Wet-Dry Tropics*, eds. M.G. Ridpath and L.K. Corbett. *Proc. Ecol. Soc. Aust.* 13: 203–214.
18 Hallam, S.J., 1975. *Fire and Hearth.* Aust. Inst. Abor. Stud. 58, Canberra.

3

THE SOUTHWEST PROVINCE AND SOUTHWESTERN INTERZONE

In the remaining chapters of this book the vegetation will be illustrated region by region from south to north, taking the natural regions shown in Fig 2.5 in their appropriate provinces, the Southwest Province in this chapter, the Eremaean Province in Chapter 4 and the Northern Province in Chapter 5. This will take the reader on a progression from the most humid forest country of the far south through into the desert and on north into the partly more humid Kimberley.

The Southwest Province comprises the relatively well watered southwestern portion of the State which experiences a mediterranean-type climate with winter rain and a dry summer. It is the only part of the State to have been closely settled and in consequence much of the original native vegetation has been cleared for agriculture, industry and urban use. Sixty-five per cent of the land in the province has been transferred to private freehold and has thus mainly been cleared or is liable to be. Clearing has affected the wet, forested areas less than the drier interior which is suited to cereal cultivation, and ranges from 31% of the karri forrest sub-region to 93% of the wheatbelt region.

The vegetation of the province varies from tall forest ranking among the tallest in the world, on the deepest soils under highest rainfall, to heath on the driest and poorest sandy soils. Forests, woodlands and mallee are eucalypt-dominated while the shrublands are heterogeneous with Proteaceae and Myrtaceae providing the most conspicuous elements. There is a complete absence of natural grasslands.

3.1 *Southwest Forest Region*
DARLING BOTANICAL DISTRICT
This region typically consists of forest country with related woodlands, in the southwestern part of the province. It is divided into four subregions or botanical subdistricts.

3.1A *Karri Forest Subregion*
WARREN BOTANICAL SUBDISTRICT
Tall forest of karri *(Eucalyptus diversicolor)* on deep loams, forest of jarrah-marri *(E. marginata-E. calophylla)* on the leached sands. Extensive paperbark *(Melaleuca)* and sedge swamps in valleys.

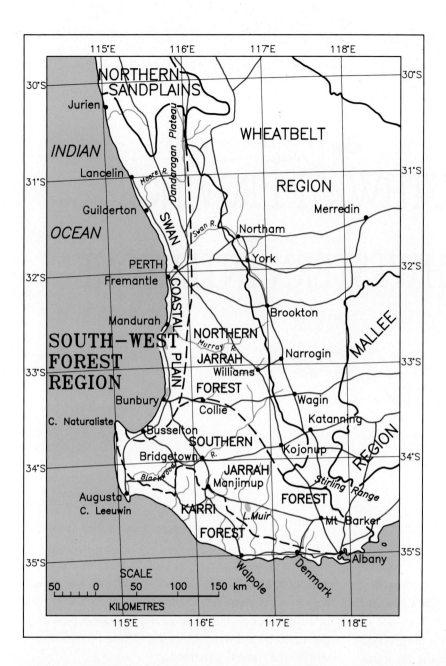

Climate: Moderate mediterranean; winter precipitation ranges from 650 to 1500 mm per annum, essential feature is short dry season of only 3–4 dry months.
Geology: Archaean granite and infolded metamorphic rocks of the Yilgarn Block.
Topography and soils: Dissected undulating country of small relief, hardsetting loamy soils alternating with leached sand soils.
Area: 8323 km², 31% cleared.
Boundary: The northern boundary is drawn where *E. diversicolor* ceases to be a significant component.

Although the vegetation of this subregion reaches its climax in the magnificent karri forest, which gives its character to the whole region, karri (*Eucalyptus diversicolor*) is by no means universal and in fact has a very patchy distribution. The reason is that forest trees of such size and luxuriance can only grow on the most favourable soils, called 'karri loams'. In Chapter 2 we have seen that types of soil occur linked in catenary sequences, and karri loams only occupy a portion of the catena. Karri forest country has been subject to clearing for agriculture only to a minor extent, fortunately for the forest, because the soils are agriculturally unproductive. They have a very good physical structure, which favours tree growth, but have been shown to require additions of phosphorus and zinc to support productivity under cultivation[1]. Attempts were made to found soldier-settlements in the forest after both world wars but were unsuccessful. The accompanying extract (Fig. 3.1) from the vegetation map of F.G. Smith[2] covering the Manjimup-Pemberton area shows the pockets of agricultural land uncoloured and also reveals the very patchy mosaic of karri forest (purple) and jarrah forest (brown).

Typical karri soil has a reddish-brown loam or sandy loam surface horizon changing gradually to red clay at about 50 cm depth. Other soils in the area unfavourable to karri are more leached, often expressed by paler colour, and may be more sandy or contain ironstone gravel. Poor internal drainage and waterlogging are also unfavourable factors. Because of these limitations, karri has a central node of abundance between the Donnelly and Shannon Rivers in an east-west direction, and from behind the coastal dunes inland for some 50 km. The Donnelly River approximately marks the line of the Darling Fault. West of it the geological formation changes to sedimentary rocks, and resulting soils are unfavourable to karri which is absent from the Blackwood Plateau but reappears in small patches in the hilly area of Margaret River-Augusta. Most of these patches are on old, highly weathered coastal dunes, some are on alluvium or on young soils along creeks. A large patch at Karridale was logged at the turn of the century and converted to farmland, except for a portion of Boranup which regenerated successfully and has recently been made into a national park. Karri occurs occasionally along the south coast on the dune systems, provided the soil is old and highly weathered, but many of these patches are actually of *Eucalyptus cornuta* (yate) which looks similar. East of the Shannon the concentration of karri declines rapidly. Much of the country is swampy and covered with reeds and paperbarks. Here the karri is joined by the two tingles, yellow tingle *(Eucalyptus guilfoylei)* mainly inland from Broke Inlet, and red tingle *(E. jacksonii)* in the

FIG 3.1

Reproduction of part of vegetation map of the Pemberton area by F.G. Smith, 1972. Agricultural and other cleared land is uncoloured, karri forst is coloured purple and jarrah forest brown. These are two principal types of forest in this area of the lower southwest which has a high rainfall and is more suited to forestry than farming. Karri is *Eucalyptus diversicolor*, and jarrah *E. marginata*. These are the forest dominants but do not always form pure stands. Marri *(E. calophylla)* and smaller trees are frequently present in addition.

FIG 3.2

Karri *(Eucalyptus diversicolor)* is the tallest and most magnificent of the forest species. Trees may reach 80 m in height, rivalling the Californian redwoods and other giant eucalypts of Victoria and Tasmania as the world's tallest trees. Such outstanding growth is only possible on the soils of best physical properties, known as karri loams, not necessarily fertile in the agricultural sense. Jarrah and other forest types occur on the poorer soils.

FIG 3.3

Karri has a smooth flaky bark giving it a dappled appearance, and will grow to about 2 m in diameter and about 400 years of age. Such trees have a considerable aesthetic and tourist value, and outstanding specimens are marked out for visitors. This row of four splendid stems is called 'The Four Aces'. Their occurrence in a straight line is attributed to having started life long ago along the burnt trunk of a fallen tree after a bush fire.

Fig 3.4

Forestry is the principal activity in this region where almost all forested land is State Forest and under the control of the Department of Conservation and Land Management which allocates coupes for felling annually according to a calculated quota system in balance with the productive capacity of the forest.

Fig 3.5

In the felling operation, certain selected stems are left as seed trees. After removal of millable logs to the sawmills, and waste logs for woodchipping, the coupe is burnt and the heat of the fire causes release of seed from capsules held on the seed trees. This rain of seed germinates the next rainy season to provide a new crop of trees for the future. This replicates the natural process of regeneration after fire which has been going on for thousands of years.

Fig 3.6

A crop of seedlings quickly establishes itself and grows on into sapling and pole stages. This young crop is 9 years old and was estimated to have numbered initially 45 000 seedlings per hectare. The crop thins itself by natural mortality. Understorey plants regenerate satisfactorily at the same time.

Arthur Fairall

Fig 3.7

Karri loam has a good appearance, and early settlers were led to believe that magnificent forest trees must indicate magnificent soil for farming. Attempts were made to found soldier settlements in the forest after both world wars, the land being cleared by the traditional method of ring-barking. Some desolate relics of this process are still to be seen. The soldier settlements were a failure, but the resulting cleared land in the district has since become more productive with the increase in scientific knowledge and improvement in methods.

Fig 3.8

Other giant trees occur in mixture with karri in the south from Broke Inlet to Nornalup. These are the tingles, red tingle seen here *(Eucalyptus jacksonii)* and yellow tingle *(E. guilfoylei)*. The white-barked tree in the rear of this photograph is a karri, the rest are tingle. These trees were probably more widespread in the distant past when rainfall was more plentiful elsewhere.

FIG 3.9
The Gloucester Tree at Pemberton. In the early days the forest officers were faced with a problem in fire suppression, as lookout towers to watch for and locate fire outbreaks in the dry season could not be built of sufficient height to rise above the canopy of the giant karri trees. The problem was solved by using selected trees on high points as lookouts, by climbing them and installing a peg ladder up the trunk and a cabin on the top. Aerial surveillance has since replaced this system, but the Gloucester Tree survives as a local attraction.

FIG 3.10
The top of the Gloucester Tree with its lookout cabin. The tree was originally stripped to a bare pole but has sprouted epicormic growth. Until recently the tower was manned throughout the dry season, reporting the compass bearing of any tell-tale smoke to forestry headquarters in Manjimup.

FIG 3.11
Another localised tree of the south coast is the highly ornamental red-flowering gum, *Eucalyptus ficifolia*. It occurs naturally on coastal dunes near the mouth of the Bow River but has been widely cultivated in many countries of the world. In Perth it tends to suffer from disease, but it has become one of the most popular street trees in Cape Town and elsewhere.

Walpole area. These are also giant trees, distinguished by their rough bark, having a very restricted distribution. It seems probable that these are relict species, formerly more widespread when climate was more favourable. Similarly, red-flowering gum *(E. ficifolia)* has a restricted habitat on the south coast on dune sands. Going eastward, karri remains common on the hills around Denmark and Torbay, but the flatter country is unfavourable to it, and only small patches are seen. However karri reappears in the Porongurup Range, adding much beauty to the national park; and there are two patches on a plain further east to the north of Mt Manypeaks. All these outliers are within the proper climatic zone for karri but restricted by the lack of suitable soil. The distribution of karri and karri loams in the landscape is shown diagrammatically in Fig. 3.12. Catena A applies in the central node from the Donnelly River to Walpole while Catena B applies to the country east of Walpole. The diagrams are not literal representations from measured profiles but show a typical arrangement in each case. The segments of the profile are not to scale and may not occur precisely in this manner on every slope. In Catena A the sequence begins with reed swamp in the bottomland or alternatively paperbark low forest, or there may be a creek lined with Warren River cedar *(Agonis juniperina)*. As the soil becomes better drained, karri appears, often towering up at the edge of the swamp, and persists up slope until on middle to upper slopes it becomes mixed with marri *(E. calophylla)*. On upper slopes if the soil becomes more sandy, or laterite appears, karri drops out and jarrah *(E. marginata)* becomes co-dominant with marri. On summits where the soil may be leached sand overlying clay and is poorly drained the jarrah declines to open stands of short crooked trees, giving way in extreme cases to shrubland of *Pultenaea reticulata*.

Catena B is to be seen east of Walpole where karri occurs either on alluvium or on red loam surrounding a granite outcrop but typically not on the lower and middle slopes where sandy and lateritic soils are found. These carry jarrah forest or jarrah-casuarina *(C. fraseriana)*, or in extreme cases on leached white sand, casuarina low forest. *Eucalyptus megacarpa* appears between the karri forest and the rock outcrop. This catena is particularly well developed in the Porongurup National Park.

Karri forms tall forest with a general canopy level of about 70 m. Below this at about 10 m is a scattered layer of *Agonis flexuosa, Casuarina decussata* and banksia and at about 3 m a continuous stratum of soft-leaved plants such as *Trymalium floribundum, Chorilaena quercifolia, Hovea elliptica* and *Acacia* spp. The lack of sclerophylly in the understorey of the karri forest is one of its most striking features. The ground cover consists of many shrubs and creepers and a very light cover of the grass *Tetrarrhena laevis* together with some mosses, liverworts and occasional epiphytic ferns.

A list of the karri forest flora below is taken mainly from McArthur and Clifton, Table 4.[1]

Canopy trees *Eucalyptus diversicolor, E. calophylla, E. marginata.*

Trees 10–30 m *Casuarina decussata, Eucalyptus megacarpa.*

Small trees < 10 m *Agonis flexuosa, Banksia grandis, B. verticillata, Persoonia longifolia.*

Tall shrubs 2-3 m *Acacia pentadenia, Paraserianthes (Albizia) lophantha, Bossiaea aquifolium, Chorilaena quercifolia, Pimelea clavata, Trymalium floribundum.*

Small shrubs 1-2 m *Acacia browniana, A. divergens, A. myrtifolia, A. pulchella, A. urophylla, Bossiaea linophylla, B. ornata, Chorizema ilicifolium, Crowea angustifolia, Hakea amplexicaulis, Hibbertia amplexicaulis, H. grossulariifolia, H. cuneiformis, H. serrata, Hovea elliptica, Hypocalymma cordifolium, Leucopogon capitellatus, L. propinquus, L. verticillatus, Phyllanthus calycinus, Podocarpus drouynianus, Sphaerolobium medium, Thomasia quercifolia, T. triloba, Tremandra stelligera, Xanthosia* spp.

Cycad, to 2 m *Macrozamia riedlei.*

Grass tree, to 3 m *Xanthorrhoea preissii.*

Herbaceous *Anigozanthos flavidus, Dampiera hederacea, D. linearis, Lepidosperma longitudinale, Lomandra* spp., *Opercularia hispidula, Orthrosanthus laxus, Patersonia umbrosa, Pteridium esculentum, Scaevola auriculata, S. striata.*

Creepers *Cassytha glabella, Clematis pubescens, Chorizema diversifolium, Hardenbergia comptoniana, Kennedia coccinea.*

Other vegetation besides karri in this region includes principally the jarrah forest which replaces karri forest on the poorer soils and originally probably covered about half as much area as karri. To avoid duplication the jarrah forest will be dealt with in the next section under the Southern Jarrah Forest Region. Other communities cover relatively minor areas. Jarrah is a very variable and plastic species in both habitat and form.

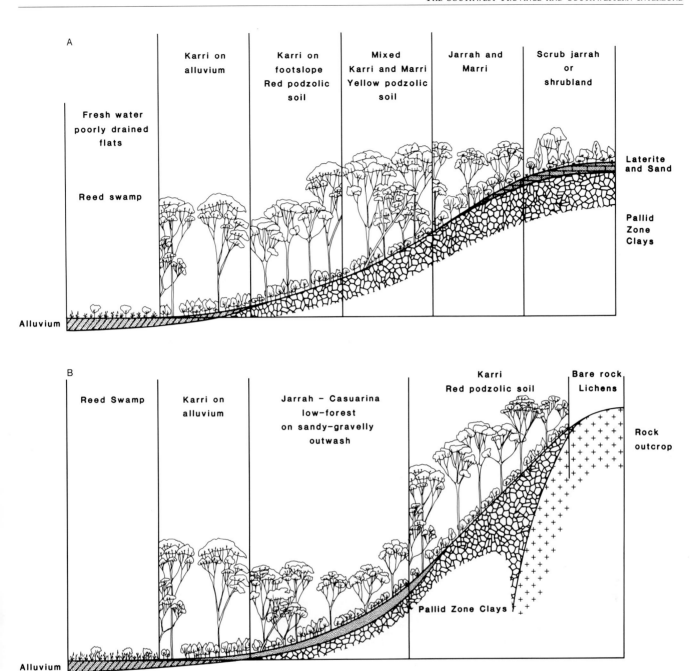

FIG 3.12
Diagrammatic catenary sequence showing relation between karri and other vegetation units. A—Central Manjimup–Pemberton area; B—Eastern area, east of Shannon River.

It grows on a wide range of sandy and lateritic soil types and ranges in height from a giant tree in association with the karri forest to a mallee in the Stirling Range with a suite of intermediate forms in forest, low forest and low woodland. As low woodland it occurs in Catena A on poorly drained sands on watersheds, and in association with *Banksia* spp. on sand ridges in swamps. As low forest it may occur on low ground and sandy soil in Catena B.

The paperbark tree *Melaleuca rhaphiophylla* forms a low forest in deep swamps, either completely covering the swamp or occurring as a ring round the edge. There is an understorey of rushes. Another paperbark, *M. preissiana*, appears as a small tree, scattered or in small groups, covering large areas of open flats of leached sand subject to seasonal flooding. The swamp banksia, *B. littoralis*, stunted jarrah, and *Nuytsia floribunda* may also be present, and blackboys (*Xanthorrhoea preissii*). The ground layer consists of heathy shrubs, merging into reed swamp in wetter parts. Reed swamps occur principally close to the coast where drainage has been obstructed by the coastal dune systems.

FIG 3.13

This view taken looking north from the summit of Mount Frankland shows the catenary organisation of the country. The pale coloured unforested swamps in the valley bottoms show clearly, surrounded by low forest on somewhat better drained land. Karri forest on the lower and middle slopes shows distinctly on the ridge in the left middle ground. Part of the granite outcrop of the summit of Mt Frankland shows in the foreground.

FIG 3.14

The paperbark tree *Melaleuca rhaphiophylla* forms a low forest in deep swamps, either completely covering the swamp or occurring as a ring round it. There is an understorey of rushes (Restionaceae) and sedges (Cyperaceae).

FIG 3.15

Another paperbark, *Melaleuca preissiana*, appears as a small tree, scattered or in small groups, covering large areas of open flats of leached sand subject to seasonal flooding. The ground layer is similar to that of the open reed swamps (Fig 3.17). Denbarker area.

FIG 3.16

A common type of swamp is the 'heath swamp' where shrubs are dominant. It occurs on sandy, seasonally wet ground where reeds are mixed with shrubs such as the Albany bottlebrush (*Callistemon speciosus*) flowering in left foreground, and *Agonis marginata* with white flowers. In summer the red-flowering *Beaufortia sparsa* will be conspicuous. Denbarker area.

FIG 3.17

Reed swamps occur where water-logging is more or less continuous. These consist principally of dense Cyperaceae and Restionaceae with few woody plants, though blackboys are persistent and occasional shrubs of *Agonis*, *Melaleuca* and so on are seen. The pitcher plant (*Cephalotus follicularis*) occurs in these swamps. Among the reeds the restionaceous *Leptocarpus tenax* seems to be generally dominant with species of *Anarthria* and *Restio*, and sedges such as *Lepidosperma* and *Mesomelaena*. Northcliffe near Mt Chudalup.

The swamps consist principally of dense Cyperaceae and Restionaceae with woody plants present in various communities. One phase, probably perennially wet, has scattered paperbark trees *(Melaleuca preissiana)* and small teatree *(M. densa)*, another has clumps of jarrah mallee *(Eucalyptus marginata)* or of scattered shrubs *(Beaufortia sparsa, Callistemon speciosus)* or blackboys *(Xanthorrhoea preissii)*. There is also a 'heath swamp' association on sandy, probably seasonally wet ground, in which the reeds are mixed with *Adenanthos obovatus*, *Acacia myrtifolia*, *Agonis flexuosa*, *A. juniperina*, *A. marginata*, *Andersonia caerulea*, *Beaufortia sparsa*, *Boronia spathulata*, *Callistemon speciosus*, *Cosmelia rubra*, *Grevillea brevicuspis*, *Kunzea ericifolia*, *Leucopogon obovatus*, *Oxylobium lanceolatum*, *Pultenaea reticulata*. Many of these shrubs are very showy in flower, especially *Beaufortia sparsa* which is summer flowering. Among interesting smaller plants in these swamps is the pitcher-plant *Cephalotus follicularis*. Among the reeds *Leptocarpus tenax* seems to be generally dominant; others identified are *Anarthria prolifera*, *A. scabra*, *Evandra aristata*, *Lepidosperma persecans*, *Leptocarpus scariosus*, *Lyginia barbata*, *Mesomelaena tetragona* and *Restio tremulus*.

The geography of the south coast is controlled by

FIG 3.18
Jarrah forest of *Eucalyptus marginata* occupies the poorer sandy and lateritic soils in the region. Under optimum conditions jarrah can be a giant tree like karri but is generally smaller and may be reduced to scrub or mallee in extreme cases.

granite outcrops which rise as low domes and form capes and headlands. From one headland to the next the coast extends in shallow curving bays backed by sand dunes. During the last glacial period which ended about 12 000 years ago, sea level was much lower than now, the coastline lay further off shore and the rivers cut down valleys accordingly. With the onset of the current warmer interglacial period sea level gradually rose drowning the river valleys. On reaching a position of approximate still-stand some 4000 years ago large quantities of sand were deposited on the strand line and blown into sand ridges which now occupy belts 2-3 km wide. In addition to forming dunes on low country this sand also mantles the granite headlands whose summits only may peep through it. With the passage of time most of the sand has become stabilised and vegetated but some is still mobile either due to continued accretion as along the shore of William Bay or due to disturbance or to long-term failure to stabilise on high ground, e.g. on Tower Hill and the Flinders Peninsula. In the former case the bare area contains the so-called 'Petrified Forest' where solidified root-channels are exposed by deflation and give the impression of being fossilised roots. Ponding of the rivers by the sand dunes has created lakes and swamps behind them.

Most of the vegetation is developed on a mantle of recently consolidated sand which is little weathered, poor in nutrients, and does not constitute a true soil.

Fig 3.19

The bark of jarrah is fibrous and is not shed annually, and is readily distinguished from that of karri (Fig 3.3). In eastern Australia jarrah would be classed as a stringybark; it closely resembles *E. obliqua* of Victoria.

Fig 3.20

Leaves and fruit of jarrah *(Eucalyptus marginata)*.

Fig 3.21

The common associated tree marri or red gum also has a fibrous persistent bark but is divided into small sections instead of long strips. In the general Australian classification this bark type is referred to as a bloodwood. Whereas jarrah produces a prime quality timber, marri is unfortunately useless for sawing. It has a tough timber of a light brown colour but the logs are full of circular gum pockets called ring shakes which limit its structural value. Rather than see marri wasted in the forest during logging operations, the logs are now converted into woodchips for export.

Fig 3.22

Leaves and fruit of marri *(Eucalyptus calophylla)*. The large size of the gum nuts is a particular feature.

Fig 3.23

On poor badly drained sites jarrah is reduced in size to form low forest of crooked branchy trees. The understorey in flower here consists mainly of *Agonis marginata*. Walpole.

FIG 3.24

On low-lying sites where the soil consists of pure white sand, the tree cover is of *Casuarina fraseriana* forming low forest. There is commonly rather little understorey, in general a selection of swamp species. Between Shannon and Walpole.

FIG 3.25

The granite outcrops which occur frequently throughout the region are interesting. With karri on their lower slopes and a fringe of *Eucalyptus megacarpa* around the rock, scattered plants grow on the outcrop rooting in crevices. Here on Mt Chudalup near Northcliffe, blackboys *(Xanthorrhoea preissii)* are common. Shrubs *Bossiaea linophylla, Eutaxia obovata, Leucopogon pulchellus* and *Verticordia plumosa* have been identified and the creeper *Kennedia glabrata*.

FIG 3.26

The ground layer in the karri forest is regularly swept by fire. Karri itself is relatively fire tender and may be damaged or killed by severe fires, but is normally unhurt by gentle ground fires. After a fire there is a flush of growth in the understorey in response to reduction of competition and release of nutrients, accompanied by a profusion of flowering. Here the scarlet flowered *Kennedia coccinea* is conspicuous. Warren National Park.

Fig 3.27

Flowers and foliage of the creeper *Kennedia coccinea*.

Fig 3.28

Another creeper conspicuous after fire is *Clematis pubescens*.

Fig 3.29

A common undershrub of the karri forest conspicuous when in flower is the legume *Hovea elliptica*. October is the best month for wild flowers in the forest; spring comes later in the south than it does for example in northern sandplains.

Fig 3.30

Yellow-flowering legumes are a conspicuous element at this time. *Bossiaea linophylla* is common in both karri and jarrah forest. Manjimup.

Fig 3.31

Flowers and foliage of *Bossiaea laidlawiana*. This species has broader leaves than *B. linophylla* and is a component of the karri forest. Pemberton.

Fig 3.32

A charming understorey plant is the southern cross, *Xanthosia rotundifolia*. The white flowers are arranged in groups of four in the form of a cross.

Fig 3.33

An understorey shrub which is of great botanical interest but has no showy flowers is *Podocarpus drouynianus*, because it is the last survivor of the Tertiary rainforest which once covered Western Australia. *Podocarpus* found today in all southern continents are normally great forest trees, but this one has survived by shrinking to a subshrub which resprouts from its rootstock. The plant is a conifer and after a bushfire produces small male and female cones. After pollination the female cone-stalk becomes fleshy and turns into a plum-coloured edible fruit, alas quite flavourless.

Fig 3.34

The dominant plant of coastal dune sands all along the south coast from Cape Leeuwin to Bremer Bay is the peppermint tree, *Agonis flexuosa*. If free from fire long enough to develop to maturity it forms a woodland of handsome trees. Bremer Bay.

Fig 3.35

More commonly the peppermint is regularly burnt and reduced to a sort of mallee form which may form dense groves or be scattered in heathland. A range of understorey shrubs has been recorded and is mentioned in the text. William Bay.

Fig 3.36

Cape Leeuwin is a projecting tongue of granite marking the meeting point of the Indian and Southern Oceans, the southwestern extremity of the Australian continent. The lighthouse is on a spray-swept rock, but the slopes behind exhibit wind-pruned coastal vegetation in which the grey-leaved *Olearia axillaris* is conspicuous but the wattles *Acacia littorea* and *A. cyclops* are dominant, or near the sea *Scaevola crassifolia* and *S. nitida*.

Fig 3.37

The south coast is an alternation of bold granitic headlands and sandy bays, all backed with the same windswept coastal vegetation and fronted by a sea of very deep blue. *Pimelea ferruginea* (centre) provides brilliant colour against the grey foliage of *Olearia axillaris*. Albany.

Fig 3.38

Certain plants are amenable to adapting themselves to the windswept coastal environment by submitting to dwarfing, notably *Banksia grandis*, elsewhere a forest tree but frequently seen along the coast reduced to a shrub. As is usual in these cases the floral parts remain of normal size. William Bay.

FIG 3.39

Burning of coastal shrubland may lead to a successional stage in which *Pimelea ferruginea* is conspicuous, giving colour effects worthy of the best of kwongan in other regions of the southwest. William Bay.

FIG 3.40

Apart from strand colonisers, the first permanently established plants above the beach are likely to be the blue-flowered scaevolas, *S. crassifolia* and *S. nitida*, adapted to this habitat by thick, fleshy leaves with a protective coating which serves both to reduce desiccation and to prevent penetration of salt. William Bay.

When this vegetation is burnt it does not regenerate rapidly and most commonly exists as heath and scrub-heath, depauperate shrubland communities which do not fully express the potential of the good rainfall. The climax on this calcareous sand appears to be *Agonis flexuosa* low woodland but this is only seen in low-lying or protected places. Inland of the sand mantle on soils of greater age there is an immediate change to forests, tall, medium and low, of karri and jarrah. Such forests may occur on older coastal sands where these have become well weathered.

Agonis flexuosa is dominant in a range of structural types from scrub to low woodland. More generally burning has reduced the woodland to an early successional stage of shrubs only. Such scrub heath contains scattered large shrubs of *A. flexuosa* and *Banksia grandis* with clumps of *E. angulosa* mallee. Smaller shrubs include *Acacia littorea*, *A. divergens*, *Adenanthos cuneatus*, *A. sericeus*, *Andersonia simplex*, *Anigozanthos flavidus*, *Anthocercis viscosa*, *Casuarina humilis*, *Dryandra sessilis*, *Hakea elliptica*, *H. oleifolia*, *H. prostrata*, *Hibbertia cuneiformis*, *Jacksonia horrida*, *Leucopogon parviflorus*, *Lysinema ciliatum*, *Melaleuca acerosa*, *Olax phyllanthi*, *Pimelea clavata*, *Senecio lautus*, *Spyridium globulosum* and there is a large reed *Loxocarya flexuosa*.

In the Torndirrup National Park the association described above is given a distinctive aspect by the presence of scattered large shrubs or small trees of

Banksia ilicifolia. These only occur on the sheltered north side of the main ridge.

Close to the sea and on mobile sand dunes and blow-outs the shrubland is characterised by the grey-leaved shrub *Olearia axillaris*. Other constituents are *Acacia cyclops, A. littorea, Calocephalus brownii, Exocarpos sparteus, Helichrysum cordatum, Hibbertia cuneiformis, Isolepis nodosa, Pimelea clavata, Spyridium globulosum*. *Scaevola crassifolia* and *S. nitida* may become very common on some sites close to the sea.

3.1B *Southern Jarrah Forest Subregion*
MENZIES BOTANICAL SUBDISTRICT
Jarrah forest on duricrusted plateau and on loam soils of valleys therein: marri-wandoo woodlands on drier laterite-free soils. Understoreys similar to those in the Karri Forest Subregion.
Climate: Warm mediterranean; winter precipitation 600-1200 mm, with 5-6 dry months per year.
Geology: Archaean granite of the Yilgarn Block.
Topography and soils: Duricrusted plateau on the Yilgarn Block, surfaced with ironstone gravels, dissected towards the east with hard-setting loamy soils.
Area: 26 572 km², 61% cleared.
Boundaries: W—Darling Scarp. N—arbitrary line across the jarrah forest. E—inland limit of marri-wandoo woodlands. S—bioclimatic boundary between dry and moderate mediterranean zones.

The subregion covers the southern part of the Darling Plateau and dissected country further inland as far as these are covered by marri-wandoo woodlands. It extends southeast to Cranbrook and reaches the sea between Albany and Bald Island.

The plateau is dissected by the Collie, Preston, Blackwood, Donnelly and Frankland Rivers, which cut across it from inland, but more and more towards the east it comes to form a broad level surface of poor drainage with one large permanent lake, Lake Muir, and innumerable small freshwater lakes and swamps. At the same time ironstone gravels become steadily less in evidence, being buried beneath sands which become paler and more leached. There is an area of neutral and acidic red soils around Bridgetown in response to gneissic country rock, but otherwise acidic yellow-mottled soils containing variable amounts of ironstone gravel are dominant. Towards the southeast the mosaic contains progressively larger areas of deep, leached sands overlying ironstone gravels and clays.

The most characteristic formation is jarrah forest, followed by woodlands of marri and wandoo *(Eucalyptus wandoo)*. It is estimated that jarrah forest originally covered three quarters of the whole area but agricultural settlement has made inroads into it, mainly in valleys which are free of laterite, leaving the forest intact on the lateritised higher plateau levels. Jarrah occurs in pure stands on laterite but is joined by marri where there is more superficial soil. These trees are able to survive by putting their roots down through the surface duricrust into the deep pallid zone beneath which acts as a water storage reservoir. Within the Darling Plateau jarrah and marri also grow on the valley soils — or did before clearing — so that the forest covers virtually the whole catena, being absent only from rock outcrops, stream banks, and some sand-filled valley bottoms on higher parts of the plateau. On the more favourable valley soils blackbutt *(Eucalyptus patens)* joins the forest dominants. On streambanks the river gum *(E. rudis)* is characteristic. In the upper reaches of valleys on the plateau bullich *(E. megacarpa)* forms woodland usually in narrow strips along the valley bottom. Its pale bark and straight stem gives it the appearance of young karri but it is readily distinguishable by its large fruits.

The trees in jarrah-marri forest vary in height from 20 to 30 m. There is a lower layer of small trees at about 7 m and a sclerophyll shrub understorey of larger and smaller shrubs reaching 1–2 and 0.5 m respectively. Leguminous creepers develop in open spaces, for example, where large trees have fallen, and are very prolific after fire. A species list is as follows[3].

Canopy trees *Eucalyptus marginata, E. calophylla*.

Small trees <10 m *Banksia grandis, Casuarina fraseriana, Nuytsia floribunda, Persoonia longifolia, P. elliptica, Xylomelum occidentale*.

Shrubs 1-2 m *Acacia browniana, A. extensa, Agonis marginata, A. hypericifolia, Bossiaea linophylla, B. ornata, Dryandra formosa, Hakea amplexicaulis, H. oleifolia, H. varia* (in wet places), *Hemigenia divaricata, Hibbertia amplexicaulis, Hovea elliptica, Hypocalymma angustifolium, Isopogon dubius, Leucopogon propinquus, L. verticillatus, Oxylobium* sp., *Petrophile diversifolia, P. serruriae, Podocarpus drouynianus*.

Cycad, grass trees *Macrozamia riedlei, Kingia australis, Xanthorrhoea preissii*.

Fig 3.41

The jarrah forest in its northern and southern regions is distinguished more by the nature of the understorey than by variation in the forest itself. In the southern region the understorey more resembles that in the karri region, in the northern it possesses a more strongly sclerophyll character. The picture above is of a rare example of virgin, unlogged jarrah forest at Jarrahdale in the Darling Range. Undergrowth in virgin forest is relatively sparse.

Fig 3.42

The soil profile beneath jarrah forest consists typically of a surface layer of massive laterite 1–2 m thick, seen here as brown pebbles, overlying a clay layer known as the pallid zone which becomes whitish at depth. Roots of jarrah are capable of penetrating the laterite to tap a moisture reservoir in the pallid zone. The whole profile however is deficient in nutrients and the habitat is a very harsh one. Brookton Road.

FIG 3.43

Rainfall declines inland from the coast so that the forest is progressively reduced in stature from about 30 m tall in the west to about 20 m in the east. Jarrah forest has a lower layer of sparse small trees of *Banksia grandis, Casuarina fraseriana* and *Persoonia longifolia*. There is a rich flora of understorey plants, which studies by the Forests Department have shown can be classified into associations related to site, but these are not readily recognised in the field. North Bannister.

Small shrubs <1 m *Acacia alata, Pimelea lehmanniana, Verticordia habrantha, Xanthosia rotundifolia.*

Herbaceous *Anarthria prolifera, Conostylis* sp., *Johnsonia lupulina, Lepidosperma angustatum, Pteridium esculentum.*

Creepers *Clematis pubescens, Hardenbergia comptoniana, Kennedia coccinea, K. prostrata.*

Jarrah declines to form low forest and woodland on shallow and poorly-drained soils. These become prevalent in the south and southeast of the region particularly along the watershed between the south coast rivers and the drainage of the Blackwood River, i.e. along and to the south of the Muir Highway, also on the plain between Mt Barker and Albany.

These poorer jarrah areas are commonly a mosaic of jarrah forest with poor, crooked trees up to 15 m and lower stands with thinner more crowded stems and increasing mixture of *Casuarina fraseriana*. The latter comes in more and more on bleached sands and may become a virtually pure stand on deep sand. *E. marginata* drops out in favour of the related species *E. staeri* on bleached sands over laterite on high rainfall sites near the coast. The understorey communities of these associations have not been separately studied but are believed to be similar. A general list for the mosaic was made in Reserve 18739 known as the Millbrook Road Reserve, near Albany[3].

Trees & small trees *Eucalyptus marginata, E. decipiens, E. staeri, Banksia attenuata, B. grandis, B. ilicifolia, B. verticillata* (in valley), *Casuarina fraseriana, Nuytsia floribunda, Persoonia longifolia, Xylomelum occidentale.*

Grass trees *Kingia australis, Xanthorrhoea preissii.*

Shrubs >1 m *Adenanthos cuneatus, A. obovatus, Agonis marginata, A. hypericifolia, Banksia brownii, B. coccinea, B. quercifolia, B. sphaerocarpa, Dryandra carlinoides, D. squarrosa, Hakea amplexicaulis, H. ferruginea, H. oleifolia, Isopogon cuneatus, I. formosus, I. sphaerocephalus, Kunzea ? recurva, Lambertia inermis, Melaleuca striata, M. thyoides, Persoonia teretifolia, Persoonia* sp., *Petrophile rigida.*

Small shrubs <1 m *Actinodium cunninghamii, Andersonia* sp., *Banksia goodii, B. repens, Beaufortia anisandra, Bossiaea linophylla, Burtonia scabra, Conospermum caeruleum, C. flexuosum, C. petiolare, Darwinia vestita, Franklandia fucifolia, Grevillea synapheae, Leucopogon verticillatus, Pimelea spectabilis, P. suaveolens, Sphaeolobium macranthum, Stirlingia tenuifolia, Synaphea* sp.

Herbaceous *Anarthria scabra, Cyathochaeta avenacea, Dasypogon bromeliifolius, Drosera* spp., *Velleia macrophylla.*

The marri-wandoo woodlands occupy the drier, laterite-free soils both east and west of the Darling Plateau. On the west they cover the steep slopes of the Darling Scarp. On the east, where the rainfall declines, they appear on the valley slopes while jarrah persists on duricrusted summits for some distance eventually giving way to wandoo and brown mallet *(E. astringens)*. In the valley bottoms one finds York gum *(E. loxophleba)* on the better-drained red soils and swamp yate *(E. occidentalis)* on grey alkaline clays. Most of this country has been cleared for farming and if it has not the understorey has been eaten out by sheep. As a result it is somewhat difficult to give more details of original plant associations. Stands of wandoo and brown mallet on laterite cappings are generally intact but normally have little or no understorey. In marri-wandoo woodland proper, jam *(Acacia acuminata)* and *Casuarina heugeliana* are present as occasional small trees, and may become locally abundant. *Banksia grandis* may also be seen. The most typical undershrubs seem to be species of *Gastrolobium* and *Oxylobium* belonging to the category of poison plants. Smith[4] listed the following shrubs from various sites:

Acacia nervosa, Beaufortia bracteosa, Calothamnus planifolius, Dryandra armata, D. nivea, Eucalyptus foecunda, E. incrassata, Gastrolobium bilobum, G. spinosum, Hakea lehmanniana, H. lissocarpa, Hypocalymma angustifolium, Hypolaena exsulca, Isopogon teretifolius, Leptospermum erubescens, Oxylobium capitatum, O. parviflorum, Petrophile squamata and *Trymalium ledifolium.*

Small woody plants such as *Lechenaultia formosa* and *Hibbertia* spp. may also be seen, while spring annuals are a feature of this association not seen in the jarrah forest. They include *Podolepis canescens, Ursinia anthemoides* and *Waitzia citrina.*

3.1C *Northern Jarrah Forest Subregion*
DALE BOTANICAL SUBDISTRICT
Jarrah *(E. marginata)* forest on ironstone gravels, marri-wandoo *(E. calophylla-E. wandoo)* woodlands on loamy soils, sclerophyll understoreys.

Climate: Somewhat drier than the Southern Jarrah Forest.

Geology, topography and soils: Essentially as in the Southern Jarrah Forest.

Area: 19473 km², 57% cleared.

Boundaries: E—where *E. accedens* and *E. astringens* replace *E. marginata* on lateritic residuals, *E. calophylla* retires from woodlands of mid slope, and *E. loxophleba* and/or *E. occidentalis* appear on lower slopes. S—ill-defined line where ironstone gravels become less prevalent and the understorey changes. W—Darling Fault.

This subregion occupies the northern portion of the Darling Plateau, east of the Darling Scarp. It overlies Archaean granite and metamorphic rocks and has an average elevation of about 300 m. The plateau is an ancient erosion surface capped by an extensive lateritic duricrust which has been dissected by later drainage. The general level is broken here and there by granite hills of unusual elevation. The principal summits are: Mt Gungin 410 m, Mt William 482 m and Mt Curtis 415 m close to the scarp; Mt Saddleback 575 m and Mt Wells 547 m near the Hotham River; a chain of hills in the centre including Mt Cooke 582 m and Mt Randall 525 m; and Mt Dale 548 m in the northeast. The plateau is dissected by streams rising locally, and by rivers originating in the interior which cut across it from east to west. The latter comprise the Swan-Avon System and the Murray River. In the east the plateau becomes more and more deeply dissected and eventually is broken up into isolated remnants.

The dominant soils are lateritic gravels consisting of up to 5 m or more of ironstone gravels in a yellow sandy matrix, and related lateritic podzolic soils with ironstone gravels in a sandy surface horizon overlying a mottled yellow-brown clay subsoil. These materials frequently overlie a pallid zone up to 30 m or more in thickness. Massive ironstone pavements are common on ridge tops and some slopes. The mid-slope gravels are those currently being mined as bauxite.

This region gradually tapers towards the north due to declining rainfall, and finally pinches out between New Norcia and Moora. For the same reason the jarrah forest which remains typical of the duricrusted plateau throughout becomes of progressively lower stature towards the north. North of the Great Eastern Highway it was mainly not considered of commercial quality for timber production and, except for one portion, was not included in State Forest, being surrendered to the demands of agriculture. However, portions are preserved in the John Forrest, Walyunga and Avon Valley National Parks. At its limits jarrah is joined by wandoo and powderbark wandoo *(Eucalyptus accedens)*. Jarrah forest terminates at Bakers Hill on the Great Eastern Highway and just short of the Calingiri turnoff on the Great Northern Highway. North of the Serpentine River, jarrah ceases to occupy major valleys incised into the plateau which then carry woodland of marri and wandoo. The latter as before covers the slopes of the Darling Scarp, and dissected, laterite-free country on the inland side of the plateau. Marri and wandoo are thus more widespread relative to jarrah than in the southern jarrah subregion. The major catena in the northern subregion comprises:

Alcoa of Australia

FIG 3.44

The lateritic gravels which underlie the northern jarrah forest are sufficiently rich in alumina to be mined for bauxite. At present two mining companies are active, and Western Australia has become the world's largest alumina producer. Stripping of the surface gravels is confined to richer pockets, but destroys the forest in the process. Government environmental controls however ensure that after mining the pits are rehabilitated and replanted with native trees and smaller plants. Cautious experiments have proved that it is possible to restore the original forest composition, and Alcoa of Australia now uses only indigenous jarrah forest species in its rehabilitation. This photograph shows a three-year old stand of mixed jarrah *(Eucalyptus marginata)* and marri *(E. calophylla)* in the foreground against a background of untouched forest and a mine pit which has been restored and already seeded with native species, at the Huntley mine site. In time, due to removal of the inhospitable surface laterite, site quality will be enhanced by this process.

FIG 3.45

The plateau which supports the northern jarrah forest is dissected by broad flat valleys filled with winter-wet sandy soil. These carry an open low woodland of the paperbark *Melaleuca preissiana*. Associated undershrubs are *Astartea fascicularis, Hakea ceratophylla, H. varia, Hypocalymma angustifolium* and *Pericalymma ellipticum*, with the reeds *Leptocarpus scariosus* and *Mesomelaena tetragona*.

a the open vegetation of granite rock outcrops which protrude through the laterite mantle;
b jarrah forest on the laterite plateau and screes descending from its edges;
c marri-wandoo woodland on the younger red soils of the scarp and the slopes of the deeply excavated small valleys;
d river gums and paperbarks along the watercourses in these.

There are also minor catenas on the plateau itself in the undulations of its surface running from the highest ground into the shallow valleys and these are separated into those of a higher-rainfall western sector and a drier eastern sector.

Rock outcrops

In the wetter western sector these may often have no special vegetation other than lichens on the rocks which are interspersed with jarrah and marri trees. Elsewhere the rocks are more open; characteristic species are lichens, tussocks of *Borya nitida*, shrubs *Grevillea bipinnatifida, Hakea elliptica, H. undulata*, small trees *Casuarina huegeliana*.

Outcrops of granite and gneiss are common on the scarp, and as with granite outcrops that occur on steep slopes within the Darling Plateau, these outcrops support a variety of mosses and lichens. Mosses on the granite surface include *Campylopus bicolor, Hedwigia ciliata, Bryum* sp. and *Tortella* sp. *Borya nitida* is often seen with the mosses. A lichen *(Siphula coriacea)* is also associated with the moss *Campylopus bicolor*. Other lichens that occur on the granite surfaces are *Cladia aggregata, C. ferdinandii, Cladonia zanthoclada, Parmelia* aff. *cheelii, P. conspersa* group, *P.* ? *dichtoma* and *P. subnuda*. The combined effect of extra water shedding and the dearth of tree shade leads to a proliferation, both in density and floristics, of low shrubs and herbs. *Calothamnus quadrifidus, Cryptandra arbutiflora, Grevillea bipinnatifida, G. endlicheriana, Thomasia glutinosa* and *Verticordia acerosa* often form dense shrub thickets.

Jarrah forest

The jarrah forest is one of the only two forest formations in Western Australia and is composed of trees averaging 25 to 30 m tall in the western sector, about 4 m less in the eastern. Stem density is 125 to 150 per hectare. Today nearly all stands have been logged so that many or most of the largest trees have gone and been replaced by a larger number of younger stems. Whereas the virgin forest orginally contained mainly large mature trees most stands now contain smaller immature trees. In addition many stands have been thinned by dieback disease caused by *Phytophthora cinnamomi*. Jarrah is the principal dominant tree, normally accompanied by marri in proportions varying from 50% downwards, and by blackbutt on valley soils and by wandoo and powderbark wandoo on northern and eastern drier sites. There is frequently a lower layer of small trees of 10-15 m including *Banksia grandis, Casuarina fraseriana* and *Persoonia longifolia*; and a ground layer of sclerophyll shrubs 1-2 m tall averaging 185 individuals to the hectare. J.J. Havel working for the Forests Dept published[5] a study of understorey associations in the forest based on a mathematical analysis of 320 sample plots. He found that 21 understorey associations could be distinguished. Three of these occur in swampy bottomlands and two on or around granite outcrops. These five give recognisable photo-patterns and could be mapped using aerial photography. The other 16 are not recognisable from the air and are even difficult to distinguish on the ground except by looking carefully for indicator species, though they are definitely associated with topographic positions. The swampy bottomland types, which are readily recognisable, are as follows:

Type A Winter-waterlogged leached acid sands over hardpan in broad heads of valleys. Scattered *Melaleuca preissiana, Banksia littoralis, E. calophylla, E. patens*, or largely treeless. Shrubs *Astartea fascicularis, Hakea ceratophylla, H. varia, Hypocalymma angustifolium, Pericalymma ellipticum*. Reeds *Leptocarpus scariosus, Mesomelaena tetragona*.

Type B Winter-wet leached acid grey sands in upland depressions. Jarrah, marri, scattered *Banksia grandis*. Shrubs *Conospermum stoechadis, Dasypogon bromeliifolius, Hibbertia polystachya, Leucopogon cordatus*. Reeds *Lepidosperma angustatum, Leptocarpus scariosus, L. tenax, Mesomelaena tetragona*.

Type C Moist to wet sandy loams along creeks and swamp margins. Mainly blackbutt with bullich, marri, jarrah, occasionally river gum and swamp banksia. Shrubs *Acacia alata, Agonis linearifolia, Astartea fascicularis, Grevillea diversifolia, Hypocalymma angustifolium*. Reeds *Lepidosperma angustatum, L. tetraquetrum, Leptocarpus scariosus, Mesomelaena tetragona*.

The other site types may be used to make up a general list of diagnostic understorey plants of the northern jarrah forest.

Shrubs 1-2 m *Acacia browniana, A. extensa, A. urophylla, Adenanthos barbigerus, Baeckea camphorosmae, Banksia attenuata, Bossiaea aquifolium, Casuarina humilis, Chorizema ilicifolium, Conospermum stoechadis, Daviesia decurrens, Diplolaena microcephala, Gastrolobium calycinum, Grevillea wilsonii, Hakea ceratophylla, H. cyclocarpa, H. lissocarpha, H. ruscifolia, Isopogon dubius, Lasiopetalum floribundum, Leptomeria cunninghamii, Pericalymma ellipticum, Leucopogon cordatus, Styphelia tenuiflora, Trymalium floribundum*.

Small shrubs >1 m *Dampiera alata, Dillwynia* 'sp.A', *Hibbertia lineata, H. polystachya, Hovea chorizemifolia, Hypocalymma angustifolium, Leucopogon capitellatus, L. oxycedrus, L. propinquus, L. verticillatus, Phyllanthus calycinus, Stirlingia latifolia, Synaphea petiolaris, Trymalium ledifolium*.

Cycads & grass trees *Kingia australis, Macrozamia riedlei*.

Climbers *Clematis pubescens, Kennedia coccinea*.

Herbaceous: *Caustis dioica, Lepidosperma angustatum, Leptocarpus scariosus, L. tenax, Mesomelaena tetragona, Patersonia occidentalis, P. rudis, Pteridium esculentum*.

Marri-wandoo woodland

Marri-wandoo woodland replaces jarrah forest on soils of the scarp and valleys where the deep-weathered zone has been stripped off so that they provide less water storage than the plateau.

FIG 3.46

As in the karri forest a ground fire encourages a burst of flowering activity. Species vary from place to place and in this case the display is provided by *Hibbertia polystachya*. As in the karri forest the creeper *Kennedia coccinea* can be prominent. Brookton Highway.

FIG 3.47

The most typical flowering plant associated with the jarrah forest is the blue leschenaultia, *Lechenaultia biloba*. Unnecessary confusion in spelling has been created by the decision of botanists that the botanical name should be spelt *Lechenaultia* even though the spelling Leschenaultia had been in use for years and become the common name of the plant, has been adopted in place names (Lake Leschenaultia, Leschenault Inlet), and was the way the original botanist that it is named after, Leschenault de la Tour, spelt his own name! There are at least 20 species of *Lechenaultia* in W.A., of a great variety of flower colours. Armadale.

FIG 3.48

The State's floral emblem, the kangaroo paw *Anigozanthos manglesii*, is a perennial herbaceous species which occurs casually throughout the Southwest Forest Region and can become locally abundant after bushfires or other disturbance. In recent years after clearing of the site, the cemetery at Gingin has become home to a colony of them which is an attraction for visitors.

FIG 3.49

The strange flowers of the kangaroo paw are adapted to be pollinated by birds. A visiting bird perches on the flower stalk and reaches for nectar held in a cup at the base of the flower. In so doing, the bird's forehead brushes the style and anthers which are presented at the top of the floral structure.

Fig 3.50
Granite rock outcrops occur frequently in the jarrah forest. Bare rock is covered with lichens making blotched patterns but mats of moss are able to build up on flat places or depressions, surviving in a dormant state from one rainy season to another. When wet these mats may be home to small flowering plants such as bladderworts *(Polypompholyx* and *Utricularia)*. Karagullen.

Fig 3.51
Tiny pink and white *Polypompholyx* in an oozing moss mat create nature's rock garden for a brief season in the spring. Karagullen.

Fig 3.52
During the last fifty years the jarrah forest has unfortunately become subject to attack by a root fungus, *Phytophthora cinnamomi*, which causes the malady known as 'jarrah dieback'. The trees stagnate and eventually die. Poorly drained sites underlain by an impermeable laterite hardpan are the worst affected and it now seems unlikely that the whole forest is at risk. Research on the disease is in progress. Brookton Highway.

Such sites carry a lower and more open formation than the forest. On the west side of the Darling Range tree height is 20-25 m, and density approx. 100 stems per ha. The two trees marri *(E. calophylla)* and wandoo *(E. wandoo)* constitute the woodland with occasionally *E. laeliae*. *Nuytsia floribunda* occurs as a smaller tree. The grass tree *Xanthorrhoea preissii* is common and conspicuous with 600 to 1000 plants per ha. *Macrozamia riedlei* is less common. Tall shrubs (1.2 to 1.5 m) include *Daviesia horrida*, *Dryandra sessilis*, *Hakea cristata* and *H. trifurcata*. Small shrubs of 45 to 60 cm typically include *Acacia pulchella*, *Dryandra nivea*, *Hibbertia hypericoides* and *H. montana*.

On the east side of the Darling Range the woodland tends to be lower and more open, and both *Acacia acuminata* and *Casuarina huegeliana* appear as small trees, sometimes locally abundant.

Below the 500 mm rainfall line jarrah disappears on laterite capped residuals, replaced by powderbark wandoo and brown mallet *(E. astringens)*. The understorey beneath these consists of sparse shrubs which include *Gastrolobium microcarpum* and *G. spinosum*, *Dryandra cirsioides* and other *Dryandra* spp., *Calothamnus quadrifidus*, *Leptospermum erubescens* and blackboys *(Xanthorrhoea preissii)*. At the poorest extremes this woodland opens out to a mallee-heath with *E. drummondii* conspicuous, while on the more favourable sites there is taller woodland of wandoo with occasional marri.

Gravelly slopes bear a woodland of wandoo and

Fig 3.53

Where the laterite cover has been eroded away on the Darling Scarp and in dissected country to the east of the main forest belt, the wandoo, *Eucalyptus wandoo*, replaces jarrah as the dominant tree, still usually in association with marri, *E. calophylla*. Wandoo has a white smooth bark peeling annually, placing it in the classification of gum trees. Contrary to popular belief, a gum tree is not just any eucalypt but one of the above specific bark type, and the early settlers correctly called wandoo white gum. Undergrowth is relatively sparse.

Fig 3.54

Bark of wandoo *(Eucalyptus wandoo)*. Otherwise called white gum, this species is a true gum tree in having smooth bark the outer layers of which peel annually, renewing the fresh pale colour beneath. Partial peeling gives a mottled appearance.

Fig 3.55

Leaves, buds and fruits of wandoo. Botanical classification in the eucalypts is based on buds and fruit, both of which may be necessary for accurate identification.

Fig 3.56

In the eastern part of the Darling Range, with declining rainfall, woodlands of marri and wandoo replace the jarrah forest. Trees are shorter and more branchy, more open, and the composition of the understorey changes. The shrub layer is less tall and more scattered, and there may often be a predominance of poisonous legumes of the genera *Gastrolobium* and *Oxylobium*, whose leaves are toxic to grazing animals. However not all have been shown to be poisonous and the small yellow-flowered species here, *Oxylobium drummondii*, appears to be innocent.

Fig 3.57

The grass tree or blackboy, *Xanthorrhoea preissii*, can be particularly common in marri-wandoo woodlands. Blackboys occur throughout Australia and there are several species in the west. They can be regarded as primitive lilies, being originally placed in the lily family, Liliaceae, and only recently accorded a family of their own, Xanthorrhoeaceae. Flowering is stimulated by bush fires, when the enormous inflorescence is covered by hundreds of tiny white flowers. The trunk is a cylinder with a central pith, surrounded by a mass of dead leaf bases.

Fig 3.58

Easily confused with wandoo and occurring in mixture with it on some soils is the powderbark wandoo *Eucalyptus accedens*. Usually a smaller tree than true wandoo and with different buds and fruit, it can be distinguished by an orange tinge to the bark. However the most positive distinction is that the bark rubs off as a white powder, when rubbed with the hand.

Fig 3.59

A tree seen east of the Darling Range in farming country of the Great Southern is the flat-topped or swamp yate, *Eucalyptus occidentalis*. It originally formed woodlands in pure stands or in mixture with other species on heavy alkaline clay soils. It has a very black fibrous bark and a spreading crown.

powderbark; the principal understorey species here is *Oxylobium parviflorum*. On less gravelly slopes there is marri and wandoo with smaller *Acacia acuminata* and *Casuarina huegeliana*. The principal undershrub may originally have been *Gastrolobium microcarpum*. York gum *(E. loxophleba)* normally replaces marri and wandoo in valley bottoms.

3.1D *Swan Coastal Plain Subregion*
DRUMMOND BOTANICAL SUBDISTRICT
Mainly *Banksia* low woodland on leached sands with *Melaleuca* swamps where ill-drained; woodland of tuart *(Eucalyptus gomphocephala)*, jarrah *(E. marginata)* and marri *(E. calophylla)* on less leached soils.

Climate: Warm mediterranean; winter precipitation 600-1000 mm with 5-6 dry months per year.

Geology: Mesozoic to recent sediments of the Perth Basin.

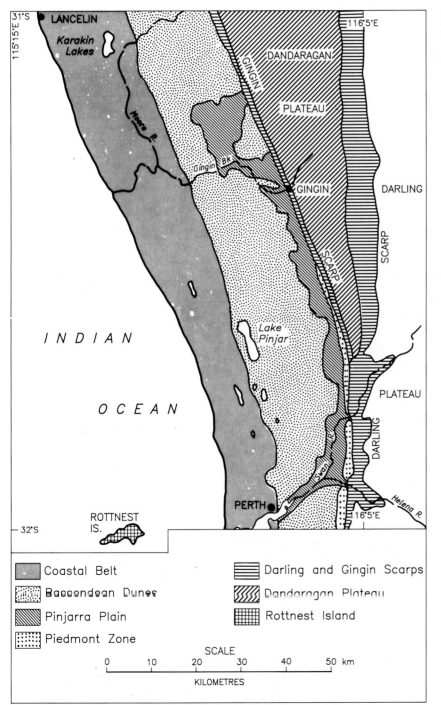

FIG 3.60

Physiographic diagram of the Swan Coastal Plain.

FIG 3.61

The Swan Coastal Plain is essentially a low-lying sandy plain of Quaternary deposits at the foot of the Darling Scarp, although it also includes the Dandaragan Plateau in the northwest. The most widespread and characteristic plant formation is the banksia low woodland comprising three species of *Banksia* with *Nuytsia* and *Eucalyptus todtiana*. It is at its most beautiful stage in November–December in those areas, principally between Perth and the Moore River, where the yellow morrison, *Verticordia nitens*, occurs. *Banksia attenuata* will be in flower at this time, as well as the occasional tree *Nuytsia floribunda* (W.A. Christmas tree) seen flowering in the centre of the above photograph. Moore River National Park.

Topography and soils: A coastal plain, low-lying, often swampy, with sandhills; soils mainly recent sands or swamp deposits. Also dissected country rising to the duricrusted Dandaragan plateau on Mesozoics, mainly yellow sandy soils.

Area: 14 637 km², 78% cleared.

Boundaries: N—northern limit of *Banksia* low woodland as above. E & S—geological boundary of the Perth Basin, mostly coinciding with the Darling Scarp but including a portion of the Dandaragan plateau in the north.

In addition to the Swan Coastal Plain itself this subregion contains a portion of the Dandaragan Plateau rising behind the plain in the northeast with an inland boundary along the Darling Fault where the geological formation changes from sedimentary to granite. It is thought that at one time the Dandaragan Plateau was probably continuous with the Blackwood Plateau in the south, south of Busselton, and that marine erosion has cut away a huge crescent shaped embayment. In the centre this erosion reached the Darling Fault and was halted there by the harder rocks, creating the Darling Scarp. Later, due to continual rises and falls of sea level during the Quaternary (roughly the last 2 000 000 years) the present coastal plain was formed by deposition of sediments on the eroded embayment. Successive periods of deposition have created some topographic complexity, clarified in Fig. 3.59 which is from the Geological Survey's notes on the Perth Geological Sheet[6]. The divisions shown apply equally to the whole Coastal Plain and are useful for understanding changes in vegetation.

1 Rottnest Island, together with Carnac and Garden Islands, has a rolling, sand-dune topography and reaches a maximum elevation of about 45 m. There are no permanent drainage channels though a number of swamps form in low inter-dune areas during winter and a few lakes occur towards the eastern end of the island. Benches around the coastline and around the margins of the lakes at 0.5, 1.5 and 3.0 m above mean sea level are believed to have resulted from higher sea levels of recent date. During the last glacial period the islands were joined to the mainland and have only been cut off for about 5000 years.

2 The **Coastal Belt** consists of two Quaternary dune systems, the Quindalup and Spearwood Dune Systems[7]. The Quindalup is the younger and is formed of both fixed and mobile sand dunes which trend north-south, while the Spearwood consists of dunes lithified to limestone and without any recognisable trend patterns. The dunes are piled up to heights of 60 m. There are some permanent lakes tending to occur in chains parallel to the coast which may represent old lagoons cut off by foredunes from a prograding shoreline. Salinity of these is very variable. Lake Richmond is permanent and fresh, Lakes Coolongup and Walyungup are intermittent and saline, Lake Preston and the chain of small lakes between it and Lake Clifton are saline, but Lake Clifton is fresh to slightly brackish. Lakes situated along the boundary between the coastal belt and the Bassendean dunes are all fresh.

The Swan River estuary and the Peel Inlet (with the Harvey estuary) are drowned valleys flooded by the same rise in sea level following the last glacial period which cut off the islands. The Peel and Harvey Inlets with the Murray and Serpentine Rivers form the largest estuarine system in the southwest, covering 115 km². They have a very small daily tidal range, a maximum of only 5% of the ocean tide. River flow is relatively small and salinity seldom drops below 10% for more than a brief period and may become slightly hypersaline in summer.

3 The **Bassendean Dunes** occur in a belt 15 km wide and consist of low, vegetated hills of quartz sand with numerous interdunal swamps and lakes. There is no organised drainage except where the various rivers cross the plain.

4 The **Pinjarra Plain** is an alluvial tract lying between the Bassendean Dunes and the Darling Scarp. It is from 1.5 to 25 km wide.

5 The **Piedmont Zone,** better known as the Ridge Hill Shelf, is a narrow strip of country 1.5 to 3 km wide along the foot of the Darling Scarp. It consists of coalescing alluvial fans deposited by streams losing grade at the bottom of the scarp and remnants of marine terraces.

6 The **Darling Scarp** rises steeply from the Ridge Hill Shelf to the Darling Plateau.

7 The **Darling Plateau** comprises the whole area to the east of the Darling Scarp and has been dealt with in the previous sections on the jarrah forest. It overlies Archaean granite and metamorphic rocks and has an average elevation of about 300 m.

The vegetation of Rottnest and Garden Islands is interestingly different from that of the adjacent mainland (or was prior to settlement). As these islands could not be visited by Aborigines who, in this part of Australia, had no canoes, they were not regularly burnt and therefore carried a climax vegetation of cypress pine *(Callitis preissii)* with some subclimax communities on special soils, areas exposed to the sea and so on. Owing to the use of Rottnest Island as an Aboriginal prison settlement from 1838 the cypress pines have been virtually exterminated but survive on Garden Island where they form a low forest of dense slender trees 5-6 m tall with little undergrowth. Destruction of the forest on Rottnest led to a population explosion of quokkas whose numbers had originally been limited by lack of food in the dense forest, and later by hunting. After closure of the prison settlement in 1923 they became so numerous that after each successive fire regeneration of palatable species was eaten out. The plant cover of the island today has regressed to a grassy scrub in which the semi-succulent shrub *Acanthocarpus preissii* and the grass *Stipa variabilis* are the dominants with sedges and some unpalatable shrubs such as *Thomasia cognata, Guichenotia ledifolia* and *Phyllanthus calycinus*. Some regeneration of trees has been effected artificially by planting in netted exclosures. On the mainland relics of cypress pine suggest that it represents the climax on stabilised dunes, but as it is tender to fire only fragments remain and it is replaced by sub-climax thickets of *Acacia rostellifera*.

Coastal vegetation on the mainland has been described by Dr G.G. Smith in various publications[8].

The vegetation is graded from the first colonisers of sand above high-water mark through sandbinders to more mature communities of stabilised dunes. Two species of strand plants colonise the beaches, *Cakile maritima* and *Arctotheca populifolia* (both introduced) but their action is ephemeral. The first perennial colonisers of the foredunes are the grasses *Spinifex hirsutus* and *S. longifolius* with *Ammophila arenaria*, a naturalised species. *Cakile* and *Arctotheca* and *Tetragonia decumbens* are also common in these open grass communities.

Sheltered hollows behind the foredunes favour more species of maritime plants. The most common species are *Tetragonia decumbens, Isolepis nodosa, Calocephalus brownii, Carpobrotus aequilaterus* and *Spinifex longifolius*. Less abundant are *Angianthus cunninghamii, Pelargonium capitatum, Trachyandra divaricata* and *Senecio lautus*. More rare and confined to this zone are *Actites megalocarpa* and *Atriplex isatidea*.

Towards the crests of these dunes the plant cover becomes closer and mainly consists of the larger shrubs *Myoporum insulare, Scaevola crassifolia, Olearia axillaris, Acacia cyclops* and the spreading clumps of *Lepidosperma gladiatum*. This low, salt-pruned shrubbery becomes taller and more luxuriant on the sheltered landward side of the mobile dunes, where dense thickets of *Pelargonium* and *Lepidosperma* and various shrubs fill the hollows in the transition area between these dunes and the scrub of the stable dunes.

On the windward slopes of the stable dunes a low dense thicket vegetation about 1.25 m tall is maintained by salt-pruning with an even canopy broken only by emergent plants of the resistant *Olearia axillaris*. *Acacia lasiocarpa* and *Melaleuca acerosa* are the characteristic dominants of this association. Lists of 65 associated species of native plants, 13 exotics and three mosses were given by Smith[8]. On sheltered slopes of the dunes taller thicket, even low forest, communities constantly develop and are as constantly destroyed by fire. The climax is probably *Callitris preissii* low forest but there are now only a few isolated remnants. *Acacia rostellifera* thicket is now the commonest community, containing *A. cochlearis, A. cyclops, Melaleuca huegelii* and *Dodonaea aptera*. Two creepers, *Clematis microphylla* and *Hardenbergia comptoniana* are common components. *Agonis flexuosa*, here at the northern limit of its range, occurs in the above but also dominates patches capable of developing into woodland.

Vegetation of the older dunes of the Spearwood System varies from north to south according to rainfall.

The terrain consists of ridges of calcarenite disposed more or less parallel to the coastline and mantled with yellow sand which becomes more bleached at the surface and less calcareous from west to east. The principal component of the vegetation is eucalypt woodland but numerous lakes of fresh, alkaline water occur in a chain along the eastern boundary from the Swan to the Murray River. They are partly or entirely overgrown by vegetation and both sedge swamp and sedge fen formations can be recognised which are similar both structurally and floristically to those in other Australian States and overseas. These formations are bordered by woody communities of *Melaleuca* or *Banksia*. The System is interrupted by the tidal Peel and Harvey Inlets. Further south it covers parallel ridges of calcarenite with long narrow lakes between them, of which those nearest to the sea are salt but Lake Clifton is fresh or more or less so.

North of Lancelin the plant cover is a mosaic in which scrub-heath is the dominant member with *Banksia prionotes* as character species among the taller shrubs, and *Dryandra sessilis, Calothamnus quadrifidus* as dominants in the lower heath layer. The mosaic includes patches of *Acacia spathulifolia* heath on shallow soil, of banksia low woodland on deep white sand and of stunted *Eucalyptus gomphocephala* in depressions. From Lancelin to Yanchep the banksia woodland becomes the general vegetation with scrub-heath limited to limestone ridges and occasional small patches of stunted eucalypts. The banksia woodland has the same dominants as on the Bassendean sands further inland but its understorey associations are different.

From Yanchep southward the Spearwood Dunes are taken over by eucalypt woodland, with the banksia community persisting as the small tree and ground layers. On less weathered dune ridges nearest the coast, tuart *(E. gomphocephala)* is the characteristic dominant, joined by jarrah on the more leached sands further inland.

The tuart woodland consists of open stands of stout, spreading trees 20-25 m tall, rarely 30 m. *E. gomphocephala* (tuart) is frequently the sole tree, only rarely with *E. calophylla* which appears on some red soils in swales, and *E. decipiens* which is patchy on shallow limestone. A lower layer of small trees under 10 m comprises *Agonis flexuosa, Banksia attenuata, B. menziesii* and *Casuarina fraseriana*. Also present are the grass tree *Xanthorrhoea preissii* and the cycad *Macrozamia riedlei*, the latter always trunkless. Shrubs of 1 to 3 m include *Acacia*

Western Australian Herbarium

Fig 3.62

Beaches of the west coast are slowly building up as sand is added by wave action, and blown inland by the prevailing wind. Above highwater mark the sand is colonised by special plants which act as sand binders and colonisers. Chief of these are the two species of *Spinifex*, *S. longifolius* (above) which grows on the west coast and round the north of Australia, and the creeping *S. hirsutus* which is a southern species. These are not the same as the 'spinifex' of the desert interior which belongs to the genera *Triodia* and *Plechtrachne*. A third important sand binder is an introduced plant of the daisy family, *Arctotheca populifolia*.

Gordon Smith

Fig 3.63

With the passage of time, other plants add themselves to the developing community; the grey-leafed *Olearia axillaris*, succulent plants such as *Salsola kali*, *Tetragonia decumbens* and *Cakile maritima*, herbaceous species such as *Isolepis nodosa*, *Trachyandra divaricata* and *Lepidosperma gladiatum*. An introduced weed from California, *Oenothera drummondii*, with yellow flowers, is common on Perth beaches. In some areas a saltbush, *Atriplex isatidea*, a tall woody plant with broad silvery leaves, forms thickets.

Fig 3.64

The final stages of colonisation of dunes take place on the lee side, sheltered from the wind. Taller shrubs such as *Scaevola crassifolia*, *Myoporum insulare* and *Acacia cyclops* invade and form thickets, leading on to dominance by *Acacia rostellifera*, the dominant dune shrub all up and down the coast. Such vegetation is very readily burnt however, and then reverts to earlier stages, recovering very slowly owing to the lack of humus and of nitrogen in these sands. Under natural conditions, with fire only a rare occurrence, the *Acacia* thicket would develop to a climax of *Callitris preissii* low forest.

saligna, *A. cyclops*, *Anthocercis littorea*, *Dodonaea aptera*, *Dryandra sessilis*, *Grevillea vestita*, *Hakea prostrata*, *Jacksonia furcellata*, *J. sternbergiana*, *Logania vaginalis*, *Melaleuca huegelii*, *Myoporum tetrandum* and *Templetonia retusa*. Characteristic low shrubs under 1 m include *Acacia dilatata*, *A. pulchella*, *Calothamnus quadrifidus*, *Casuarina humilis*, *Diplopeltis huegelii*, *Dryandra nivea*, *Grevillea crithmifolia*, *G. thelemanniana*, *Helichrysum cordatum*, *Hibbertia hypericoides*, *H. racemosa*, *Leucopogon parviflorus*, *Melaleuca acerosa*, *Petrophile serruriae*, *Phyllanthus calycinus*, *Rhagodia baccata*, *Scaevola nitida*, *S. thesioides* and *Synaphea polymorpha*. Creepers include *Clematis microphylla*, *Kennedia prostrata* and *Hardenbergia comptoniana*. Herbs are *Caladenia latifolia*, *Lagenifera huegelii*, *Ranunculus colonorum*, *Sowerbaea laxiflora*, *Trachymene coerulea*.

In the mixed stands with *E. marginata* the more calcicolous plants such as *Anthocercis*, the *Melaleuca* spp., *Myoporum*, the *Scaevola* spp. and *Templetonia* drop out, plants tolerant of acid sands become emphasised, e.g. *Banksia attenuata*, *B. menziesii* and *Casuarina fraseriana* which form a more definite understorey, while others of this class come in such as *Adenanthos cygnorum*, *Banksia grandis*, *Conospermum* and *Isopogon* spp., *Stirlingia latifolia* and other species of *Petrophile*.

Changes in composition occur from north to south. In the southern half *Agonis flexuosa* is the principal understorey tree in the woodlands, being not only more common but a larger tree than further north. *Banksia* spp. and *Casuarina* become rare but *Xylomelum occidentale* comes in.

The Bassendean System of old weathered dunes lies inland of the Spearwood System. It is probable that no trace of the original coastal dunes remains, the dunes of today being the result of reworking by wind of the sands of the plain in various arid periods. The sands are yellow at depth, bleached white at the surface. The overall cover is banksia low woodland but dune swales tend always to be swampy due usually to drainage obstructed by hardpan of various kinds. Heath communities, teatree, paperbark and reed swamps are very numerous under these conditions.

The trees of the woodland reach 6-8 m in height and comprise *Banksia attenuata*, *B. menziesii* and on wetter sites *B. ilicifolia*, besides lesser numbers of *Eucalyptus todtiana* and *Nuytsia floribunda*. *Casuarina fraseriana* comes in south of the Moore River, while from Perth southward scattered taller jarrah trees may be seen. N.H. Speck in 1952 listed understorey plants to the number of 28 species of tall shrubs, 130 of low shrubs and 182 ground plants including many Restionaceae, Cyperaceae and other herbaceous monocotyledons. Understorey communities were studied by J.J. Havel in 1968 by mathematical analysis, a prelude to his study in the northern jarrah forest described in the previous section. He was able to define 11 communities of which 5 occurred on the Spearwood Dune System, 3 on the Bassendean System, 1 on the transition between the two, and 2 on swampy sites which might occur anywhere in the region. *Verticordia nitens* becomes a very colourful and conspicuous understorey element when in flower in December, in areas between Perth and the Moore River.

South of the Serpentine River the Bassendean System becomes more swampy and there is an intricate mosaic of vegetation controlled essentially by drainage. Banksia low woodland is an important component occupying the highest and driest sands. On moister but still well-drained sites it merges into jarrah-marri woodland which varies from open trees with a thick banksia understorey to a well-formed woodland with scattered smaller banksia and casuarina. Freshwater swamps between the dunes occupy a large proportion of the area. If there is open water it is bordered by the paperbark *Melaleuca rhaphiophylla* with a belt of sedges and/or bullrushes extending into the water. Seasonally flooded swamps are covered by paperbark trees of *M. preissiana* with *Banksia littoralis* and some *Eucalyptus rudis*. Deeper swamps may consist of reeds only such as *Gahnia* spp. or *Baumea articulata*, or have been colonised by the introduced bullrush *Typha orientalis* as in Benger Swamp. A slightly different zonation is found at the margins of salt-water inlets such as Peel Inlet and Leschenault Inlet where a belt of the salt-tolerant reeds *Juncus kraussii* and *Baumea juncea* fronts open water, followed by the salt-tolerant paperbark *Melaleuca cuticularis* then the less salt tolerant species *M. preissiana* and *M. rhaphiophylla*, finally *Eucalyptus rudis* and the vegetation of dry ground. Near the mouth of the Leschenault Inlet there are small populations of the mangrove *Avicennia marina*, the most southerly on the Western Australian coast.

South of Gingin the Coastal Plain is bordered at the foot of the scarp by a belt of heavier alluvial soil constituting the Pinjarra Plain unit, and this takes up a higher and higher proportion of the Coastal Plain itself in a southerly direction, pinching out the Bassendean sands. As this unit contains the best soils for pasture development and irrigation, there is no virgin vegetation left. Even relict wooded areas must have been selectively

Fig 3.65

The climax vegetation of stabilised coastal dunes of the Swan Coastal Plain is low forest of Rottnest pine, *Callitris preissii*. However this is tender to fire and few examples remain. Burning by Aborigines had already reduced it to a few remnants on the mainland by the time of European settlement, but it was still dominant on Rottnest and Garden Islands. The latter is still its principal refuge, and it must be regarded not as a threatened species but as a threatened plant community.

Westralian Sands Ltd

Fig 3.66

In many places the sands of the Swan Coastal Plain contain heavy minerals deposited along former shore-lines, which are now being mined. Stringent conservation laws require restoration of mined areas and one solution adopted as shown here has been to convert them into wetlands. Over 70% of the natural wetlands of the Plain have been lost to development, so that new wetlands can provide valuable refuges for wildfowl. The concept includes islands created for nesting sites, sandbars for loafing areas, and adjacent grazing, in a perennial, fresh water body, a 'friendly' environment for water fowl throughout the year.

Fig 3.67

On Rottnest Island clearing, cutting and burning over many years opened up the vegetation so much that it greatly encouraged the population of quokkas, which grazed off regeneration of palatable species. Most of the island is now covered by a selection of unpalatable plants, the grass *Stipa variabilis*, the semi-succulent shrub *Acanthocarpus preissii* and in the background *Olearia axillaris*. The tuart tree *(Eucalyptus gomphocephala)* on the right is the result of attempted reafforestation.

culled for timber during the earlier years of settlement, and their original shrub and ground layers have been replaced by pasture plants. It has been suggested that the dominant vegetation, occurring on the better drained soils, would have been marri forest or woodland with a wandoo component, and some jarrah on higher ground. Lower lying wetter areas would have supported *Eucalyptus rudis*, or paperbark swamp on the worst-drained sites of all. Banksia low woodland may still be seen on occasional sand ridges. In the most northerly section, north of Bullsbrook, there are saline swampy areas dominated by *Casuarina obesa*.

The last unit to consider in this subregion is the Dandaragan Plateau, a triangular-shaped section lying between the Gingin Scarp and the Darling Fault and consisting of sedimentary rocks. The plateau surface has been dissected by the Moore River and lesser streams but in general is fairly even and thickly mantled with laterite and sand. The uppermost sedimentary rocks however are calcareous, consisting of chalk and greensand, and while weathered and lateritised they afford better soils able to support eucalypt woodland while the rest of the plateau being more sandy supports only banksia low woodland and kwongan.

Beginning in the extreme south at Bullsbrook where the sedimentary rocks pinch out against the Precambrian, the original cover was woodland of marri and wandoo with patches of jarrah forest. After crossing the Great Northern Highway the expanse of the plateau becomes very sandy and is covered by banksia low woodland of similar composition to that of the Bassendean sand on the plain below but with scattered taller trees of marri and jarrah, thickening locally to forest of these two species. Dissected country around Gingin has woodland of marri which seems always to have been the sole tree species. North of Gingin to Red Gully the plateau is still covered by the banksia low woodland with scattered trees as before but north of Red Gully this begins to become restricted to the deeper sands, giving way to scrub heath and heath on the shallower soils.

North of the Moore River the plateau becomes distinctly divided into a western and an eastern half. The western half, stretching north through Dandaragan almost to reach the Dinner Hill road has good agricultural soil — seen in profiles, a yellowish-red to brown sand or sandy loam with a layer of pea ironstone at depths of 1.5 to 2 m — and has almost all been cleared. The general cover appears to have been marri woodland reaching 20 m in height. There is no indication that jarrah was ever present, but there is some York Gum in the north.

The following understorey species have been listed:

Small trees *Acacia saligna, Banksia grandis*.
Cycads *Macrozamia riedlei* (locally abundant).
Grass trees *Xanthorrhoea preissii*.
Shrubs *Acacia pulchella, Boronia scabra, Bossiaea* sp., *Daviesia divaricata, D. preissii, Dryandra sessilis, Hakea prostrata, H. ruscifolia, H. trifurcata, Hibbertia hypericoides, Hybanthus calycinus, Hypocalymma angustifolium, Jacksonia sternbergiana, Lechenaultia biloba, Malleostemon roseus, Pimelea* sp., *Petrophile linearis, P. serruriae, Stirlingia latifolia*.
Herbaceous *Conostylis* sp., *Orthrosanthus* sp., *Sowerbaea laxiflora, Ptilotus polystachyus*.

The eastern half of the Dandaragan Plateau here is by contrast extremely sandy so that the vegetation is banksia low woodland for the most part, with dryandra heath in a southern section which has a soil of bleached sand overlying laterite.

The floristic composition of the banksia low woodland shows a mingling of elements of the *Banksia-Xylomelum* alliance from further north with those of the *Banksia-E. todtiana* alliance of the Swan Coastal Plain. The following list was taken in the area northeast of Badgerabbie Hill.

Trees *Actinostrobus arenarius, Banksia attenuata, B. burdettii, B. prionotes, Eucalyptus todtiana, Nuytsia floribunda, Xylomelum angustifolium*.
Shrubs *Acacia blakelyi, Adenanthos cygnorum, Banksia sphaerocarpa, Calytrix ? strigosa, Calothamnus quadrifidus, Casuarina humilis, Chamaelaucium drummondii, Conospermum stoechadis, Daviesia preissii, Eremaea pauciflora, Grevillea biformis, G. eriostachya, Hakea costata, H. obliqua, H. prostrata, H. ruscifolia, H. trifurcata, Isopogon divergens, Lambertia inermis, Lachnostachys eriobotrya, Lechenaultia biloba, Lomandra hastilis, Petrophile ericifolia, Stirlingia latifolia, Synaphea petiolaris, Verreauxia reinwardtii, Verticordia grandiflora, V. patens, V. picta*.
Herbaceous *Anigozanthos humilis*.

In the heath community, *Dryandra* spp. are dominant except in valleys on deeper soil where *Hakea obliqua* or *Banksia burdettii* may become dominant in patches, or locally *B. attenuata, B. menziesii* and *Eucalyptus todtiana*.

A floristic list in the heath was made as follows:

Andersonia sp., *Astroloma microdonta, Calectasia cyanea, Calothamnus* sp., *Calytrix* sp., *Casuarina humilis, Conostephium* sp., *Conospermum stoechadis, Dryandra bipinnatifida, D. carlinoides, D. nivea, D. shuttleworthiana, Eremaea* sp., *Eriostemon spicatus, Gompholobium* sp., *Hakea baxteri, H. conchifolia, H. incrassata, H. prostrata, H. ruscifolia, Hibbertia hypericoides, Isopogon ? scabriusculus, Lambertia inermis, Leptospermum erubescens, Lechenaultia biloba, Leucopogon* sp., *Lysinema ciliatum, Petrophile media, Synaphea petiolaris, Verreauxia reinwardtii, Verticordia grandiflora, V. patens, V. polytricha, Xanthorrhoea drummondii,* Cyperaceae and Restionaceae spp.

The xanthorrhoea are small and trunkless and are not a conspicuous element as they are nearer the coast. The dryandras are dominant.

FIG 3.68

Estuary margins such as those of the Peel Inlet and Blackwood Estuary, and originally the Swan River, have a salt-tolerant swamp vegetation with species of *Juncus* and *Chenopodium* fringing the water, backed by trees of the salt-water paperbark *Melaleuca cuticularis* and *Casuarina obesa*. Behind these come the freshwater paperbarks *M. preissiana* and *M. rhaphiophylla*, river gums *Eucalyptus rudis*, and finally the vegetation of dry ground. Shore of Harvey Estuary.

FIG 3.69

An interesting relict community occurs in the Leschenault Inlet at Bunbury—mangroves. A single species *Avicennia marina* grows on tidal mud bordered by rushes of *Juncus kraussii*. These are the most southerly mangroves on the Western Australian coast, their nearest relatives being in the Abrolhos Islands.

FIG 3.70

Most of the sand piled into coastal dunes is calcareous, consisting of pieces of powdered shells. In time this weathers and consolidates to limestone on which soil is formed, and this habitat is the home of the tuart *Eucalyptus gomphocephala*. It is at its best in the south, seen here near Busselton with a lower storey of peppermint trees *Agonis flexuosa*. Tuart woodland extends as far north as Yanchep and the tree itself patchily as far as Jurien.

FIG 3.71

Inland of the coastal dunes, much of the Swan Coastal Plain is swampy. Deep swamps with open water are bordered by groves of paperbarks, *Melaleuca rhaphiophylla* as seen here, with undergrowth of sedges and bullrushes. Seasonally flooded swamps are more likely to contain the other species *M. preissiana*.

Fig 3.72

The most typical formation of the Swan Coastal Plain is the banksia low woodland which covers the extensive poor leached sands of the Bassendean dune system. Moisture storage in these sands is insufficient to permit larger trees to survive the dry season. Scattered larger trees of first casuarina and then jarrah appear more and more towards the south as rainfall increases. Three species of banksia are represented, *B. attenuata* and *B. menziesii* with *B. ilicifolia* in low-lying places. The undergrowth consists of intensely sclerophyllous small shrubs, sedges and Restionaceae.

Fig 3.73

Flowering cone and foliage of *Banksia attenuata*. A banksia cone carries hundreds of small individual flowers, only a small number of which are fertilised to produce seeding follicles. It would in fact be physically impossible for fertile follicles to be produced from all flowers. The leaves show typical sclerophylly: stiff, brittle, and edged with sharp points.

Fig 3.74

In early summer *Verticordia densiflora* is dominant on an open swampy flat in the woodland while scattered *Nuytsia* trees add their Christmas colour. Fig 2.24 shows this tree which is one of the finest flowering trees in the world and has many botanical peculiarities. It belongs to the mistletoe family and is a root-parasite.

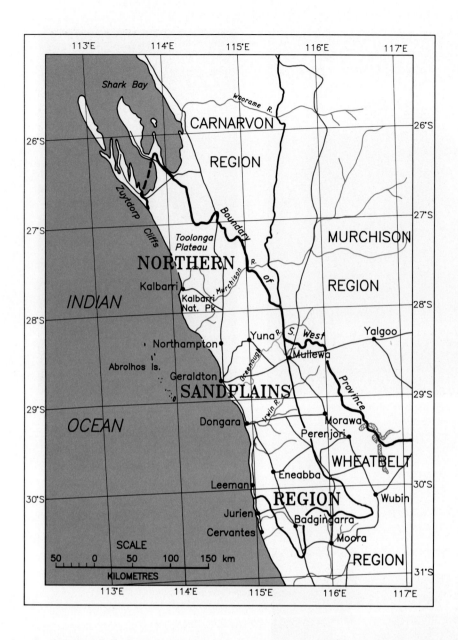

3.2 Northern Sandplains Region
IRWIN BOTANICAL DISTRICT
Scrub heath on sandplains near the coast; *Acacia-Casuarina* thickets further inland. *Acacia* scrub with scattered trees of *Eucalyptus loxophleba* on the hard-setting loams.

Climate: Dry warm mediterranean; winter precipitation 300-500 mm, with 7-8 dry months per year.
Geology: Mainly sedimentary basins exposing Permian to Cretaceous sediments; horsts of Proterozoic rocks.
Topography and soils: Prior land surface forming extensive lateritic sandplain, locally dissected especially near the coast. The sandplains are covered with leached sandy soils near the coast, and yellow sands with an earthy fabric further inland, both overlying laterite. Soils developed after stripping of the sandplain are mostly hard-setting loams with red clay subsoils.
Area: 39 656 km², 59% cleared.
Boundaries: NE—provincial boundary. E—vegetational expression of the Darling Fault, with a lobe towards Dalwallinu and Pithara occasioned by a major sheet of aeolian sand. S—northern limit of *Banksia* low woodland.

This region comprises the northernmost extension of the Southwest Province west of the Darling Fault, so that it is underlain almost entirely by sedimentary rocks of a mostly siliceous nature. The principal

FIG 3.75

North and east of the Southwest Forest Region where rainfall is lower, tree growth is confined to the deeper and heavier valley soils and then in the more open form of woodland instead of forest. Drier sandy soils carry the characteristic shrub formation known as kwongan or scrub-heath. Kwongan is from an Aboriginal word in the southwestern Nyungar language. Kwongan is typical of sandplains, extensive stretches of sandy soil which cap the higher ground on the interior plateau or occur more widely upon sandstone rocks. As the latter are common in the Perth Basin, the Northern Sandplains Region lying between Jurien and the Murchison River contains a high proportion of both sandplains and kwongan vegetation. Typical kwongan, Kalbarri National Park.

exception is a block of Proterozoic metamorphic rocks with some granite, between the Greenough and Murchison Rivers. The sedimentary rocks form a series of plateaux at the same level as the Yilgarn Plateau of the interior east of the Darling Fault (200-300 m elevation), the Dandaragan Plateau from the south up to the Irwin River, the Victoria Plateau from the Irwin to the Murchison River, and the Toolonga Plateau north of the Murchison. On the Toolonga Plateau the inland boundary of the region, which here is also the boundary of the Southwest Province, ceases to follow the Darling Fault and gradually moves across until it meets the shore of Shark Bay. These plateaux have been eroded by the sea on the west and dissected by rivers, but substantial stretches of the plateau surfaces are still preserved and form extensive monotonous sandplains. Sandy soils occur throughout except upon the Proterozoic rocks where red loams are found. An isolated portion of the Victoria Plateau forms the flat-topped Moresby Range inland of Geraldton.

Along the coast the same features are found as on the Swan Coastal Plain further south. A plain of marine denudation forms the sandy Eneabba Plain and reappears in the alluvial Greenough Flats further north. Above Port Gregory it narrows to the coast and disappears so that the sea is fronted by high cliffs. Known as the Zuytdorp Cliffs north of Kalbarri after a Dutch ship wrecked there in the 17th century, these become very impressive and reach a height of 224 m. The coastal plain has two distinct dune systems piled upon it as it does further south, older lithified dunes corresponding to the Spearwood System and younger unconsolidated dunes corresponding to the Quindalup System. Where there is no coastal plain, calcareous sand has piled on top of the cliffs and materially increased their height. For the whole length of the coast, the two dune systems form hilly sections.

Owing to the predominance of sandplains and other sandy soils it is not surprising to find that various types of kwongan have been mapped as having originally covered 81.5% of the region. The next most common vegetation type, acacia scrub with or without scattered trees, which occupies heavier soils, covered 14.3%. Eucalypt woodlands were only 2%. Only brief details of vegetation types can be given in this chapter. Fuller information is contained in Vegetation Survey Publications[9] including flora lists for the Burma Road Reserve east of Walkaway, the Chapman River Reserve at Geraldton and the Kalbarri National Park.

At the edge of the sea, the sandbinders and colonisers are as described under the Swan Coastal Plain, with the *Spinifex* grass the most conspicuous element, except that the saltbush *Atriplex isatidea* appears locally and forms dense thickets along the beaches at Geraldton for example. Unlike most saltbushes it is a tall shrub of 2 m with broad silver-tomentose leaves. Stabilised dunes of the more recent unconsolidated series appear to have a climax vegetation of thickets of *Acacia rostellifera, A. xanthina, Casuarina lehmanniana, Alyogyne huegelii, Melaleuca cardiophylla* and *M. huegelii*. These dominants are joined locally by the mallee *Eucalyptus obtusiflora (E. dongarraensis)* particularly in the vicinity of Cliff Head. However, fire in these thickets is very common, causing a regression to a pioneer stage of low heath of *Acacia lasiocarpa, Melaleuca acerosa* and a *Scaevola* sp.

Consolidated dunes or 'coastal limestone' have a varied vegetation. In the south from Jurien to the Arrowsmith River, in a belt of rough limestone country known as the Beekeepers' Reserve as the native vegetation has not been cleared and supports an active bee-keeping industry, the plant cover is scrub-heath with occasional groves of the small tree illyarrie or *Eucalyptus erythrocorys*. This is not a mallee, being fire-tender, and seems to be confined to rocky patches where inflammable ground vegetation is at a minimum. The scrub-heath mostly consists of shrubs of under 1 m among which *Beaufortia squarrosa* and *Banksia sphaerocarpa* are generally the most conspicuous, though *Acacia spathulifolia* may be locally dominant. Occasional banksias and nuystia provide larger shrubs/small trees.

North of the Arrowsmith River the illyarrie continue to occur on the coastal limestone but less conspicuously, and are embedded in *Acacia-Casuarina-Melaleuca* thickets as described above. North of Geraldton to the Hutt River the limestone tends to be covered with red sandy soil, in which case a more open shrubland dominated by *Acacia rostellifera* and *Banksia prionotes* is found with several other *Acacia* and *Banksia* species. North of the Hutt River there are dunes with *Acacia rostellifera* thicket pinching out against cliffs which continue to the mouth of the Murchison River. A low heath pruned by wind and spray is found on the cliff tops, featuring cushion plants of *Baeckea pachyphylla* and *Scaevola crassifolia*. At Kalbarri the slopes facing the sea are covered with thicket of *Acacia rostellifera* with *A. scirpifolia, A. acuminata, Jacksonia cupulifera, Hakea stenophylla, Alyogyne hakeifolia* and an understorey of *Melaleuca megacephala*. Elements of the Kalbarri sandplain also appear sporadically, especially *Grevillea leucopteris*.

The Zuytdorp Cliffs north of the Murchison are not in fact vertical cliffs but steeply sloping bluffs with a low vegetation containing many halophytes. Thickets of small *Diplolaena dampieri* are common containing also *Acacia ligulata, Olearia axillaris, Scaevola crassifolia, Anthocercis littorea* with small ground plants of *Atriplex, Sclerolaena, Frankenia* and *Carpobrotus*. Low heath covers the hill tops.

The Eneabba Plain was formed during a high sea level epoch of the Pleistocene when mineral-rich beach sands were deposited on the shoreline of the time and are now being mined. The plain is very sandy, in places with small longitudinal ridges, and with numerous small lakes and swamps in depressions. The mature structure of the sandplain community consists of scattered small trees up to 5 m tall, an open layer of tall shrubs of 1-3 m and a closed layer of small heath-like shrubs of under 1 m. Owing to frequent fires this mature structure is not often seen, and more commonly there are scattered trees of *Eucalyptus todtiana* (which is not killed by fire) and low shrubs with emphasis on smokebush *(Conospermum)*. Dominants include *Banksia attenuata, B. menziesii, B. prionotes* and *Xylomelum angustifolium*. An interesting small shrub is the dwarf banksia, *B. candolleana*.

Sand ridges are the home of the beautiful *Banksia hookeriana* with woody pear *Xylomelum angustifolium*.

The Greenough Flats south of Geraldton consist of two alluvial plains, the 'front flats' and the 'back flats'. As the area has been intensively farmed for over a century no trace of the original vegetation remains even on roadsides but it is believed from analogy that

Fig 3.76

Kwongan consists of both large and small shrubs and herbaceous plants, the large shrubs belonging mainly to the family Proteaceae and the smaller to the Myrtaceae. The latter are of a type called *ericoid shrubs* by Diels, from their resemblance in form to the heath plants of the family Ericaceae common in his European homeland. The ericoid shrub distinguishes kwongan from other types of shrubland. It is from 30 to 60 cm tall (taller if long unburnt), much branched and with small evergreen leaves, and generally flowers brilliantly in spring. This example is *Pimelea ferruginea*, in cultivation.

Fig 3.77

Flowers and foliage of *Pimelea ferruginea*, a typical ericoid shrub, in cultivation.

Fig 3.78

Coastal dune formations in the Northern Sandplains Region are somewhat similar to those of the Swan Coastal Plain: the same sand binders and colonisers, similar scrub on stabilised dunes. Here, at Leeman, *Acacia rostellifera* is the dominant element with *A. xanthina, Casuarina lehmanniana, Alygyne huegelii* (seen flowering at right), *Melaleuca cardiophylla* and *M. huegelii*. The mallee *Eucalyptus obtusiflora* (formerly *E. dongarraensis*) may be common locally.

Fig 3.79

Burning of the coastal dune scrub reduces it to a pioneer stage of *Acacia lasiocarpa* (foreground), *Melaleuca acerosa* and a *Scaevola* sp. Here an annual daisy *Podolepis* and small shrubs of *Pimelea* are in flower. Leeman.

the front flats were covered with dense acacia and melaleuca thickets with occasional trees of *Eucalyptus camaldulensis,* while the more sandy back flats had patches of York gum *(E. loxophleba)* and banksia scrub. Surviving *E. camaldulensis* today are all deformed into extraordinary shapes by the predominant south wind and are a source of wonder to visitors.

On and around the various plateaux of the interior, woodlands as we have noted are quite rare. South of Eneabba small patches of marri, wandoo or powderbark wandoo may be seen on heavier soils or laterite. River red gum *(E. camaldulensis)* appears on alluvial flats on river banks and around small freshwater lakes on the Eneabba Plain. York gum occupies small alluvial patches on the Arrowsmith, is common in the Irwin Valley, and occurs in slight depressions on the Victoria and Toolonga Plateaux.

From Geraldton to the Murchison River much of the country is underlain by Proterozoic rocks which form red loam soils carrying *Acacia* scrub with scattered trees of York gum and river red gum, and in the north particularly *Casuarina huegeliana*. As the soils are relatively young and are not deep-weathered, there is insufficient soil depth and moisture reserve to support actual woodlands. The acacias are *A. acuminata* (general) and *A. rostellifera* (near the coast) with *Hakea preissii, H. recurva* and *Jacksonia cupulifera*. Remnants of overlying younger sandstones occur in this country forming flat-topped hills such as the Moresby Range, and on their steep slopes there is a unique kwongan community — one of the very few not associated with sandplains. This is known as the *Melaleuca-Hakea* thicket, up to 2 m tall, with *Melaleuca megacephala* and *Hakea pycnoneura* as dominants. There is an occasional mallee, *E. redunca;* other shrubs, especially *Gastrolobium spinosum, G. oxylobioides* and *Grevillea pinaster;* and a ground layer of *Melaleuca scabra, Verticordia chrysantha* and other ericoids. This community may be seen at Howatharra Gap on the Geraldton-Northampton Highway, where *V. chrysantha* makes a brilliant show in spring.

The predominant sandplains cover a range of rainfall, declining from south to north and from the coast inland, and a range of soil types. Three major types of soil may be distinguished — dense laterite, sand overlying laterite or pea ironstone and deep sand >1 m in depth. Shallow sands are normally grey in colour, and deep sand yellow changing to red in the northeast close to the boundary of the Southwest Province. The vegetation of the sandplains is always kwongan, but it varies structurally and in composition according to variations in climate and soil.

In the south, for example in the Badgingarra National Park, three communities may be seen: the blackboy heath, the *Hakea obliqua* scrub-heath and the banksia scrub-heath. The blackboy heath occupies dense laterite. Blackboys *(Xanthorrhoea drummondii)* up to 2 m tall are common and conspicuous, rising from a low heath of typically *Dryandra* and *Hakea* spp. with numerous sedges and restioids. *Hakea obliqua* is common and conspicuous, acting as character species for a taller scrub-heath on sand over laterite, of very mixed and rich composition. Banksia dominated scrub heath occuring on deep sand is an attenuated form of the banksia low woodland of the Swan Coastal Plain further south, and contains the same dominants.

North and east of Eneabba up to the Greenough River, there is still low heath on dense laterite with *Hakea auriculata* and *Dryandra fraseri* but without the blackboys. On lateritic sand *Hakea obliqua* drops out and the scrub heath is very heterogeneous without any obvious dominance or character species. A list of 140 component species is given in (9a). On deep sand a well-marked community known as the *Banksia-Xylomelum* alliance appears. The larger shrubs normally reach about 3 m in height but may reach 6 m if long unburnt. Character species are banksias — in this area *B. attenuata, B. burdettii* and *B. prionotes* and woody pear *Xylomelum angustifolium,* together with the small cypress-like tree *Actinostrobus arenarius* and *Grevillea leucopteris,* and in this area *Adenanthos stictus* and *Eucalyptus pyriformis*. In the lower shrub layer coppercups *Pileanthus peduncularis* and several *Verticordia* spp. provide massive colour effects in spring. This community may be seen in the northern half of the Watheroo National park or along the coast road from Coorow.

On the Victoria Plateau northeast of Geraldton the *Banksia-Xylomelum* alliance formed the original plant cover, except for occasional patches of mallee and York gum on red soil, as far inland as the 350 mm rainfall line. Beyond this there is a gradual patchy change in soil from incoherent yellow sand to red sand containing a higher proportion of earthy material. The *Banksia-Xylomelum* alliance continues to occupy the yellow sand with some species changes from further south. *Banksia sceptrum* replaces *B. burdettii, B. ashbyi* replaces *B. prionotes* and *Grevillea gordoniana G. leucopteris. Adenanthos sticta* and *Eucalyptus pyriformis* are no longer present, the latter replaced by several sandplain mallees. On the red sand

FIG 3.80

Flatter expanses on the coastal limestone tend to form sandplains on which a low form of kwongan appears. In this scene the white flowered smokebush *Conospermum stoechadis* contrasts with the coastal wattle *Acacia spathulifolia*. In spring, at Leeman.

FIG 3.81

Coastal sandplains from Green Head to Dongara contain scattered groves and individual trees of *Eucalyptus erythrocorys*, the illyarrie, a local species confined to this habitat. It is a striking ornamental tree when in flower, with scarlet bud caps and large yellow flowers. As it is tender to fire, its distribution within its habitat is controlled by the amount of inflammable undergrowth, and in general it is found growing on rocky places where undergrowth is at a minimum.

FIG 3.82

North of the Arrowsmith River the illyarrie continues to occur but is embedded in thickets of *Acacia rostellifera, Melaleuca huegelii* and other species. Much of this thicket remains uncleared on the more stony ground where limestone is close to the surface, as may be seen on the highway between Dongara and Geraldton.

FIG 3.83

The *Acacia–Melaleuca* thickets, improved perhaps to the form of low forest, are believed to have been the original vegetation of the fertile Greenough Flats south of Geraldton, with the addition of scattered trees of the river gum *Eucalyptus camaldulensis*. As the Flats were settled early and have been intensively farmed, the eucalypts are the only surviving trace of the original plant cover. Exposure to the strong south winds has caused them to be bent over into recumbent shapes growing away from the wind. This is not because the wind has actually bent them but because it destroys the buds and growing tips on the exposed side, and growth occurs on the sheltered side.

FIG 3.84

North of Port Gregory high cliffs replace the coastal dunes. Plant growth on the top is dwarfed by the wind for some distance back and consists of small domed shrubs. Here *Baeckea pachyphylla* is in flower, interspersed with *Scaevola crassifolia*. Kalbarri National Park south of Red Bluff.

there is a change from scrub-heath to thicket (i.e. to a denser form of shrubland) belonging to the *Acacia-Casuarina-Melaleuca* alliance. This consists of a closed tall shrub layer about 2.5 m in height with scattered emergent larger shrubs and small trees reaching about 5 m, and a sparse understorey at about 1 m of mainly ericoid type shrubs. Principal emergents here are *Banksia ashbyi*, *Eucalyptus eudesmioides* and other mallees, and *Grevillea gordoniana*. The dominant thicket layer includes *Acacia acuminata*, *A. longispinea*, *A. stereophylla*, *Casuarina acutivalis*, *C. campestris*, *Eremaea pauciflora* and *Melaleuca uncinata*. The ground layer includes *Baeckea pentagonantha*, *Beaufortia elegans*, *Melaleuca cordata*, *Pileanthus peduncularis* and *Verticordia etheliana*.

A curious feature of this area is the frequent presence of narrow strips of mallee along the boundary between red and yellow soil types, perhaps reflecting a seepage zone. The mallee may be *E. eudesmioides*, *E. oldfieldii* or *E. redunca*. It has an understorey either of spinifex (*Triodia* sp. undescribed) or of the reed *Ecdeiocolea monostachya* giving a similar superficial appearance. Mallee and spinifex is normally a vegetation type of the desert!

The Kalbarri National Park contains a very large expanse of sandplain traversed for 40 km along the main road into Kalbarri. Although this overlies a different geological formation, the Tumblagooda Sandstone, which is older and so hard that the Murchison River has cut a deep gorge into it, the sandplain is essentially the same as elsewhere and zoned according to rainfall.

FIG 3.85

North of the Murchison River mouth the cliffs become extremely high and steep in a long unbroken line called the Zuytdorp Cliffs after a Dutch ship once wrecked there. These have a thin vegetation containing many halophytes because of the exposure to the sea. Thickets of small *Diplolaena dampieri* are common, containing *Acacia ligulata*, *Olearia axillaris*, *Scaevola crassifolia* and *Anthocercis littorea* with small ground plants of *Atriplex*, *Sclerolaena*, *Frankenia* and *Carpobrotus*.

FIG 3.86

In close-up, the slope of the Zuytdorp Cliffs is a stony scree with scattered succulents.

The coastal slope of these hills as we have noted above is covered by *Acacia* thicket with *A. rostellifera* and *A. acuminata*. This spills over the main ridge above Kalbarri onto the eastern slope where it will be seen to terminate along a definite line giving way to the scrub heath of the sandplain. Nearer the coast the soil is principally bleached sand overlying laterite but changes patchily to deeper yellow sand inland. On the former, *Adenanthos cygnorum*, *Banksia attenuata*, *B. menziesii* and occasional blackboys *Xanthorrhoea drummondii* are characteristic species, while on the latter the typical *Banksia-Xylomelum* alliance takes over, with *B. sceptrum* and *B. prionotes*, *X. angustifolium*, *Actinostrobus* and *Grevillea leucopteris*. The latter is a pioneer species and frequently forms a spectacular 'hedge effect' alongside the main road.

A more comprehensive account of the Kalbarri sandplain and a list of 200 species for the National Park will be found in reference 9c. Both the structure of the vegetation and its superficially visible composition vary considerably, as everywhere in kwongan, according to the time since the last fire. In the early stages it has the appearance of a low heath, but if long unburnt the taller elements outgrow the smaller to form a layer of scattered large shrubs or small trees, which may often attain 4 m in height. The biggest banksias and woody pears ever seen by the writer had attained 5 m in height and 30 cm in diameter at the base.

The Murchison Gorge country is very rugged and consists of precipitous cliffs, steep rocky slopes and more gentle sand-covered benches. The vegetation of the most

FIG 3.87

Behind the coastal belt, at Eneabba there is a sandy plain corresponding to the Bassendean Dunes of the Swan Coastal Plain, covered with an attenuated form of the banksia low woodland. Sand ridges in the Eneabba plain are the home of the beautiful small banksia *B. hookeriana*, seen here in flower, and restricted to this area. They also mark the first appearance of the woody pear *Xylomelum angustifolium* which becomes important further north.

FIG 3.88

The more hilly country of the Badgingarra–Eneabba areas is developed on sandstones which produce poor sandy soils, and almost the whole of this country is covered with kwongan. Scattered small trees of *Nuytsia floribunda*, which flourish because they are robbing surrounding plants of moisture and nutrients, frequently stand isolated in low heathy kwongan. Acacias and smokebush (*Conospermum*) are in flower in this spring scene. Badgingarra.

FIG 3.89

Distinct associations can be recognised in the Badgingarra sandplains. Deep sandy soil is dominated by banksias, the same *B. attenuata* and *B. menziesii* as on the Swan Coastal Plain further south, but reduced in stature and now regarded as kwongan. Such deep sand is normally confined to depressions.

Fig 3.90

On the other hand some soils on higher ground may consist of little but dense laterite, and this carries the bizarre 'blackboy heath' given character by scattered *Xanthorrhoea drummondii* as the only tall plants in a low heath of extremely hard, sclerophyllous plants, typically *Dryandra* and *Hakea* species with numerous sedges and restioids.

Fig 3.91

In close-up the ground surface consists of laterite pebbles in a sandy matrix, in which the heath plants are sparingly rooted. This is one of the most inhospitable environments in the State.

Fig 3.92

A third readily recognisable community in the Badgingarra sandplain is one dominated by *Hakea obliqua*, distinguished by its strange spindly habit. It has very prickly cylindrical leaves, and showy white flowers in spring. The small tree in the background is a *Banksia*. This community grows in sand of moderate depth over laterite.

Fig 3.93

In the Northern Sandplains Region the climatic regime is capable of supporting woodland but in practice areas of woodland are very restricted and limited to heavier soils. Wandoo occurs patchily in the Herschel Range near Badgingarra, and both marri and powderbark wandoo are seen locally as well. York gum *(E. loxophleba)* appears in the Irwin valley and occurs locally in patches as far north as Nerren Nerren beyond the Murchison River.

Fig 3.94

In the hills north of Geraldton known as the Moresby Range, illustrated in Fig 2.12, an unusual shrub community covers the slopes, the *Melaleuca-Hakea* thicket seen here. The dominants are *Melaleuca megacephala* with large yellow flower heads in the picture; and *Hakea pycnoneura* not illustrated. The flowering shrub in the centre is *Melaleuca scabra*. This community is a form of kwongan but occurs only in this habitat. It is one of the few kwongan types which do not occur on sandplains. Howatharra.

Fig 3.95

Flowers and foliage of *Melaleuca megacephala* ('large headed'). The terminal flowerheads 3 cm in diameter are unusually large for a *Melaleuca* and also exceptional in shape as this genus usually favours the 'bottlebrush' type of inflorescence.

Fig 3.96

A common component of the *Melaleuca-Hakea* thicket is the shrub *Gastrolobium spinosum*, 'prickly poison'. This is one of the most widespread of the poisonous legumes which are highly toxic to stock. This particular form belongs to the variety *triangulare* which according to Gardner and Bennetts 'is apparently restricted to the Northampton-Geraldton area. It is a highly toxic form and is common on the sandstone hills, especially around White Peak.'

FIG 3.97

The footslopes of the Moresby Range and the undulating country around Northampton are occupied by an *Acacia* scrub with large bushes up to 3–4 m of *A. acuminata* and *A. rostellifera* with *Hakea preissi*, *H. recurva* and *Jacksonia cupulifera*. Eucalypts are quite rare, occasional trees only of York gum and river red gum. Originally there was a ground layer of native spring annuals, but these are being replaced more and more by introduced weeds such as wild oats.

FIG 3.98

The occurrence of eucalypts in the country between Geraldton and the Murchison River is limited to occasional patches and individual trees of York gum (*E. loxophleba*) with a mainly herbaceous understorey of spring annuals, occupying deep moist soil. River red gum is also seen as individuals on hillsides occupying small drainage lines.

rocky places consists of *Acacia acuminata* scrub with *Jacksonia cupulifera* and small trees of *Casuarina huegeliana*. *A. acuminata* with scattered *Eucalyptus camaldulensis* occupy lower flats in the gorges, and the river itself is lined by *E. camaldulensis* and *Casuarina obesa*, with occasional *Callistemon phoeniceus* in the river bed.

Further inland as seen on the main highway going north from the Murchison River, we have the *Banksia-Xylomelum* alliance on yellow sand changing patchily to *Acacia-Casuarina-Melaleuca* thicket on red sand, as previously described. There are some patches of York gum woodland on this stretch on red soil in depressions which seem to represent former river channels.

At the northern limit of the Southwest Province where it reaches the shores of Shark Bay and for some few kilometres south of this, there is a red sandplain with a remarkable formation which has been labelled 'tree heath'. It may be seen along the side road to Tamala Station and Useless Loop. The soil is an incoherent and structureless red-brown sand, everywhere swept up into sand ridges mostly small and of a confused pattern. The tree heath consists of herbs and grasses, small and large shrubs, and small trees up to 6 m, all very irregular and open at all levels. Perhaps due to this openness the formation seems to be immune to fire. Floristically it is related to the red sand thickets but contains a number of local endemic species notably *Adenanthos acanthophyllus*, *Eucalyptus beardiana*, *E. roycei*, *Grevillea rogersoniana* and two species of *Melaleuca*. The principal character species are *Banksia ashbyi*, *Calothamnus chrysantherus*, *Eremaea pauciflora*, *Grevillea gordoniana* and a coarse grass *Plectrachne danthonioides*.

Arthur Fairall

Fig 3.99

This photograph was taken in 1962 in the Northampton area in the *Acacia* shrub country (in this case with some large *Nuytsia* trees) to show the carpet of spring annuals which was common in those days. The pink-flowering *Schoenia cassiniana* provides most of the show of colour. It is no longer possible to take a photograph of this kind owing to weed invasion.

Fig 3.100

In spring 1988, in wattle scrub of the Northampton area, weed invasion has progressed to a point where a few native annuals (yellow *Podolepis* at left) are fighting a losing battle against wild oats, Paterson's curse and lupins. Widespread agricultural clearing in the area has led to the invasion of native vegetation by agricultural weeds.

Fig 3.101

The largest and richest remnant of the original kwongan of the Northern Sandplains is preserved in the Kalbarri National Park, on a sandstone plateau near the mouth of the Murchison River. The kwongan has a number of different aspects, according to time elapsed since the last fire, local soil differences, and the season of the year. In this September scene colour is provided by *Conospermum stoechadis* (smoke bush, white), *Melaleuca scabra* (reddish), *Verticordia grandiflora* (yellow) and the taller shrub *Grevillea biformis* (yellow).

FIG 3.102

Also in September at Kalbarri, *Conospermum stoechadis* and *Melaleuca scabra* are generally conspicuous. The large shrub in the foreground is *Isopogon divergens* while in the background the small cypress-like conifer *Actinostrobus arenarius* appears in its typical manner, in dense colonies.

FIG 3.103

A common element in early spring is the star flower *Calytrix brevifolia*, a typical 'ericoid shrub'.

Don Bellairs

FIG 3.104

Later in October in the Kalbarri sandplain major colour is provided by *Verticordia monadelpha* (pink) and *Pileanthus peduncularis*, coppercups (orange). The reeds are *Ecdeiocolea monostachya*. In the middle ground *Grevillea biformis* is still in flower.

FIG 3.105

An aspect in the Kalbarri National Park in the portion north of the Murchison River has *Actinostrobus arenarius* dominant with shrubs *Conospermum stoechadis* (white) and *Grevillea dielsiana* (red). A sand dune in the background carries mainly *Banksia sceptrum* and *Xylomelum angustifolium*.

FIG 3.106

The woody pear *Xylomelum angustifolium* is a characteristic small tree of northern kwongan on the deeper sands, and gives its name to the *Banksia-Xylomelum* association. It can become a small tree of 4–5 m. The white flowers are borne in summer and are thus not often seen by visitors, but the pear-shaped woody fruits are conspicuous. The tree belongs to the Protea family, and the fruits like those of *Hakea* contain two winged seeds.

FIG 3.107

Banksia ashbyi is one of the most conspicuous species in the northern part of the Northern Sandplains Region, replacing *B. prionotes* which it somewhat resembles. It can become a spreading tree of 5–6 m in height and the flowerheads are showy of a rich golden colour. It does particularly well in cultivation.

Fig 3.108

When kwongan is long unburnt, the taller shrubs grow up into small trees, suppressing the lower layer of ericoids. In this case we have an example of the *Banksia-Xylomelum* association, with *Banksia sceptrum* on the left of the vehicle and *Xylomelum angustifolium* on the right. Most areas of kwongan however do not reach this stage, contrast Fig 3.101 and 3.102. The picture above was taken in a remote part of the Kalbarri National Park.

Fig 3.109

Although most kwongan plants are woody, herbaceous elements are represented. This *Conostylis* is one of a genus of two dozen species in the southwest, all with yellow flowers, and perennials growing in clumps. They belong to the family Haemodoraceae and are small relatives of the kangaroo paws.

Fig 3.110

One of the most showy shrubs in the Kalbarri sandplain, when in flower, is *Grevillea leucopteris*, called 'old socks' from the bad smell of the flowers. Never pick them and take them home! It is a pioneer species regenerating after fire or roadside disturbance, and frequently forms a 'hedge effect' along the main road through the park. Like a number of the grevilleas, the bush develops special floral branches whose function is to bear the flowers and fruits waving high in the air above the bush itself.

Fig 3.111

A feature of the interior sandplains north of Geraldton is the occurrence of a belt of mallee with a spinifex-like understorey along the margin of the sandplain. The mallee may be *Eucalyptus eudesmioides, E. oldfieldii* or *E. redunca*, and the ground layer may be composed of an undescribed species of *Triodia* or the reed *Ecdeiocolea monostachya*. Yuna East Reserve.

FIG 3.112

One of the most colourful occupants of kwongan in the more interior sandplains is *Hakea bucculenta*, 'red pokers', a large shrub, usually in thickets of *Acacia–Casuarina–Melaleuca*. Yuna.

FIG 3.113

Immediately to the south of Shark Bay and extending up the base of the Peron Peninsula is a unique community known as tree-heath. The sand here is red, rainfall is lower, the whole community more open. This openness apparently saves it from fire, and it is an irregular mixture of small trees, shrubs and grasses. In this photograph we see *Banksia ashbyi, Calothamnus chrysantherus* and in the ground layer *Plectrachne danthonioides*. The latter is not strictly a spinifex. The community contains a number of endemic plants such as *Adenanthos acanthophyllus, Eucalyptus beardiana, E. roycei, Grevillea rogersoniana* and two species of *Melaleuca*.

FIG 3.114

The Murchison River has incised a spectacular gorge into the hard Tumblagooda sandstone which underlies the Kalbarri sandplain, and provides one of the finest scenic attractions of the national park. *Acacia acuminata* (jam tree) with scattered river red gums occupies the lower flats in the gorges and the river bed itself is lined by the latter *(Eucalyptus camaldulensis)* and *Casuarina obesa* with occasional *Callistemon phoeniceus* in the river bed.

3.3 Wheatbelt Region

AVON BOTANICAL DISTRICT

The typical sequences of vegetation comprise scrub-heath on sandplain, *Acacia-Casuarina* thickets on ironstone gravels, woodlands of York gum *(Eucalyptus loxophleba)*, salmon gum *(E. salmonophloia)* and wandoo *(E. wandoo)* on loams, halophytes on saline soils.

Climate: Dry warm mediterranean; winter precipitation 300-650 mm per annum, with 7-8 dry months.

Geology: Archaean granites with infolded metamorphics of the Yilgarn Block.

Topography and soils: Undulating plateau, mostly with disorganised drainage. Remnants of prior land surface are preserved, giving rise to catenary sequences of soils, typically yellow earths on sandplain, with ironstone gravels peripheral to same, hard-setting loam soils on slopes and bottomlands, and saline soils in depressions.

Area: 93 520 km², 93% cleared.
Boundaries: NE—provincial boundary as defined. SE—boundary of the mallee formation. W—Darling Fault, approximately, in the north; south of Moora the boundary of the SW Forest Region.

The Wheatbelt occupies a large portion of the drier interior below the 500 mm rainfall line. It is distinguished from the Northern Sandplains Region by overlying the granite and metamorphic rocks of the Yilgarn Plateau, and from the Southwest Forest Region and the Mallee Region (Fig. 2.5) by changes in vegetation accompanying changes in climate and soils. The Wheatbelt extends inland as far as the boundary of the Southwest Province which approximately coincides with the 300 mm rainfall line. It occupies a plateau 200-300 m above sea level rising gently towards the east, whose surface is gently undulating and well organised into ridges and valleys. The erosion which shaped this landscape however occurred very long ago and the rivers are no longer active except in a narrow belt on the western margin from Narrogin to Moora. Elsewhere the valleys are occupied by chains of salt lakes which normally absorb any drainage and only flush out in years of unusually heavy rainfall.

The Wheatbelt Region is the largest in the southwest, almost a third of the province, and the most intensively occupied for agricultural purposes. Ninety-three per cent is alienated land but there are many small reserves which have been intensively studied in recent years. Also most farmers preserve some uncleared bush if only along fence lines or as isolated trees in paddocks, and with roadside vegetation often largely intact it is normally easy to interpret original vegetation. Maps showing original vegetation of the Wheatbelt were in fact confidently constructed[10].

Owing to the underlying geological uniformity the landscape of the region is reasonably uniform but the vegetation upon it changes gradually from west to east with declining rainfall. The key to understanding the vegetation of this region more than any other is the catenary sequence as described in Chapter 2 and illustrated for this region in Fig. 3.115. The soil catena is essentially the same throughout though individual segments will vary in the proportions they occupy, and the vegetation upon each segment varies with the rainfall. Soil and vegetation are absolutely interrelated, and are topographically controlled. There is a major division between valley soils which are red loams developed on the pallid zone of the deep-weathered profile, and upland or sandplain soils which are sandy and contain varying proportions of ironstone gravel. These are usually separated by a relatively narrow band of duplex soil (which has contrasting horizons) where sand has washed down slope above the loam. Woodland is characteristic of the valley soils and kwongan of the sandplain, with mallee as intermediate. At the foot of the slope alluvial and colluvial material underlies the salt lakes and their associated dunes and flats. The regularity of the catena is broken by occasional outcrops

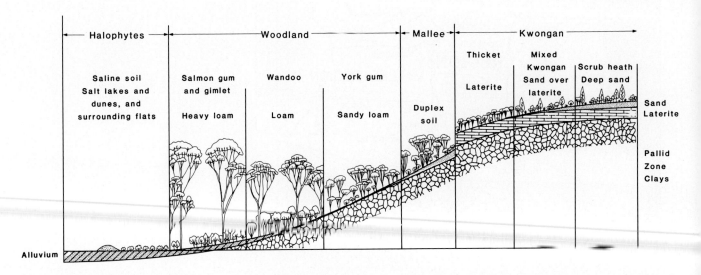

Fig 3.115

Diagrammatic catenary sequence for the Wheatbelt Region showing relationship of vegetation and soils.

FIG 3.116
The Wheatbelt Region lies inland on the Yilgarn Plateau and is the most intensively farmed of any of Western Australia's Regions, with 93% of the land alienated for this purpose. It is one of the world's greatest grain-growing areas, with 18 million hectares under cultivation for wheat in rotation with pasture. The landscape undulates very gently and is given a pleasant park-like aspect by the trees which have been left uncleared in the paddocks. In the pasture the Cape weed *Arctotheca calendula* often seems more common than grass but is tolerated as it can be grazed. Near Brookton.

of unweathered granite of varying size from small tors to massive domes, which may be surrounded by shallow grey sandy soils. Rarely, sheets of transported sand lie in the valleys and form 'low level sandplains'. An extensive sandplain of this kind, now completely cleared, stretches from Bolgart to Meckering. There is another smaller example, still vegetated, adjacent to the township of Quairading. The proportions of valley and sandplain soils vary considerably. The western wheatbelt, being in the lower part of the catchments, has relatively little sandplain whereas sandplains tend to predominate in the eastern part. Thirty-four per cent of the region as a whole is (or was) covered with kwongan on sandplain soils.

On the western margin where the Avon and other streams are still active, the salt flats at the foot of the catena are of course replaced by well-drained valleys with woodlands of York gum, and with river gum *(E. rudis)* and sheoak *(Casuarina obesa)* lining stream banks.

Salt flats occur elsewhere throughout the system. The lowest sites, seasonally flooded, dry in summer and covered with salt crystals, are devoid of vegetation. Bordering these or covering extensive flats of saline mud and less subject to inundation are samphire communities, bordering these again on sandy rises thickets of *Melaleuca uncinata* and other *Melaleuca* species. Other communities reported in this situation have been *Hakea preissii* thicket, *Callistemon phoeniceus* scrub and *E. redunca* open low woodland over *Melaleuca* thicket. Samphire communities consist of low succulent shrubs scattered to give an estimated 30–70% cover, mainly *Halosarcia halocnemoides* with *H. indica* on slightly raised areas within the salt flat. *H. pergranulata* was listed for reserve 24897, and *A. lylei* for 25112 'where water flow is greatest'. Few or no other species are normally present. *Sclerolaena* aff. *diacantha*, *Cotula coronopifolia*, *Disphyma crassifolium*, *Maireana brevifolia* and *Polypogon monspeliensis* have been recorded.

FIG 3.117

The trees left by farmers when clearing land, for shelter or for ornament, have not only given character to the landscape but enable reconstructions of original natural vegetation to be made. York gum *(Eucalyptus loxophleba)* is one of the more easily recognised trees due to its habit of branching from near the base so that it looks like an overgrown mallee. Eventually of course such trees will die of old age, and it is unusual for young ones to come up as the conditions for natural regeneration no longer apply. This situation can be taken care of by planting, and awareness of the problem is now growing. Pingelly.

In *Melaleuca* thickets *M. hamulosa* is the commonest species associating with *M. uncinata* and *M. thyoides*. These thickets attain 2 to 5 m in height with an estimated density of 30-70%. Few or no other species are normally present. *Acacia colletioides, Halosarcia halocnemoides, Carpobrotus* sp., *Disphyma crassifolium, Hakea preissii, Einadia nutans* and *Rhagodia spinescens* have been recorded.

Woodlands in salt areas consist of *E. loxophleba, E. loxophleba* and *E. gracilis* or *E. gracilis* and *E. salubris*. The understorey where *E. loxophleba* is present consists usually of teatree — *Melaleuca uncinata* but also *M. hamulosa* and *M. cymbifolia* — with *Templetonia sulcata* and a ground layer of the same succulents as listed above.

The principal types of red loam valley soils and their associated woodland dominants are as follows:

1 A fine incoherent clay over calcareous hardpan at depth 5-15 cm. Valley bottoms, usually adjacent to saline areas. Cover — morrell *(Eucalyptus longicornis)*.

2 A brown to red-brown sandy loam increasing in texture to a clay within 7.5 cm, overlying pallid zone at >90 cm, sometimes with *kunkar* (lime concretions) in the profile. Valley bottoms. Cover — salmon gum *(E. salmonophloia)* and gimlet *(E. salubris)*.

3. A brown to red-brown coarse sandy loam to gritty clay. Middle to upper slopes. Cover — wandoo *(E. wandoo)*.

4 A brown to red-brown sandy loam increasing in texture with depth overlying pallid zone at 60-90 cm. Topographic position indiscriminate. Cover — York gum *(E. loxophleba)*.

York gum and wandoo tend to predominate in the western wheat belt, salmon gum and gimlet further east, where the similar tree *E. capillosa* replaces *E. wandoo*. Salmon gum is the tallest tree, reaching over 20 m in height with gimlet somewhat smaller. Wandoo may also reach 15-20 m but York gum is a smaller tree with a low-branching habit and rarely exceeds 12 m. Since the soil most favourable to each species forms a scale with varying sand content and depth over clay, and soil types merge into one another, these eucalypts may form mixed stands, most commonly York gum with wandoo, York gum with salmon gum and salmon gum with gimlet. Other associated trees are few, *E. gracilis,* and *E. kondininensis* confined to salt country, *E. transcontinentalis* and *E. erythronema* in drier eastern areas. The woodland may be very open and irregular, with trees as much as 60 m apart. Jam *(Acacia acuminata)* may be frequent as a smaller understorey tree, mainly with York gum and wandoo. *Casuarina huegeliana* is another smaller tree but usually favours rocky patches and may form dense stands around granite outcrops. Occasional mallees of *E. redunca* and *E. transcontinentalis* are also seen.

Shrub and ground layers in these woodlands are normally sparse due to the competition of the trees and species diversity is likewise reduced. The flora is not as rich as that of the kwongan, none the less long lists of species for different areas are available in (10) compiled from the work of various authors. Space does not permit these to be repeated here. Shrubs are predominantly of a sclerophyll character. There is a substantial component of herbaceous species both annual and perennial. With few exceptions it does not appear to be possible to

FIG 3.118

In the Wheatbelt Region the lowest parts of the landscape are occupied by salt lakes and flats in which drainage water collects and evaporates, as the rainfall is not high enough to operate active river systems. Such saline areas carry salt-tolerant vegetation. In the background are some small lakes with beds of bare salt surrounded by samphires, *Halosarcia* spp. (See Figs 1.28 and 1.29.) On slightly higher ground woody plants appear. The trees are *Eucalyptus melanoxylon*. Johnston Lakes.

FIG 3.119

Around the larger lakes, such as Lake Moore in this case, the influence of salt is spread around a wide marginal belt since salt particles are blown by the wind from the dry lake bed. A zonation may then occur with samphires (darker plants on right) closest to the lake and saltbush (lighter, grey-coloured plants). Saltbush is *Atriplex*, a world wide genus, and there are several Australian species. The ground in between is occupied by annuals which are grazed by sheep.

FIG 3.120

Even salt vegetation can produce a show of colour! The pigface *Disphyma crassifolium*, belonging to the *Mesembryanthemum* family, makes gay show here among saltbushes, contrasting with sombre samphires on the right. Samphires have very inconspicuous white flowers. Salt flat at Jibberding.

Fig 3.121

Upslope from the saline depressions in the Wheatbelt come the woodlands on the red loam valley soils, and chief among the woodland trees is the salmon gum, *Eucalyptus salmonophloia*, often in association with gimlet *(E. salubris)*. Salmon gum is perhaps the most handsome tree in the State, tall, elegant, beautifully shaped, with shining leaves. After the bark has peeled each summer it is shining copper colour. The tree's name is derived from the inner bark which when cut or blazed resembles tinned salmon (so they say). Gimlet on the other hand has a twisted trunk like a carpenter's gimlet. Understoreys vary, and may be of a saltbush, broombush or sclerophyll type. Beacon North area.

Fig 3.122

Wandoo *(Eucalyptus wandoo)* is an important wheatbelt tree, mainly in the west, becoming more rare in an easterly direction. It occupies paler, more siliceous loam soils and is distinguished by its white bark and dull, bluish leaves. There is frequently rather little understorey in wandoo, consisting of small sclerophyll shrubs. East of Kellerberrin *E. wandoo* is replaced by the superficially similar species *E. capillosa*, seen here near Westonia and in Fig 2.19.

FIG 3.123

York gum *(E. loxophleba)* may occur in pure stands or associate with either wandoo or salmon gum. The soil is typically a red loam sandy at the surface. York gum country was considered to offer the best prospects for farming and there is now little of it left so that it is difficult to find a virgin stand to photograph. Apart from the large shrubs such as raspberry jam tree, *Acacia acuminata* (because the dry heartwood smells of raspberry jam), York gum stands typically have a herbaceous ground layer. Although this appears beautifully grassy at a distance, however, most of it is sedges and rushes (Restionaceae) mixed with spring annuals. New Norcia.

FIG 3.124

Mallee (eucalypts in a short, many-stemmed form) is not a well-developed formation in the Wheatbelt, occupying usually narrow bands between the woodlands of the valley soils and the kwongan of the sandplains on higher ground. Mallee has an intermediate soil with a sandy wash over loam, and an intermediate flora with many understorey plants shared with the woodland on one hand and the kwongan on the other. *E. redunca* seems to be the most consistent mallee species. Dongolocking Reserve, near Dumbleyung.

FIG 3.125

High ground in the Wheatbelt is normally occupied by lateritic soils consisting of ferruginous laterite and sand. These soils are agriculturally very poor but can be cultivated if trace elements are given to the crop. Natural vegetation is no more than scrub—kwongan, as it is called—due it is believed more to the poor moisture capacity of such soils than their infertility. Laterite can be seen ploughed up at the front of this photograph. Where plants are growing directly on laterite in this way vegetation consists of thickets, of *Casuarina campestris* in the western Wheatbelt or of *Acacia neurophylla* as seen here in the drier interior. Beacon North area.

distinguish between understorey communities associated with particular dominant eucalypts and it may be that as elsewhere in Australia the understorey varies independently of the overwood, responding to different ecological factors. This is illustrated by the fact that on saline bottomlands under whatever eucalypt is present, the understorey may change to saltbush of *Atriplex paludosa* and other halophytes (*Sclerolaena, Enchylaena, Maireana, Rhagodia* spp.). However, quite frequently under salmon gum the 3 m shrub boree *(Melaleuca pauperiflora)* may become virtually the sole understorey species.

Mallee is not extensive in the Wheatbelt, consisting of relatively small patches scattered both in kwongan and woodland or forming a transition between the two, wherever the necessary sand-over-clay soil is to be found. Height varies from 2 to 12 m averaging 5–7 m, and density is most commonly estimated at 10–30% cover. *Eucalyptus redunca* is the most consistent species but at least a dozen others can be enumerated (see Section 3.4). Understoreys vary from very open to very dense, the most typical being a dense layer of *Melaleuca* spp. but other types include mixed shrubs or a thick growth of the reed *Ecdeicolea monostachya*.

As in the Northern Sandplains Region, sandplain soils vary and may be conveniently grouped into three main types:

1 Shallow: Yellow-brown sand over massive laterite or ironstone gravel from surface to 45 cm.
2 Moderate: Yellow-brown sand over sandy loam over ironstone gravel at 45-90 cm.
3 Deep: Yellow-brown sand >90 cm deep.

These soils are not necessarily topographically related though the 'shallow' type may occur towards the edges of a sandplain especially if, as frequently in the east, a small laterite scarp or 'breakaway' forms the boundary. It is evident that wind has moved the surface sands in the past so that depth is a haphazard factor. Accordingly sandplain vegetation may often be very patchy and form intricate mosaics.

Vegetation varies with rainfall and may be roughly distinguished into communities of the moister west and the drier east. Taking first the western sector, on 'shallow' soil the formation is thicket 1–2.5 m tall with *Casuarina campestris* as the sole tall shrub, densely crowded together. Some smaller shrubs up to 1 m manage to crowd in, of *Acacia, Hakea, Melaleuca. Grevillea petrophiloides* can be conspicuous. Smaller ericoid shrubs e.g. *Baeckea crispiflora* will be present, various sedges — *Gahnia, Lepidosperma* — and the reed *Ecdeiocolea*. It is possible in some cases for this thicket to occur with scattered wandoo trees.

On soil of moderate depth the typical cover belongs to the *Acacia-Casuarina-Melaleuca* alliance and is known as mixed kwongan. Height and structure vary with the time since the last fire. The typical aspect is of scrub-heath with an open upper layer 2-3 m tall in which *Casuarina acutivalvis* is the character species, and a denser lower shrub layer. The flora is richer than in the dense *Casuarina campestris* thickets, the smaller shrub layer is usually dominant and while Restionaceae and sedges are present they do not form a conspicuous ground layer. Composition of mixed kwongan varies considerably as it is very widespread and long lists of species are available in (10). Several species of *Acacia* and *Melaleuca, Casuarina campestris* and *C. corniculata* are normally present among the taller shrubs to characterise the alliance with *M. cordata* as the most common small species. A variant is thicket dominated by *Melaleuca uncinata* which appears to favour winter-wet places on sandy clays. This community is dense and poor in species but is noteworthy as the home of the underground orchid *Rhizanthella gardneri. Ecdeiocolea monostachya* is always present and may be abundant.

On deep sand in the western sector we have the *Banksia-Xylomelum* alliance as in the Northern Sandplains with the typical *Actinostrobus arenarius, Banksia attenuata, B. prionotes* and *Xylomelum angustifolium*. Characteristic smaller shrubs are *Eremaea pauciflora* and *Grevillea pritzelii*. Associated flora has been listed in (10).

In the drier eastern sector, on shallow soil thickets of *Acacia neurophylla* replace those of *Casuarina campestris*. As before there are few other species. On the moderately deep soil we have mixed kwongan of the same general character as before with *Casuarina acutivalvis* conspicuous, but with species changes taking place in the associated flora. On the deep sands a profound change takes place to sole-dominant thickets of *Acacia resinomarginea*, which is a small tree. Height varies with age of the stand and is usually 3-5 m but may reach as much as 10 m. In the dense thicket formed there are few other species but there are usually some small ericoid shrubs such as *Thryptomene australis, Eriostemon thryptomonoides* and *Phebalium tuberculosum*, and *Ecdeiocolea* also.

It remains to speak of granite outcrops which are common in some areas. Low outcrops may expose slabs of rock covered at the sides by shallow sandy soil, on

Fig 3.126

Where sand overlies laterite, vegetation improves to a formation called 'mixed kwongan' dominated by species of *Acacia, Casuarina* and *Melaleuca*. With increased depth of sand, height of the vegetation improves further, in the western Wheatbelt to an open community dominated by *Banksia* spp. and *Xylomelum angustifolium*. In the drier interior one finds single-dominant stands of *Acacia resinomarginea* as seen here. It is a paradox that this community is taller and denser than the former in spite of the lower rainfall. On rabbit-proof fence north of Bonnie Rock.

Fig 3.127

In the Wheatbelt kwongan for the most part is not such as to give spectacular displays of wildflowers in the spring, consisting as it does of thickets rather than of more open scrub-heath. There are exceptions however, one of the best known being the Reynoldson's Reserve near Wongan Hills. This has an interesting history. It was cleared and cultivated in 1953, cropped in 1954, and then abandoned to revert to bush. Dense regeneration mainly of *Verticordia* spp. then took place with a few other heath plants, which are still dominant to this day although other taller kwongan species are gradually invading. Here *Verticordia brownii* (white) and *V. monadelpha* (pink).

Fig 3.128

Wheatbelt thickets have their interesting plants, but they often have to be searched for. One of the most remarkable is the wreath leschenaultia, *Lechenaultia macrantha*, found in the area of Wubin and Mullewa beneath *Acacia-Casuarina* thickets. It sprouts annually from a rootstock and grows to a diameter of 50 cm, flowering around the outer perimeter as it does so. Many people think that leschenaultias must be blue, but this is by no means the case. Near Mt Singleton.

FIG 3.129
After burning of kwongan thickets the regeneration contains a number of pioneer plants which disappear as the thicket becomes re-established, and many of these are soft leaved and unlike the climax sclerophyll species. Standing here among the dead skeletons of the thicket we see small trees of the native poplar, *Codonscarpus cotinifolius*, already senescent; and among young regeneration, *Glischrocaryon aureum* (yellow) and *Keraudrenia integrifolia* (purple).

FIG 3.130
The granite rock habitat is a common one in the Wheatbelt, where there are many outcrops, some very extensive, and mostly still in a natural state. On Mt Caroline near Tammin shrubs have taken root in crevices or where soil has accumulated on ledges, while exposed rock is covered with lichens or mats of moss, or the pin grass *Borya nitida*. Among the shrubs, *Thryptomene australis* is seen in flower here with jam, *Acacia acuminata*. Mt Caroline is a locality for one of the rare mallees, *Eucalyptus caesia*, which are only found on such outcrops. *E. caesia* is very ornamental and has become well known in gardens.

which *Casuarina campestris* thickets again appear, this time even into the drier eastern sector where however the thicket is more open so that *Calothamnus gilesii* and the succulent shrub *Calycopeplus ephedroides* join the dominants. The outcrop is frequently fringed by *Thryptomene australis,* and the rock itself bears clumps of 'pin grass' *Borya nitida.*

Larger outcrop areas are a mosaic of bare granite slopes with shallow or deep soil pockets and are surrounded by shrublands or low woodlands on detrital material receiving run-off from the rock. As a typical example, on the Yorkrakine Rock bare areas tend to be on steeper slopes and have only blue-green algae stains or occasional lichens, particularly *Rhizocarpon* sp. or *Caloplaca irrubescens.* Flatter areas have *Grimmea* sp. moss or *Parmelia conspersa* lichen or colonies of *Borya nitida.* Shallow soil pockets have *Kunzea pulchella, Dodonaea viscosa* or *Diplolaena microcephala* shrubs. Deeper pockets contain *Melaleuca elliptica, Kunzea pulchella, Scaevola spinescens, Dodonaea viscosa, Stypandra imbricata* or *Lepidosperma costale* with occasional *Acacia lasiocalyx* trees. Still deeper and more extensive soil pockets develop thickets of *Leptospermum erubescens* containing also the last-mentioned group of species and with still more root room small trees of *Casuarina huegeliana* appear, and even *Acacia acuminata* and *Eucalyptus loxophleba* as may be seen on Mt Caroline. Typically however the last-named three species dominate associations peripheral to outcrops. *Casuarina huegeliana* occurs on coarse sandy soil, and the others on loam, but they can intermingle. *Casuarina* stands vary from open to very dense and from 5 to 10 m in height, sometimes more, sometimes less. The understorey varies with the density of the overwood. *Leptospermum erubescens* is normally conspicuous in the shrub layer, and large tussocks of *Lepidosperma gracile* or *L. tenue.* Other associates are likely to be the shrubs mentioned above and annual grasses and forbs such as *Avena* spp., *Stipa elegantissima, S. hemipogon* and *Waitzia acuminata* which appear in winter.

Large, high rocks tend to be more sparsely vegetated than this, with lichens and *Borya nitida, Kunzea pulchella, Diplolaena microcephala* and the 'blind grass' *Stypandra imbricata* rooting in crevices. These last are confined to granite rocks and it is interesting that several mallee eucalypts are also confined to rocks in certain areas, e.g. *E. caesia, E. crucis, E. kruseana, E. orbifolia* and *E. websteriana.* They are closely related botanically and as they have showy flowers and attractive form make excellent garden subjects in spite of their restricted natural habitat.

FIG 3.131

One of the most showy and attractive granite rock plants is the shrub *Kunzea pulchella*. In nature it is absolutely confined to rock crevices on granite outcrops, yet it is most successful in gardens and easy to grow. This specimen was taken in cultivation.

Fig 3.132

Eucalyptus orbifolia is one of the group of mallees confined to granite rock outcrops and is distinguished by its silvery leaves. Others, besides *E. caesia* mentioned under Fig 3.130, are *E. crucis*, *E. kruseana* and *E. websteriana*, all of them decorative in cultivation to which they are very amenable. All share in a curious feature of having similar bark which peels in a reticulated fashion exposing paler brown bark beneath. Yorkrakine Rock.

Fig 3.133

The Wongan Hills near the township of that name are botanically interesting in a number of ways and have been written up in a Field Naturalists Club handbook. From the point of view of vegetation they are interesting in being one of the few localities where kwongan occurs other than on a sandplain. The slopes shown here are covered, apart from the scattered trees of *Eucalyptus transcontinentalis*, by shrubland of *Melaleuca undulata* as virtually the sole species. This community may be compared with that in the Moresby Range near Geraldton (Fig 3.94) where *Melaleuca megacephala* is the principal species and again represents kwongan on hill slopes.

3.4 *Mallee Region*

ROE BOTANICAL DISTRICT

The general cover is mallee with *Eucalyptus eremophila* the most consistent species. Patches of eucalypt woodland occur on lower ground, and scrub heath and *Casuarina* thickets on residual plateau soils.

Climate: Dry warm mediterranean; winter precipitation annual total 300-500 mm, with 7-8 dry months.

Geology: Archaean and Proterozoic granites overlain in the east by early Tertiary sediments.

Topography and soils: Gently undulating country of low relief with duplex mallee soils i.e. sand overlying clay.

Area: 78 957 km², 44% cleared.

Boundaries: The limit of the mallee formation is the all-round boundary. On the north and west the mallee gives way to eucalypt woodland and on the south to mallee-heath.

The Mallee Region lies to the southeast of the Wheatbelt, extending from it as far as the Great Australian Bight. Like the Wheatbelt it occupies part of the surface of the Yilgarn Plateau but there is a change in the rainfall pattern exemplified by the climate diagrams for Perth and Ravensthorpe in Fig. 3.135. In these two cases the length of the dry season is the same but the total rainfall is very different. Perth, typical of west coastal districts, has a high winter rainfall capable of recharging the deep-weathered pallid zone of the soil profiles with moisture, and thus of supporting forest and woodland. In the southeast as at Ravensthorpe the lower rainfall is much less effective even though it is spread over an equivalent season, and accordingly we find no forest and very little woodland in this region where the dominant cover is still of eucalyptus, but only in mallee form.

FIG 3.134

Adjacent to the Wheatbelt, the Mallee Region is as the name implies characterised by the mallee formation which originally covered 50% of the region or more. Mallee is formed by small species of eucalypts which grow with numerous thin stems arising from a common rootstock. There is usually a dense understorey of sclerophyll shrubs, mainly *Melaleuca*, which is very inflammable. After fire the mallees resprout from the base. Various heights and densities are found, various combinations of species, and the ground layer may sometimes be of saltbush or spinifex. Between 100-mile Tank and Lake Hope.

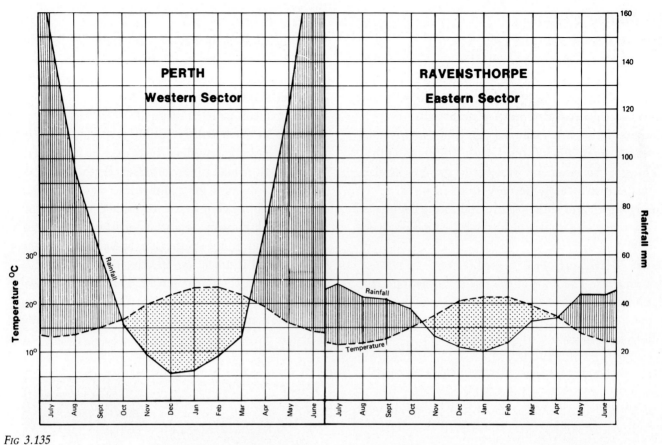

FIG 3.135
Contrasted climate diagrams for Perth on the west coast and Ravensthorpe in the Mallee Region. Summer rainfall and the length of the dry season are much the same, but total winter rainfall is very much less in the mallee.

Mallee is a shrub-eucalypt formation. Each plant has an underground rootstock about 0.03 m³ in size, from which arise numerous slender stems giving a bushy crown similar to the 'broombush' habit. Mallee is subject to frequent fires which destroy the top growth, regeneration taking place from coppice. It is not clear to what extent the mallee habit is genetically controlled or due to fire. Certainly any small *Eucalyptus* of a species having the power to coppice would automatically assume a mallee habit if frequently burnt, but very many mallee species can also be found in tree form, either moderately-sized or in the small but single-stemmed tree form known as *marlock* in Western Australia. The structure of mallee is extremely variable, its height varying with age from the last fire, and its density and associates varying also. The most typical form of mallee is a closed community of mallee habit rising to 3-4.5 m in height, with an understorey of small ericoid shrubs of the genus *Melaleuca*. The understorey may elsewhere consist of mixed shrubs belonging to the scrub-heath where there is a transition to the latter formation, of saltbush under alkaline soil conditions, or of hummock-grass on red sand. In this region the last two types occupy only small patches. The same species of mallee eucalypts may occur over different understoreys and vice-versa, and the stature of the eucalypts may vary without change of species from true mallee to marlocks and small trees.

The suggestion that fire is responsible for the mallee habit is supported by the fact that a number of eucalypt species which are obligate reseeders, not resprouters, occurs within the mallee formation, usually in colonies on certain soils. These are tender to fire and must regenerate from seed but do so successfully, growing into small single-stemmed trees 5 to 7 m in height. Such trees are not mallee but of the marlock type. These occurrences lead to the supposition that the climax vegetation of the Mallee Region in the absence of fire would be low forest.

In addition to the change in rainfall pattern in the Mallee Region there is a change in soil types which has no doubt resulted from the difference in climate. It was shown in the catena for the Wheatbelt Region (Fig. 3.115) that mallee is associated with a duplex soil of sand over clay in two distinct layers, the sand having

Fig 3.136

Not all denizens of the mallee are many-stemmed resprouting species. Some are single-stemmed small trees called marlocks which regenerate from seed, in this case *E. falcata* and *E. flocktoniae*. There are about ten of these marlock species which usually grow in colonies on specific soil types, forming a low forest. It is therefore believed that low forest is actually the climax formation in the mallee region, and that mallee as such has been degraded by burning. Southeast of Lake Magenta.

Fig 3.137

Diagrammatic catenary sequence for the Mallee Region showing relation of mallee with other vegetation units.

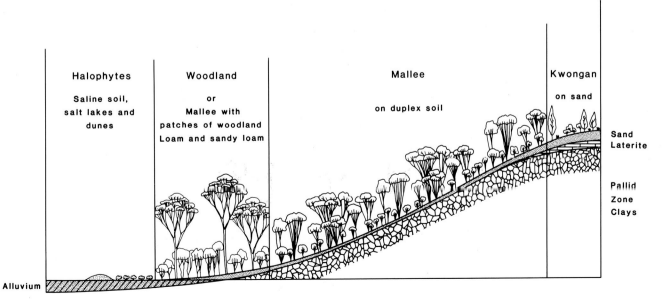

drifted downslope from the sandplain. In the Mallee Region this component of the catena expands to occupy the major proportion, and the duplex soil appears to have originated *in situ* (Fig. 3.137). Sandplains still occur on highest ground with their typical lateritic sandy soils, but merge laterally into the mallee. Salt flats and woodlands still occur in valley bottoms but to a lesser extent and again merge upslope into mallee which covers almost all the long gentle slopes of the region. The typical mallee soil is described as a sandy alkaline yellow-mottled duplex soil known technically as a *soloth*. The underlying clay layer is of acid reaction. The mallee soils of South Australia and Victoria are somewhat different.

There have been difficulties with the agricultural development of mallee soils in Western Australia so that less than half of the region has been alienated for this purpose, mainly in the western part up to Lake King and Ravensthorpe. Further east salinity becomes an increasing problem. There is no lack, therefore, of original natural vegetation. Mallee is mapped as having originally covered 50% of the region and 'mallee with patches of woodland' 25%. Kwongan on sandplains accounted for 7.5% and woodlands 10%. Mallee was therefore overwhelmingly dominant.

Salt flats and associated vegetation are essentially the same as in the Wheatbelt Region. Thickets of *Melaleuca* spp. typically surround lakes and samphire flats but much has been affected by salinisation of the bottomlands and has died or is in bad condition.

In bottomland country a typical zonation would be as follows. The bed of a salt lake if not consisting of bare mud and salt would be vegetated with samphire — *Halosarcia halocnemoides* and *Sarcocornia blackiana*. Its margin carries a zone of small, low *Frankenia*. Around the lake is an irregular stand of teatrees, mainly *Melaleuca thyoides* but also *M. scabra*, *M. lateriflora* and *M. hamulosa*, tall shrubs with dense foliage. A little further out the community is joined by *Eucalyptus kondininensis* as a taller tree, next the teatree changes to *M. pauperiflora*, and *E. salmonophloia* and *E. longicornis* come in. Furthest away from the lake there is a mixed woodland of trees and large mallees including *E. salubris*, *E. gracilis*, *E. loxophleba*, *E. oleosa*, *E. sheathiana*, *E. flocktoniae*, *E. annulata* and *E. spathulata*. A saltbush understorey to the woodland of *Atriplex hymenotheca* is often present, otherwise scattered woody shrubs of *Acacia*, *Eremophila*, *Pittosporum* and some grasses. *Eucalyptus loxophleba* occurs almost entirely in mallee form in this region; *E. capillosa* is rarely seen. In the south where grey alkaline clay is present in bottom lands *E. occidentalis* is a common tree.

In the mallee formation *E. eremophila* is the most consistent species, being nearly always present, but it has numerous associates, about three of which seem usually to be present at any one site. A full list is as follows:

E. celastroides, *E. cerasiformis*, *E. conglobata*, *E. cylindriflora*, *E. deflexa*, *E. dielsii*, *E. eremophila*, *E. foecunda*, *E. georgei*, *E. goniantha*, *E. incrassata*, *E. leptocalyx*, *E. longicornis*, *E. loxophleba*, *E. merrickiae*, *E. micranthera*, *E. oleosa*, *E. ovularis*, *E. pileata*, *E. redunca*, *E. sheathiana*, *E. transcontinentalis*, *E. uncinata*.

The understorey is most commonly dominated by one or more species of *Melaleuca* forming a more or less continuous layer in which other casual species are interspersed. The recorded composition is: *Melaleuca pungens*, *M. spicigera*, *M. viminea*, *M. urceolaris*, *M. hamulosa* (swampy places), *Acacia beauverdiana*, *A. ericifolia*, *A. fragilis*, *A. hemiteles*, *Baeckea grandibracteata*, *B. polystemona*, *Boronia ternata*, *Callitris roei*, *Cryptandra parviflora*, *Eremophila lehmanniana*, *Gastrolobium parvifolium*, *Grevillea paradoxa*, *G. petrophiloides*, *Hakea falcata*, *H. multilineata*, *Hybanthus floribundus*, *Isopogon scabriusculus*, *I. teretifolius*, *Leptospermum erubescens*, *Micromyrtus* cf. *imbricata*, *Mirbelia spinosa*, *Petrophile ericifolia*, *Phebalium filifolium*, *Pimelea suaveolens*, *Pityrodia axillaris*, *P. lepidota*, *Pultenaea capitata*.

A very small mallee, *E. grossa*, with fleshy leaves, is found around low granite rock outcrops. *E. loxophleba* is commonly seen on sandy soil around larger outcrops as at the Wave Rock, Hyden. In the far east on the limestone adjoining the Nullarbor Plain the mallees are *E. cooperiana* and *E. socialis*.

Marlock species are associated with particular soil types, e.g.:

E. falcata and *E. gardneri* on ironstone ridges
E. erythronema and *E. flocktoniae* on mallee soil
E. annulata and *E. diptera* on red clay
E. platypus and *E. spathulata* on winter-wet grey clay
E. forrestiana on gilgai country.

E. erythronema and *E. flocktoniae* occur mixed in with the mallee. *E. forrestiana* forms a particular association with *E. eremophila*. The other species are normally coloniate and may form quite dense stands singly or together, suppressing any understorey and thus acquiring some immunity from fire.

Sandplain vegetation is kwongan, similar to the

Fig 3.138

The Mallee Region is without organised drainage so that the bottomlands contain small salt lakes in the valleys trending north and east as part of the Swan–Avon basin. Samphire and saltbush communities surround the lakes as in the Wheatbelt.

Fig 3.139

Sometimes small salt lakes occur at the foot of granite rocks, absorbing the runoff from the rock, giving rise to interesting local communities. In this case *Verticordia preissii* makes a conspicuous show while the larger shrubs are *Melaleuca elliptica* which has large red inflorescences in its proper season. Emu Rock near Lake Carmody.

Fig 3.140

The larger lakes such as the Johnston Lakes complex often have wide areas of salt-affected land around them due to wind blowing salt from the dry lake bed, and salt tolerant communities develop. The trees here are mirret *(Eucalyptus flocktoniae)* with a ground layer of saltbush *(Atriplex hymenotheca)*. Lake Hope.

FIG 3.141

Less salty country may carry salmon gum (*E. salmonophloia*) and Kondinin blackbutt (*E. kondininensis*) with a normal sclerophyll shrub understorey. In the Mallee Region in general, woodlands can only occupy the moister bottomland soils, and then only in a patchy manner, because of the lower wet-season rainfall which does not afford adequate ground water recharge to support stands of large trees through the dry season.

FIG 3.142

The mallee itself which covers most of the region is of variable height and density but the above is a good average sample. The clumps of mallee are each growing from a massive woody rootstock. In the course of land development these are ploughed out by farmers and piled in heaps as the much-prized 'mallee roots' for firewood. The understorey in mallee consists typically of species of *Melaleuca*. The mallees here are *E. redunca* and *E. uncinata*. Jerramungup District.

FIG 3.143

Taller, denser mallee inland from Esperance consists of *E. oleosa* and *E. eremophila*. The latter is generally dominant inland, and *E. redunca–E. uncinata* nearer the coast. *E. oleosa* favours calcareous soils. Mt Ney area.

FIG 3.144
Along the inland boundary of the Mallee Region the mallee merges gradually into woodlands and it is sometimes difficult to distinguish between mallee and woodland which has been burnt. Here the size of the dead trees shows that woodland was present but the stand is of course at present regenerating as mallee.

FIG 3.145
Within the mallee, stands of single-stemmed small trees called marlocks occur on certain soil types, always regenerating from seed after fire as they do not resprout. Moort *(E. platypus)* occurs on patches of heavy clay soil.

FIG 3.146
Replacing *E. platypus* towards the east is *E. diptera* which has a shining copper bark and can be easily confused with small salmon gum. It rarely becomes more than a small whipstick tree however. The stand here has been recently burnt killing both trees and understorey. East of Lake King.

mixed kwongan of the wheatbelt, and varying in a mosaic fashion according to depth of sandy topsoil and amount of ironstone present. It may appear as scrub-heath with species of Proteaceae dominant or as thicket with *Casuarina* spp. dominant. In the former case

Grevillea hookeriana is a character species with *Banksia baueri, B. elderiana, B. media* and several species of *Hakea, Isopogon* and *Petrophile.* Dominant casuarinas are *C. acutivalvis, C. campestris* and *C. corniculata.* Lists of associated species will be found in (11). Plants of

FIG 3.147

On higher ground mallee is replaced by kwongan, and on upper slopes the two intergrade in a very patchy manner. The work of B.G. Muir on these communities emphasised their 'highly mosaic character' as he expressed it. In the above photograph kwongan in the foreground is seen running into mallee in the background with an outstanding clump of *E. falcata*. West Bendering Reserve.

FIG 3.148

Kwongan in this region is a rich assemblage with a lower layer of ericoid shrubs and scattered larger bushes mainly of Proteaceae. Smaller *Verticordia* in the foreground are about to come into flower. The larger shrubs are (centre) an *Isopogon* and some *Banksia audax*; scattered larger *Casuarina acutivalvis* behind. East of Lake King.

FIG 3.149

When in full flower kwongan in the Mallee Region can be as spectacular as any in the State, mainly in October–November when the verticordias are out. Three species are seen here, the white *V. roei*, the yellow *V. serrata* and the pink *V. picta*. The taller shrub at right rear is *Grevillea hookeriana* which is characteristic of these sandplains. Some taller sombre *Casuarina* also dot the landscape. New Holland's Rocks.

Fig 3.150

A smaller very attractive grevillea is G. *asteriscosa* which is only a casual species, called star-leaf grevillea from its distinctive leaf shape. It occurs between Kulin and Bruce Rock, in this case at Narembeen.

Fig 3.151

The dwarf blackboy *Xanthorrhoea nana* is found here and there in the Wheatbelt. It is trunkless and the flower spikes are quite short, coming out sideways from the apical bud and bending upward.

Fig 3.152

Various species of dwarf banksia are found in these southern regions. In response to much fire and poor soil they have retreated underground, an adaptation also found in African Proteaceae. In some species as in this *B. gardneri* the stems are prostrate, i.e. they lie on the soil, while in others they are actually subterranean, growing below the surface and pushing up their leaves and flowerheads from below. It can be distinctly startling to find a solitary banksia head protruding from the soil!

Fig 3.153

One of the most extraordinary kwongan plants in Western Australia is the compass plant, *Casuarina pinaster*. This species has separate male and female plants, and the females develop a long pointed stem which leans over always pointing towards the south. A study showed that they all point within 45° of true south, the average bearing being 10°E. The female flowers and seed cones are borne on the upper north side of the plant, which no doubt has a connection with the tendency to lean. The male bushes are normally symmetrical. The foliage in this species consists of short prickly green stems resembling pine needles, hence the name. On Hall's Track west of Lake Magenta.

FIG 3.154

Among many granite rock outcrops in the Mallee Region, the Wave Rock at Hyden is a favourite tourist attraction. The granite has weathered out into the wave shape, made more spectacular by patterns of algae growing where water runs off the rock. The rock is used as a catchment for the town water supply and part of a wall built to direct runoff into the dam can be seen at upper left. The foot of the rock is surrounded by thickets of *Casuarina huegeliana* and *Acacia lasiocalyx* grading into woodland of York gum.

FIG 3.155

Many other outcrops in the Hyden district offer scenic and botanical interest to the visitor. At The Humps (above) the rock carries some *Casuarina huegeliana* in pockets of soil while *Verticordia preissii* makes a show in the foreground.

Fig 3.156

A small mallee, *Eucalyptus grossa*, grows in the shallow soil on low outcrops of granite together with *Casuarina* and *Melaleuca* species. The name is derived from its thick fleshy leaves. This species is only a shrub normally, very rarely attaining any larger size.

Fig 3.157

In undeveloped country between Lake King and Salmon Gums stands Peak Charles here seen from 15 km to the north with its adjacent group of the Fitzgerald Peaks. It is the greatest granite rock in the State rising to 650 m above sea level, 450 m above the surrounding country. The Fitzgerald Peaks, named after Governor Charles Fitzgerald and his wife Eleanor, have been placed in a national park. Foreground vegetation is *Casuarina* thicket, a kwongan community.

Fig 3.158

Peak Charles differs rather markedly from other granite outcrops in size, height, steepness and other respects, and its vegetation has correspondingly unique features. There is an undescribed species of *Leptospermum* which is peculiar to the site and is common on the summit and slopes in thickets containing *Calothamnus quadrifidus, C. gilesii, Melaleuca fulgens, Callitris preissii* and other species. A salt lake is seen spread out below.

FIG 3.159

On Peak Charles, looking east towards Peak Eleanora.

particular interest include the small white mallee *E. albida* which is normally present, the dwarf blackboy *Xanthorrhoea nana* and the 'compass plant' *Casuarina pinaster* whose female bushes lean over at an angle pointing to the south. A low mallee of curious gnarled habit, *E. burracoppinensis,* is locally conspicuous in kwongan of the north and west.

Granite rock outcrops are locally conspicuous as they are elsewhere and include the famous Wave Rock at Hyden and further east Peak Charles, the greatest granite rock in the State, rising to 650 m. In general the vegetation of rock outcrops is much the same as described under the Wheatbelt. Peak Charles differs rather markedly from other granite outcrops in size, height, steepness and other respects, and its vegetation has correspondingly unique features. Formed of a pink granite, the rock soars some 450 m above the surrounding country. In many places the slopes are precipitous and bare of vegetation except lichens, but elsewhere patches of soil cling to the rock and scrub has developed. The summit is largely bare with a few old gnarled bushes of a *Leptospermum* (undescribed species peculiar to this site) 60 cm tall, growing in crevices. The middle slopes bear thickets 1.25 to 2 m tall containing the *Leptospermum, Calothamnus quadrifidus* and *C. gilesii, Melaleuca fulgens, Callitris preissii, Baeckea behrii, Darwinia* sp., *Hibbertia mucronata, Labichea lanceolata, Anthocercis genistoides, Drummondita ericoides, Oxylobium parviflorum. Kunzea pulchella* was not observed here, its niche being occupied by *Melaleuca fulgens*.

3.5 *Esperance Plains Region*

EYRE BOTANICAL DISTRICT

Scrub-heath and mallee-heath on sandplain with tallerack (*Eucalyptus tetragona*) as characteristic species. Mallee (*E. redunca, E. incrassata*) occupies valleys incised in the plain.

Climate: Warm mediterranean; winter precipitation, total of 500-700 mm per annum, with 5-6 dry months.

Geology: Mainly Eocene sediments with outcrops of granites and quartzites.

Topography and soils: A plain with a little-dissected prior land surface rising gently from near sea level to a height of 200 m, broken by quartzite ranges (Barren Ranges, Russell Range) and granite domes. Soils are sandy overlying clay or ironstone gravels.

Area: 28 702 km², 52% cleared.

Boundaries: The limit of the predominant mallee-heath formation is the all-round boundary. On the north it gives way to mallee and at the western end to low woodland.

The Esperance Plains Region shares the climatic characteristics of the southeast — low winter rainfall in relation to length of the dry season — which favour mallee against woodland, so that it can be said to form part of a greater mallee region. However its geological structure is distinctive as are the resulting soils, and the dominant vegetation is mallee-heath rather than mallee proper, so that this is separated as a distinct region.

The Esperance Plains are contiguous to the southern coast, and the principal features have been developed in relation to this coastline. The edge of the continental shelf is only 30 km offshore at Bremer Bay. The coastline itself shows evidence of uplift during the Tertiary, as there is a coastal strip mantled with Tertiary sediments and other indications. The modern coastline is controlled in the south from Cape Riche to Point Hood by a chain of large granite bosses — Mt George, Mt Belches, Mt Groper, Tooleburrup Hill and the Doubtful Islands — some of which rise to over 150 m above sea-level. These are responsible for the numerous bays — Cheyne Bay, Dillon Bay, Bremer Bay — and headlands — Cape Riche, Groper Bluff, Cape Knob, Point Henry and Point Hood. Inland of these features is a flat and monotonous coastal plain which rises gently inland from sea level at the coastline to about 100 m and is formed on the Tertiary sediments of the Plantagenet Group, which are sands and siltsones now generally referred to Upper Eocene age. The seaward margin of the plain is much mantled with Quaternary drift sands, some consolidated, some still in movement. The surface of the plain is marked by numerous small, often circular depressions, which fill with water and become swamps in winter. The rivers, the Pallinup and its tributary the Corackerup Creek, the Bremer River and the Gairdner River, have cut deep and steep-sided trenches into this plain.

The plain continues to the east inland of the Barren Ranges at the same 90 m level as their platform deeply entrenched by the Fitzgerald and Hamersley Rivers. Once the Barrens are passed, beyond the Phillips River, the plains slope gently up from the shore, where there is a narrow band of dunes, lakes and inlets built upon a platform slightly above present sea level and probably of Recent age. Rivers have cut deep steep-sided valleys in the plain, a development considered to have taken place during Quaternary low sea levels.

Chief soils are sandy neutral yellow-mottled soils containing variable amounts of ironstone gravel in the surface sand, alternating with leached sands, which sometimes contain ironstone gravel, and are underlain by a clay substrate at depths of 90-150 cm. Valleys incised into the Esperance Plains, with a mallee vegetation, have hard alkaline and neutral yellow-mottled soils.

What would otherwise be a relatively featureless and monotonous coastal plain is broken by the various ranges and granite hills rising from its surface. Beginning in the west, the Stirling Range is the only real mountain range in the southwest of the State. Further east come the Barren Ranges, less massive, less tall, but fronting scenically onto the coast, and close beyond them the Ravensthorpe Range. East of Esperance the plain is broken by numerous granite domes whose smooth contours rise on every hand, with a particular concentration in the Cape Le Grand National Park. Offshore where the plain itself is submerged the granite domes continue to protrude above sea level and form the islands of the Recherche Archipelago. In the far east of the plain, close to Israelite Bay, comes another quartzite range, the Russell Range, consisting of Mt Ragged, Brook Peak and some other hills. These ranges introduce floristic as well as topographic diversity.

Alienation for agriculture in this region is mostly recent and has taken place since the second world war. Agricultural difficulties have been and still are encountered, and alienation now rests at approximately half the region (52%). The principal ranges and hills are all reserved in national parks.

FIG 3.160

The Esperance Plains stretch all along the southern coast from Albany to the Bight and provide vast expanses of monotonous sandplains broken by granite outcrops and some quartzite ranges. The soil consists of sand over laterite in the west grading into sand over clay in the east. Vegetation throughout is kwongan in the form of scrub-heath on the deeper sand but more commonly as mallee-heath dominated by small mallees. *Eucalyptus tetragona* is characteristic, a low mallee of gnarled form with large blue-green leaves and a silvery bloom. It is seen here on a lateritic patch where the ground layer is dominated by *Dryandra*. *Isopogon buxifolius* in the right foreground. Esperance.

The predominant plant formation was originally mallee-heath, the typical vegetation of the plains, which covered 58% of the region. Other kwongan on sandplains totalled 14%. Mallee, mostly in valleys, was 17%, woodland only 4%. Actual area occupied by hills and ranges is relatively small.

Vegetation patterns of the plains are of a mosaic rather than a catenary character owing to their flatness. The plains are broken by small lakes and swampy depressions, and by linear sand ridges east of Esperance. Surface soil is sand of variable depth overlying clay with a band of ironstone pebbles between the two horizons. This laterite stratum is more strongly developed in the west and declines eastward to disappear altogether east of Condingup. Variable depth of sand leads to variation in the structure and composition of the kwongan as it does in other regions. Owing to the weak development of laterite a 'shallow' soil type is uncommon, but the 'moderate' and 'deep' types can be readily distinguished underlying respectively the widespread mallee-heath which is the analogue of mixed kwongan and a *Banksia-Lambertia* association which is the analogue of the *Banksia-Xylomelum* alliance.

Mallee-heath is so called because it is regarded as a special form of scrub-heath in which the taller, scattered shrubs consist principally of small mallee-form eucalypts instead of Proteaceae. It differs from mallee proper in that the mallees are shorter and more scattered, allowing the heath understorey to be the dominant layer. Moreover the commonest and most characteristic species, *Eucalyptus tetragona* (tallerack), is of extremely straggly growth. The stems are twisted and rambling, and form very open clumps, rarely attaining 3.5 m in height. The leaves are large and broad for a eucalypt and with the

FIG 3.161

The glaucous (blue-green) leaves of *Eucalyptus tetragona*, called tallerack, make it conspicuous throughout the Esperance Plains where in fact it gives a dominant character to the vegetation. Here on a sandplain adjacent to the Barren Ranges it is in an early stage after fire and not very tall. Maximum height is shown in Fig 3.160. Ground layer is made up of small ericoid shrubs, sedges and restios as elsewhere in kwongan.

twigs and fruit are notably *glaucous*, i.e. of a bluish colour and thickly covered with a silvery bloom. *E. tetraptera* is not glaucous but has a similar habit. Other associated mallees such as *E. incrassata* are of more normal mallee habit.

The principal species are *Eucalyptus tetragona* and *E. incrassata*, dominance passing from the former to the latter in a west to east direction. On the soils where an ironstone horizon is present, mainly west of Condingup, *E. tetragona* is very common and characteristic, in its typical very glaucous form. East of Condingup where the ironstone disappears, *E. incrassata* becomes the commonest mallee species although *E. tetragona* is still present in a form with green leaves and fruits, lacking the usual glaucousness. There are usually few other large shrubs except for *Hakea cinerea* and *Grevillea hookeriana*. Patches with *Nuytsia* occur in depressions.

The following species are conspicuous in this assemblage but the flora is very rich and more comprehensive lists will be found in (11).

Mallees *Eucalyptus tetragona, E. incrassata, E. redunca, E. goniantha, E. spathulata; E. cooperiana* (on patches of limestone).
Large shrubs *Grevillea hookeriana, Hakea cinerea, H. corymbosa, H. prostrata, Isopogon buxifolius, Lambertia inermis, Petrophile* sp.
Small shrubs *Agonis linearifolia, Brachysema latifolium, Casuarina* spp., *Daviesia ? teretifolia, Dryandra longifolia, D. nivea, Grevillea pectinata, Coopernookia strophiolata, Lechenaultia formosa.*

In patches where the depth of sand in the topsoil is greater, the mallees drop out and are replaced by Proteaceae and *Nuytsia*. In the west *Banksia attenuata* and *Nuytsia* form trees of 4.5 m filled in with large

FIG 3.162

On deep sandy soils of the plains the mallee-heath changes to scrub-heath as the mallees largely drop out and the conspicuous elements become large shrubs of Proteaceae, especially *Banksia speciosa* and *Lambertia inermis* (chittick) with *Grevillea hookeriana* and *Nuytsia floribunda*. This association is analogous to the *Banksia–Xylomelum* association of the Northern Sandplains in general aspect and soil type. The scene here is at Condingup east of Esperance with Condingup Hill, a granite outcrop, on the skyline.

FIG 3.163

The kwongan of the Esperance Plains has much in common with that of the Northern Sandplains. In many cases the same species occur in both, or pairs of closely related species. Aspect of the kwongan can also be very similar. The above taken near Esperance can be compared with Fig 3.92 taken at Badgingarra. At Esperance the hakea is *H. cinerea*, the banksia *B. speciosa*, and the smokebush (*Conospermum*) is also different.

FIG 3.164

One hundred and fifty kilometres east of Esperance near Israelite Bay where the above was taken, the aspect of kwongan on deep sand is just the same. *Banksia speciosa* is still the conspicuous large shrub though *Lambertia* has dropped out. Scattered mallee as in the middleground can be expected, while species of *Melaleuca* and *Verticordia* add colour in the ground layer.

Fig 3.165

Over much of the length of the south coast, its configuration is controlled by granite bosses which form headlands with low sandy bays between. The vegetation of these is often distinctive owing to shallow soil and exposure to wind, and may take the form of cushion plants such as may be expected under alpine conditions but scarcely in Western Australia. *Dryandra pteridifolia* and *Banksia dryandroides* are typical cushion-formers, however in this photograph, on a hill summit south of Bremer Bay, we have rather *Pimelea ferruginea, Isopogon formosus, Hibbertia, Leucopogon, Melaleuca* and numerous sedges.

Fig 3.166

The general vegetation of the Esperance Plains Region is mallee and mallee-heath. Woodlands are rare and are confined to marginal areas or to patches of heavier and moister soil. On the northern border of the plains east of Esperance there is a northward transition to woodland of *E. oleosa* through low woodland and mallee of increasing height. North of Mt Ney.

Fig 3.167

This stand of yate *(E. occidentalis)* is found along the Phillips River on the track from Ravensthorpe into the Barren Ranges. Yate favours a grey alkaline clay soil, and it is usual for there to be rather little undergrowth.

FIG 3.168

The monotony of the Esperance Plains is broken by various small ranges and granite outcrops, beginning in the west with the Stirling Range which falls mostly just within this region. The Stirling Range, set aside as a national park, is the most impressive array of peaks in Western Australia, culminating in Bluff Knoll, 1074 m. The range is not continuous, consisting rather of a number of isolated peaks of much eroded Precambrian rocks. This view looking east from the summit of Talyuberlup in the middle of the park shows Toolbrunup (1052 m) centre with Toll Peak (735 m) to the left and Bluff Knoll on the skyline to the right. The range is for the most part covered with mallee-heath and contains many local endemic species.

shrubs of *B. coccinea*, *B. baxteri*, *Hakea cucullata*, *H. victoria* and *Lambertia inermis*. The heath ground layer is apparently similar throughout. East of the Barren Ranges *Banksia speciosa* replaces the other banksias and the conspicuous larger shrubs become *Lambertia inermis*, *Grevillea hookeriana* and *Nuytsia*.

There are several other aspects of kwongan determined by soil depth. Sometimes the upper layer is fairly dense with numerous species of mallees and large shrubs; sometimes it is sparse and reduced to a few scattered *E. tetragona* and *Hakea corymbosa*; sometimes — on very stony patches — it reduces to heath only in which *Andersonia parvifolia*, *Verticordia habrantha*, *V. plumosa* and *Borya nitida* are dominant. Another special type known as 'heath with scattered trees' is to be found on the coastal edge of the plain between Boyadup and the Thomas River, and lies almost entirely to the south of the main road east from Esperance through Condingup. It clearly represents an area of higher rainfall, of 600 mm a year and more, and is probably controlled mainly by the excessive wetness of the topsoil in winter as the clay subsoil is very impermeable. Ironstone may or may not be present.

There is a continuous heath layer of small ericoid shrubs 40-50 cm tall and generally no larger shrubs but scattered trees of *Nuytsia floribunda*. These never exceed 3 m in height but must be termed trees as their habit is that of a tree rather than of a shrub. Excavations show that the nuytsias have a very thick and extensive root system. Locally in patches the nuytsias are replaced by *Hakea* spp. and *Xanthorrhoea preissii*, apparently where laterite is present.

The steep sides and the floors of valleys cut down into the plain are clothed mainly with mallee, *E. redunca* and *E. uncinata* being the dominant species together with *E. flocktoniae*, *E. incrassata* and *E. conglobata*. Occasional large shrubs of *Banksia caleyi*, *Hakea laurina*, *H. crassifolia* and *H. corymbosa* are seen. A dense understorey of

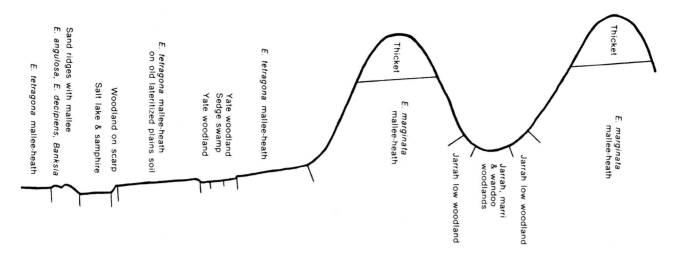

FIG 3.169

Diagram showing topographic relationship of vegetation units in the Stirling Range.

FIG 3.170

The Stirling Range contains patches of woodland in the valleys, especially towards the western end of the range where rainfall is higher. These consist of wandoo as shown here on heavy soil, and of marri where it is more sandy, sometimes with jarrah. These merge into a low woodland of jarrah-marri up slope, and this into a mallee-heath in which jarrah and marri are the principal mallee species. While normally forest trees, they are capable of existing as mallee at both extremities of their range, here and in the Northern Sandplains.

Arthur Fairall

FIG 3.171

Where valleys in the range are very sandy, banksias become dominant, the tree *B. attenuata* and the smaller species *B. coccinea*, the red-flowered banksia, which can be most spectacular if forming dense groves. A good place to see this is in Red Gum Pass at the west end of the range where this photograph was taken. *B. coccinea* has a very upright form of growth, often unbranched, with terminal heads.

Melaleuca spp. such as *M. uncinata*, *M. thymoides* and *M. subtrigona* covers the ground below. In wet and clayey ground thickets of *E. platypus*, *E. spathulata* and *E. anulata* replace the mallee.

Amid the mallee there may also be small patches of woodland of yate *(E. occidentalis)*, or on the Pallinup and Phillips Rivers York gum, and on the Pallinup only wandoo and *Casuarina obesa*. The numerous small pans which fill with water in winter usually carry stands of yate or *Melaleuca preissiana* either around the margin or across the whole area depending on whether the soil is of clay or sand respectively. A general feature of these woodlands is their lack of ground vegetation other than annual grass and ephemerals.

Along the coast there are accumulations of sand in the form of dunes and sand sheets of various ages, consolidated and unconsolidated, as there are on the west coast, but here these deposits have been built up around bosses of granite which in most places determine the coastline. The granite is often partly mantled with sand, and depth of soil upon the rock is very variable. There is therefore an intricate mosaic of plant communities upon the granite, and upon the sand dune areas also since deposit of sand is still in progress. There are considerable patches of mobile drift sand destitute of vegetation, and where the sand has become consolidated and vegetated it differs considerably in age.

Fronting the sea there are normally thickets mainly of *Scaevola crassifolia*, *Pimelea ferruginea*, *Olearia axillaris* and *Olax phyllanthi*. On the sheltered side of consolidated dunes, and west of the Barren Ranges, *Agonis flexuosa* is the dominant species usually in a mallee form up to 4-5 m in height due to the frequent fires. However it can be found in all stages up to woodland with trees of 15 m. Associated species include *Eucalyptus angulosa*, *Hakea oleifolia*, *Acacia cyclops* and other *Acacia* spp. Further east where *Agonis flexuosa* is not present the situation is somewhat different, both botanically and in the structure of the coastal belt. In the Esperance area as far east as Mt Merivale, there is a flat and swampy coastal plain bounded on the inland side by a decayed emerged sea-cliff about 30 m high and from 6 to 15 km inland of the present shoreline. There are several large salt lakes on the plain, e.g. Lake Gore, Pink Lake and Lake Warden. The seaward portion of the plain is covered by dune systems. On the seaward side of the dunes there is a low scrub in which *Scaevola crassifolia* is dominant as before. Behind this there is a scattered mallee of *Eucalyptus angulosa* with an understorey of *Melaleuca*, principally *M. pentagona*. Further inland the dunes frequently carry thickets of tall *Acacia* and *Melaleuca* or *Banksia speciosa*. The plant cover is very variable and seems particularly subject to degradation by fire from which there is no rapid recovery, so that a mature structure seldom or never develops. From the evidence of the undisturbed communities on the Recherche islands it seems that the climax would be woodland of *Callitris preissii* or *Melaleuca lanceolata*, with thickets of *Acacia cyclops* as an advanced seral stage. The other communities seen are secondary.

Eucalyptus platypus var. *heterophylla* forms thickets and low forest in depressions between the dunes. On the portion of the plain which is free from dunes, scrub-heath dominated by banksia as described above is the principal formation. Heath with scattered *Nuytsia* trees replaces this on lowlying, winter-wet ground. Actual swamps are filled with *Melaleuca* scrub and trees, sometimes with *Banksia occidentalis*. The salt lakes are associated with samphire communities *(Halosarcia)*.

From Mt Merivale to Cape Arid there is no longer any coastal plain, and the Tertiary land surface itself is over-ridden by the Quaternary dune systems. Between Mt Merivale and the massive complex of granite domes at Cape Le Grand there is an extensive belt of sand ridges, swamps and heath 20 km wide, with bare drift sand along Esperance Bay, then relatively recent dunes with *Eucalyptus angulosa*, *Acacia*, *Melaleuca* and *Banksia* as described above. Behind this zone is an area of apparently rather older sandy deposits with irregular sand ridges trending east and west with swampy hollows between them. Scrub-heath covers the ridges and *Nuytsia* heath with abundant sedges and Restionaceae the hollows. As the ground rises to the foot of the granite mountains a different heath community takes form. Much of the area is lakes, tea-tree swamps and heathy swamps, interspersed in sand heath. The granite peaks of Cape Le Grand are the most massive group in the whole State, rising to a maximum height of 345 m and covering about 25 km^2.

Further to the east around Duke of Orleans and Alexander Bays the young sand dune systems mantle the coast line and are vegetated with *Eucalyptus angulosa*, *Acacia*, *Melaleuca* and *Banksia* as before. East of Cape Arid there is an intermission in the aeolian deposits for 25 km, after which they begin again; but as the coast is now trending to the northeast and is sheltered from the prevailing wind the pattern is somewhat different. We have once more the Quaternary coastal plain, this time apparently in two levels, the lower at about present sea level, the higher some 6 to 9 m above this. The upper level is an even plain of deep sand carrying a scrub-heath with a rich flora and *Banksia speciosa* dominant. The lower level consists of a chain of salt lakes cut off from the sea by dunes. The plant cover at Israelite Bay consists of a tall scrub of *Acacia* and *Melaleuca* spp.

The vegetation of granite rocks in the higher rainfall area west of the Barren Ranges is unusual in featuring an open growth of low domed shrubs forming 'cushion plants'. However the species concerned are only those of coastal heath. Granite rocks in the plains east of Esperance are drier and resemble rocks in the interior with scattered clumps of pin grass *Borya nitida* on rock slabs, and small trees *(Casuarina huegeliana)* and shrubs growing in crevices. Some of the latter are peculiar to the Esperance area with *Kunzea baxteri* replacing *K. pulchella, Melaleuca fulgens,* and *Leptospermum sericeum* at Cape Le Grand. Where slopes are encrusted with rubble there is mallee of *E. cornuta* and/or *E. lehmannii*.

The islands of the Recherche Archipelago are granite domes of the same structure, differing only in that some of them are partly encrusted with sand and limestone, but their plant cover differs markedly from the adjacent mainland because the islands could not be reached by Aborigines and thus were not regularly burnt off. Some fires have been caused during the period of contact with Europeans, but the islands remain substantially unburnt to this day. When one goes ashore there is an immediate impression of lushness, with relatively luxuriant vegetation growing in age-old accumulations of humus, whereas on the mainland there is the usual impression of barrenness and sterility. On the granite outcrops there are as may be expected considerable exposures of bare rock but where soil has developed there are communities of trees and shrubs considerably taller and denser than any found on such outcrops on the mainland. Of these the most important is the eucalypt low forest with trees reaching a maximum height of 12 m. The principal dominants are *E. cornuta* and *E. lehmannii*, sometimes with *E. platypus, E. incrassata* or *E. conglobata*. This association is only found as mallee on the mainland. On sandy ground *Melaleuca lanceolata* and *M. globifera* mingle with the eucalypts or form woodland on their own account. These too would only be found in shrub form on the mainland. There are other equally distinctive minor communities. Where kwongan occurs on sandy plateau surfaces, the floristic composition is significantly different.

It remains to speak of the various hills and ranges which are a notable feature of this region.

The Stirling Range

The range extends east to west for 60 km, and contains Bluff Knoll (1074 m), the highest eminence in the southern half of the State. In keeping with its image as a relict geological outcrop of immense age the range consists of a series of isolated high and precipitous peaks or groups of peaks rather than a continuous chain. The formation begins in Donnelly Peak (650 m), thence through Mondurup (817 m) to Talyuberlup (784 m) with its adjacent Gog (625 m) and Magog (857 m). Next come Toolbrunup (1052 m) and Mt Hassell (848 m) and then Mt Trio (857 m) and Toll Peak (735 m). Finally the eastern part of the range is more continuous, and stretches from Yungermere (753 m) through Mt Success (750 m), Bluff Knoll (1074 m) and Isongerup (994 m) to terminate abruptly at Ellen Peak (1012 m). The peaks become steadily higher towards the east and at the same time more jagged and angular, with exposed cliffs and rock faces replacing rounded outlines.

Deep gaps between the peaks have been used as passes such as Chester Pass between Toolbrunup and Yungermere, which carries the main road from Albany to Borden, and Red Gum Pass between Donnelly Peak and Mondurup.

Principal control of vegetation is by topography and soil. Owing to the mountainous nature of the range, topography exerts a very strong influence, and the relationships of the various units can be explained in terms of a catena (Fig. 3.169), to which both geology and soils are also related. The rocks forming the Stirling Range are of sedimentary origin, consisting of quartz sandstone at the base, overlain by phyllite and muddy sandstone. As the beds are mostly fairly flat-lying, this sequence exists throughout, and there is an absence of the lithological diversity which is marked in the Barren Ranges further east and has created ecological niches favourable to local endemic species there. Broadly, the thicket of the mountain tops is associated with phyllite and the mallee-heath of the lower slopes with quartz sandstone. Woodland in the valleys is associated with colluvium brought down from the mountains, forming relatively young, undeveloped soils. Although these may contain quantities of laterite, it is in the form of transported nodules. On the other hand the *Eucalyptus tetragona* mallee-heath of the surrounding plains and

FIG 3.172
Bluff Knoll, the highest peak in the park, has an impressive precipice on the north face. It is approached by a tourist road up to a certain level, then by a good walking track which has been regraded in recent years. Pediplains around the range are covered with *E. tetragona* mallee-heath typical of the plains further east but floristically richer here, grading into the jarrah-marri mallee-heath of the mountain slopes, and this into the thicket association of the upper slopes. Snow occasionally falls on these mountain tops but does not lie for long.

FIG 3.173
The upper slopes of Bluff Knoll and of the other high peaks carry a thicket formation which is floristically quite distinctive and makes a brilliant show in a good season. This community is a form of kwongan which bears the closest resemblance of any to the *fynbos* of mountains in South Africa. Species which can be seen here are the pink-headed *Isopogon latifolius*, a local endemic which is the glory of the Stirling Range; *Dryandra formosa*, left-centre; *Oxylobium atropurpureum*, a shrub with red pea flowers; *Banksia solandri*; and *Sphenotoma* sp. (bottom left corner). The dark shrubs not in flower are *Casuarina*.

FIG 3.174
The best known local endemic plants of the Stirlings are the *Darwinia* spp. (named after Charles Darwin), the mountain bells, of which there are eight species having localised distributions within the park. The popular impression that species are localised on particular peaks is not strictly correct as they occur rather on groups of adjacent peaks. Illustrated above is *D. squarrosa* which occurs on middle slopes of the eastern peaks from Bluff Knoll to Ellen Peak at the end of the Range. In the same area *D. collina* occurs on the summits and *D. lejostyla* on the lower slopes. *D. hypericifolia*, *D. macrostegia*, *D. meeboldii* and other species occur in other parts of the park.

sediments, while it is also developed partly on colluvium, has a highly-weathered soil profile, dating probably from the Early and Middle Tertiary.

Woodland is frequently seen in the valleys of the range where available moisture is highest, and consists either of jarrah *(Eucalyptus marginata)* and marri *(E. calophylla)* on the more sandy soils or of wandoo *(E. wandoo)* on soil of heavier texture. Understorey plants in the former case feature a number of wattles such as *Acacia drummondii, A. myrtifolia* and *A. pulchella* whereas *Bossiaea linophylla* tends to be the most conspicuous beneath wandoo. Woodland trees attain 10–15 m in height, rarely 20 m. Jarrah-marri stands continue up slope with declining height and without change in composition to form low woodland of 5–10 m, and this in turn merges into a mallee-heath of 2–3 m with jarrah and marri still as the principal species. This mallee-heath formation covers all the lower slopes of the mountains throughout the range. Composition has been listed as follows (11e):

Small trees (rare) *Nuytsia floribunda*
Grass trees to 3 m *Kingia australis, Xanthorrhoea gracilis*
Mallee *Eucalyptus calophylla, E. doratoxylon, E. marginata*
Other tall shrubs *Banksia grandis, Hakea baxteri, H. cucullata, H. pandanicarpa, Lambertia ericifolia, L. uniflora.*
Smaller shrubs *Banksia petiolaris, B. sphaerocarpa, Beaufortia cyrtodonta, Boronia crenulata, Burtonia villosa, Casuarina humilis, Conospermum dorrienii, Darwinia diosmoides, Dryandra nivea, D. proteoides, Isopogon cuneatus, I. dubius, Lysinema ciliatum, Melaleuca polygaloides, Petrophile divaricata, Platytheca galioides, Sphaerolobium macranthum, Sphenotoma dracophylloides, Synaphea ?favosa, Xanthosia rotundifolia*
Herbaceous plants *Conostylis villosa, Dampiera* sp., *Diuris longifolia, Drosera* sp., *Johnsonia lupulina*
Restionaceae *Anarthria gracilis, Caustis dioica, Lepidosperma viscidum, Mesomelaena stygia, M. tetragona, Restio* sp.

Upslope the mallee-heath merges in turn into a formation of closed tall shrubs or thicket which occurs at the upper levels on all the principal peaks in the range. The stand is dense and difficult to penetrate. Examination (11e) showed a group of 15 species which were consistently present and others which were less reliable. Of the former, *Dryandra formosa, Isopogon latifolius* and *Oxylobium atropurpureum* are the most conspicuous when in flower and serve as character-species. The other consistent associates are *Acacia drummondii, Andersonia echinocephala, Banksia solandri, Beaufortia decussata, Boronia crenulata, Calothamnus* sp. aff. *gracilis, Hakea ?florida, Hypocalymma myrtifolium, Kunzea recurva* var. *montana, Leucopogon unilateralis, Sphenotoma* sp. aff. *dracophylloides* and *Xanthorrhoea gracilis.*

The surrounding plains and pediments to the range are covered with a mallee heath in which *Eucalyptus tetragona* is the character species, as described above in this section.

It has been known for a long time that the flora of the range as a whole is very rich and contains numerous endemic species but no comprehensive work on the subject has been published. One group of plants which has become well known is in the genus *Darwinia* where there are eight endemic species having localised distributions within the range. They tend to occur on adjacent groups of peaks at different levels. All of them are of outstanding floral beauty, unlike the common *Darwinia diosmoides* which is a component of the mallee heaths around the foot of the range and spreads widely in the Southwest.

The Barren Ranges

Northeast of the Gairdner River mouth the structure of the coast is controlled by the outcrop of hard massive Proterozoic quartzites, which form a platform up to 20 km in width from the coast with a general surface 90 m above sea level. The platform stretches from the Gairdner River to the Phillips River: intermediate rivers are deeply entrenched into it, forming quite impressive canyons in places, e.g. Echo Glen on the Fitzgerald. From the platform rise small abrupt mountains formed by the harder and more resistant members of the quartzites. Known in general as the Barren Ranges from the name aptly conferred by Flinders and referring to their bare appearance, these are in three groups: the West Mt Barren group between the Gairdner and the Fitzgerald comprises West Mt Barren (371 m), Mt Bland (329 m) and the low stump of Mt Maxwell; the middle group between the Fitzgerald and the Hamersley comprises Mid Mt Barren (457 m), the Thumb Peak range (about 450 m), the Whoogarup Range (about 410 m) and the off-lying Mt Drummond (309 m). The eastern group between the Hamersley and the Phillips comprises the Eyre Range with Annie Peak (about 450 m), No Tree Hill and East Mt Barren (about 275 m).

The 90 m high platform fronting onto the sea,

FIG 3.175
East of the Stirling Range, the Barren Ranges are a group of peaks of a Precambrian quartzite formation which lie adjacent to the coast between Bremer Bay and Hopetoun. They were named from the sea by Matthew Flinders owing to their barren appearance which is borne out in the above photograph of East Mt Barren. All of the peaks rise from a platform about 90 m above sea level which is thought to have been cut by the sea during the Eocene period when the land was lower. The present isolated peaks would have been eroded into separate islands at that time. Actually although the mountains appear very rocky and barren, they are well covered with kwongan thickets.

with the mountains rising behind, creates one of the most striking stretches of coastline in the south of the State.

There are four types of country whose vegetation is respectively the Barren Ranges thicket, *E. tetragona* mallee-heath, mallee and coastal scrub.

The composition of the Barren Ranges thicket varies locally. The most consistent species are the mallee *E. preissiana* and the prickly shrub *Dryandra quercifolia*, which dominate the 1.5 m thicket. Other typical components are *Eucalyptus lehmannii* (mallee form), *Banksia lehmanniana*, *Calothamnus pinifolius*, *Casuarina humilis*, *C. trichodon*, *Dryandra armata*, *Eucalyptus conglobata*, *Hakea crassifolia* and *Isopogon teretifolius*.

The effect of lithology is seen in certain distinct associations which follow the outcrop of a particular rock type. Thus, a sandstone which produces a thin yellow sandy soil is characterised by stands of the whipstick mallee *Eucalyptus sepulcralis*.

A certain very massive quartzite carries a typical association also of *Regelia velutina*, *Calothamnus validus*, *Melaleuca citrina*, *Banksia quercifolia* var. *integrifolia* and *Baeckea ovalifolia*. This quartzite, from its nature, tends to form the mountain summits, and is present in East Mount Barren, Annie Peak and West Mt Barren. However, it crops out also at lower elevations, when the identical plant association will be found with it.

The summits of the other mountains in the Barren Ranges, where not formed of this quartzite, also tend to show local peculiarities. Certain outcrops of greenstone occur locally in the ranges and carry a distinctive association, a mallee-heath, dominated by *Eucalyptus nutans* and *E. gardneri*. The Barren Ranges are well known to harbour a considerable number of

Fig 3.176

The Middle Mt Barren group looking east from Point Ann. The Barrens are in three groups. The western group comprises West Mt Barren (371 m), Mt Bland and the low stump of Mt Maxwell. The middle group comprises Mid Mt Barren (457 m), Thumb Peak and the Whoogarup Range, while the eastern group is the Eyre Range with Annie Peak (about 450 m), No Tree Hill and East Mt Barren (about 275 m). These rise from a continuous kwongan-covered sandplain (see Fig 3.161). The mountains themselves have different vegetation, often denser, varying according to the rock outcrop and containing many local endemic species.

Fig 3.177

Some of the endemic plants in the Barrens are quite bizarre. On No Tree Hill in the above photograph we see the foliage plant *Hakea victoria*, royal hakea, with an upright growth and broad leaves which begin green variegated yellow and turn red and then crimson in successive seasons. The whipstick mallee is *Eucalyptus sepulcralis*, named from its weeping habit. It has silver stems and foliage, and showy yellow flowers. Some scattered *Nuytsia* trees in the rear belie the name No Tree Hill. *E. sepulcralis* grows in a thin yellow sandy soil produced by a particular rock type. Another mallee *E. preissiana* grows on an adjacent formation, and a taxon '*E. chrysantha*', which is now recognised as a hybrid between the two, is found along the junction of the two rock formations.

Fig 3.178

Probably the most striking plant in the Barrens is the brilliantly coloured bottlebrush *Regelia velutina* which is also favoured by handsome blue-green foliage. It grows on a stratum of hard blocky quartzite which frequently (but not always) forms the summits of the peaks, rooting in rock crevices.

endemic species. The existence of these endemics is probably related to the unique habitats afforded by the differing lithology in these mountains.

On the pediments of the Barren Ranges the *Eucalyptus preissiana-Dryandra quercifolia* thicket merges gradually into *Eucalyptus tetragona* mallee-heath as the stony detritus gives way to sandy lateritic soil.

The mallee in the valleys of rivers, which have cut through the ranges, belongs to the *E. redunca-E. uncinata* association.

Coastal vegetation is similar to that in the Bremer System, except for absence of *Agonis flexuosa,* which does not spread east of the Gairdner River. Cliffs and headlands exposed to the sea carry an open stand of low cushion plants. At Point Ann these are *Banksia media, Melaleuca pentagona, Calothamnus pinifolius, Pimelea ferruginea,* and at the Hamersley River mouth *Pimelea ferruginea* and *Scaevola crassifolia.* Stabilised dune sand at the latter point is vegetated with *Eucalyptus angulosa* as mallee and an understorey of *Melaleuca pentagona.*

FIG 3.179

East of Esperance the plains are relieved by granite domes which rise through the mantle of Tertiary sedimentary rocks. The most spectacular group of these is found in the Cape Le Grand National Park visible across the bay from Esperance. These peaks rise to a maximum elevation of 345 m and cover 25 km². Vegetation is minimal on rocky slopes but includes the shrub *Leptospermum sericeum* with pink flowers and silver leaves which has become popular in cultivation since it was rediscovered by government botanist Charles Gardner in 1961. It had been originally found and described on the Recherche Islands by Labillardière in 1792, but occurs also on the mainland.

FIG 3.181

This rock formation in the Cape Le Grand National Park is not evidence that natives of Easter Island formerly visited the spot although this might be argued by a certain type of archaeologist!

FIG 3.180

One of the most interesting granite domes in the Cape Le Grand group is Frenchmans Peak, whose curious overhanging top has a cave right through it from side to side. Although most of the mountain is bare granite there is a considerable flora in patches of soil and rubble clinging to the slopes. *Leptospermum sericeum* is common on and around the top. The sandplain which surrounds the mountains had been recently burnt at the time this photograph was taken. The blackboys are *Xanthorrhoea gracilis.*

Fig 3.182
The Esperance Plains slope down to the sea and continue offshore below sea level where the granite domes now emerge as islands forming the Recherche Archipelago. These islands could not be reached by Aborigines and were not burnt in earlier times. Most of them have still not experienced bush fires and the vegetation there gives an impression of lushness, of plants growing in age-old accumulations of humus, a feeling quite different from the barrenness of the mainland. Low forest of *Eucalyptus* or *Melaleuca* is the dominant plant formation. Kwongan occurs on sandy plateau surfaces, and has a composition different from that on the mainland.

Fig 3.183
Buds, flowers and fruit of the Bald Island marlock, *Eucalyptus lehmannii*. With *E. cornuta* this species forms low forests on unburnt islands of the Recherche Archipelago (also on Bald Island near Albany) whereas on the mainland on similar granite rock habitats it is found only in mallee form. This difference is attributed to the effects of fire.

The Ravensthorpe Hills

The Ravensthorpe unit is a greenstone belt. Topographically it consists of the Ravensthorpe Range stretching southeast from Mt Short to Mt Benson, Mt Desmond (which is lower), and adjacent plains and undulating country on Archaean diorite and Proterozoic metamorphic rocks which have been stripped of sand and laterite. Ravensthorpe township appears to lie at about 260 m above sea level, and the Ravensthorpe Range has been estimated to reach about 400 m. The hills are of no great relief and of subdued outline. Mt Desmond is an elongated whaleback ridge.

The thicket of the summit ridges of the Ravensthorpe Range and Mt Desmond is similar to that of the Barren Ranges, that is to say it is dominated by the low mallees *Eucalyptus preissiana* and *E. lehmannii*, with *Dryandra quercifolia*. On the hill slopes the two most consistent and typical species in mallee are *Eucalyptus nutans* and *E. gardneri*, which in favourable places and given freedom from fire increase in stature to form low forest. At least nine other species are known to occur, including the local endemic *E. stoatei*.

Woodlands occupy broad valleys, in which the soil is at its deepest, and consist of *E. loxophleba* and *E. salmonophloia*.

The Russell Range

Situated in the midst of a limestone plain which is the southwestern extremity of the Nullarbor, the Russell Range thrusts a number of abrupt isolated peaks of a

FIG 3.184

Easternmost of the quartzite ranges of the south coast is the Russell Range comprising Mount Ragged (592 m), Brook Peak, Mt Dean, Mt Esmond and some other hills. The geological formation of the surrounding plain has changed to Nullarbor limestone, but so much grit and sand eroded from Mt Ragged has been spread out around it that the peak is surrounded by a sandplain 1–2 km wide which shares the vegetation of the Esperance Plains. Kwongan dotted with small mallees is seen in the photograph, *Banksia speciosa* in the foreground. The mountain slopes are covered with a thicket in which *Calothamnus quadrifidus* and *Dryandra armata* are the commonest species.

FIG 3.185

On Mt Ragged, looking northeast towards Brook Peak. The thicket of the mountain slopes is similar to that in the Barren Ranges and varies in height and density according to time since the last fire. On the plain below the lighter-coloured more open sandplain can be seen contrasting with the darker mallee on the limestone formation.

FIG 3.186

Where the substratum of the plain changes to limestone, the vegetation changes to mallee of this distinctive species, *Eucalyptus cooperiana*, which has a white bark, lustrous leaves and purple-red quadrangular stems. Other associated mallees are *Eucalyptus micranthera* and *E. uncinata*. This community grows in a thin red soil over limestone and the understorey is an unusual mixture of lime-loving species such as *Templetonia retusa* with typical heath plants.

quartzite formation through it, attaining 592 m at Mt Ragged. The plain around stands at about 150 m. There are three chains arranged in echelon and striking NNE—SSW. Brook Peak, Mt Dean and Mt Esmond constitute the most easterly chain, Tower Peak or Mt Ragged and the Russell Range proper the central chain 7½ km further west, while Mt Symmons, an unnamed eminence and Wolgrah Hill form the third another 5 km to the west. The third chain is much inferior in height to the others. There is yet another chain of hills of the same geological formation, 22 km south of Mt Ragged and appearing on the map as The Diamonds Hill and Hill 62, but as their vegetation is substantially the same as that of the surrounding sand plain they are not included in the Mt Ragged system. The mountain chains are formed of vertically stratified gneiss with a central band of massive quartzite whose resistance to erosion is responsible for the persistence of the ranges. A great talus of grit and sand derived from the disintegration of the rock of the ranges envelops their lower slopes and spreads out on the surrounding plain to a distance of from 1 to 2 km, carrying a heath vegetation. This evidently shares the flora of the heaths of the Esperance plains in general, the most conspicuous elements being *Eucalyptus incrassata*, *Banksia speciosa*, *B. media*, *Dryandra quercifolia*, *Isopogon buxifolius* and *Casuarina humilis*. Most of these heath plants appear to extend all over the mountain slopes also but the vegetation as a whole becomes denser so as to be considered a thicket rather than scrub, and there are changes in the dominance.

On the lower slopes of Mt Ragged *Calothamnus quadrifidus* and *Dryandra armata* become dominant — species rare to absent in the sand plain — while *E. incrassata* and banksias disappear. The dominants are associated with *Beaufortia bracteosa*, *Casuarina sclerocclada*, *Comesperma drummondii*, *Dryandra longifolia*, *Gastrolobium pycnostachyum*, *Grevillea concinna*, *Hakea* aff. *adnata*, *H. pandanicarpa*, *Hibbertia gracilipes*, *Melaleuca glaberrima*, *M. pentagona*, *Petrophile* sp. Higher up on the mountain *Agonis linearifolia*, *Acacia fragilis*, *Adenanthos sericeus* and *Leucopogon apiculatus* become conspicuous, associated with *Acacia subcaerulea*, *Boronia albiflora*, *Dillwynia pungens*, *Lasiopetalum rosmarinifolium*, *Monotoca oligarrhenoides*, *Pimelea brevifolia*, *Platysace compressa*, *Rhadinothamnus euphemiae*. On the summit ridge *Callitris preissii* (shrubs only), *Kunzea baxteri* and *Dillwynia pungens* are the most noticeable species.

3.6 *The Southwestern Interzone*

COOLGARDIE BOTANICAL DISTRICT

Predominantly eucalypt woodlands, becoming open and with saltbush-bluebush understorey on the more calcareous soils. Patches of shrub steppe adjoining the Great Victoria Desert. Scrub-heath and *Casuarina* thickets on sandplains.

Climate: Arid non-seasonal to semi-arid mediterranean; annual precipitation 200-300 mm.

Geology: Proterozoic granite and gneiss of the Fraser Range Block; Archaean granite with infolded volcanics and meta-sediments of the Yilgarn Block.

Topography and soils: Gently undulating with occasional ranges of low hills; sandplains in the western part and some large playa lakes. Principally brown calcareous earths.

Area: 126 500 km², no agricultural clearing.

Boundaries: N—climatically determined 'Eucalyptus-Acacia line' between *Acacia* woodland and *Eucalyptus* woodland on lower-slope soils, and between mallee-spinifex steppe and *Acacia* thickets on sandplains. E—edge of bluebush plains on limestone. S & W—boundary of the Southwest Botanical Province.

The Interzone all lies below the 300 mm rainfall line and must be classed as semi-desert. For this reason there has been virtually no clearing for agricultural purposes and even pastoral use of the native vegetation is minimal and localised owing to dense bush, unpalatable foliage and/or lack of water. Human population however is not inconsiderable and is engaged in the mining industry. Since the discovery of gold at Coolgardie in 1892 the region has been the principal centre of the State's gold mining. Mines are currently undergoing redevelopment after a long period of stagnation and the exploitation of nickel ore has been added in the last 30 years. At the turn of the century there was a proliferation of mines and mining settlements which involved some clearing of the woodlands, and timber was cut for use in the mines and for firewood. This activity was organised along movable tramlines laid out into the bush, which eventually extended as much as 150 km from Kalgoorlie. After cutting the woodlands were left to regenerate themselves from seed and coppice which was successful except in some stands of salmon gum which today are very open or even treeless. It is very difficult now to distinguish between virgin and cut-over woodlands.

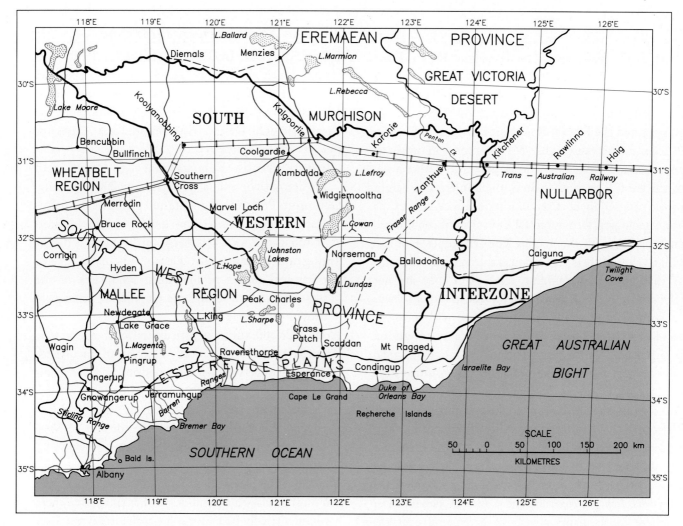

In general appearance the vegetation of the Interzone much more closely resembles that of the Southwest Province than that of the Eremaea. When travelling east from Southern Cross to Coolgardie one is scarcely conscious of any change in the sandplains and woodlands traversed whereas at Goongarrie on the road from Kalgoorlie to Menzies there is an abrupt change to mulga (*Acacia aneura* low woodland) and in the east along the railway just west of Kitchener there is an equally marked change to the bluebush country of the Nullarbor Plain. However there are species changes. Most of the woodland trees in the Interzone are different and their understoreys change progressively towards the east, with in some places the desert spinifex grass.

The region is geologically diverse with the occurrence of major greenstone belts which not only are the source of mineralisation but provide a hilly topography and heavier, less depleted soils. The greenstone belts strike NNW to SSE. One of these is situated along the boundary with the Southwest Province and extends from Mt Glass between the Johnston Lakes through the Bremer and Parker Ranges, including the gold mines of Marvel Loch and the former mines of Southern Cross and Bullfinch, into the Highclere Hills and the Die Hardy Range. North of Southern Cross beds of banded ironstone which occur sporadically within the greenstone belt and are very resistant to erosion form high abrupt ridges rising conspicuously from the plain. Of these the Koolyanobbing Range has been mined for iron ore.

Such ironstone ridges, which are very rocky, are covered with dense thickets in which *Acacia quadrimarginea* is the dominant species associated with *Casuarina acutivalvis* and *C. campestris*. On the Die Hardy Range the thicket contains the local endemic *Dryandra arborea*, the only *Dryandra* to be a small tree and to occur outside the Southwest Province. Rocky knolls on other types of metamorphic rocks carry thickets of *Casuarina campestris* and *Calothamnus asper*, or where less rocky of *Acacia* and *Casuarina* spp. On hill slopes mallee is sometimes seen, of either *E. gardneri* and *E. redunca*, or of *E. loxophleba* and *E. sheathiana*. Generally slopes and flats are clad with woodland, in which morrell

FIG 3.187
The Southwestern Interzone lies inland of the Wheatbelt, Mallee and Esperance Plains Regions, a drier, intermediate zone between the Southwest Province and the Eremaea. On the whole the vegetation has more in common with the former than the latter, consisting of eucalypt woodlands with mallee, thickets and kwongan. The general impression, seen from this breakaway on Jaurdi station, is of monotonous plains covered with woodland, however the landscape undulates with catenary sequences as elsewhere, and is varied geologically with mineralised greenstone belts between granitic areas. The greenstones tend to be hilly and to have distinctive vegetation.

(E. longicornis) is very common, with salmon gum and gimlet, locally E. corrugata and E. sheathiana (in tree form). Woodland understoreys may be either sclerophyll consisting mainly of the tall shrub Melaleuca pauperiflora, or saltbush with the glaucous soft-leaved shrubs Atriplex vesicaria and A. nummularia, in the latter case where the soil is more alkaline.

A second greenstone belt strikes NNW-SSE through Coolgardie and Kalgoorlie. It is broader and more highly mineralised. The topography is much the same, less hilly with fewer outstanding ranges. In between the two greenstone belts there is an extensive stretch of granite country with the typical alternation of sandplains, mallee and valley soils in catenary sequences. This is similar to the country in the Wheatbelt and the catena is the one illustrated as an example in Chapter 2, Figure 2.21. South of the standard gauge railway on the highest ground broad plateaux of deep yellow sand carry a rich, mixed scrub-heath association in which Grevillea eriostachya var. excelsior is usually conspicuous, forming a 'hedge effect' alongside the main road as it is a pioneer species. There is a list of associated plants in (12a). No other area has scrub-heath so far into the interior or under such dry conditions. It merges very gradually downslope into Acacia-Casuarina thickets which represent mixed kwongan. There is at first a general mixture, then small heath plants are relegated to an understorey and finally disappear. Downslope under the thicket the sand becomes shallower, and is bottomed by a lateritic hardpan which eventually comes to the surface, often with a small scarp or 'breakaway'. The shallow soil phase is under thicket of Casuarina campestris. A narrow band of mallee normally occurs below the breakaway and may comprise numerous species, typically E. eremophila, E. foecunda, E. oleosa, E. loxophleba. Below the mallee we

FIG 3.188
The mineralised belts in the Interzone became the scene of intensive gold mining activity in the 1890s but later many of the mines and townships were abandoned. This scene shows the site of Davyhurst. The pile of bricks represents the hotel. Such deserted remains are common throughout the area, while in other places new mines have opened or old ones are still active.

FIG 3.189
There are no rivers in the Interzone, all drainage being received into salt lakes large and small. There is a tendency for small lakes to migrate to the WNW under the influence of the prevailing wind when the lakes are full. Evidently wave action erodes the western end of the lake and transports the material to the eastern end where it is deposited on sandy beaches and dunes.

enter woodland and in this catena there is a distinct association on upper slopes, of *E. transcontinentalis* and *E. flocktoniae* with trees up to 20 m tall. On lower slopes and flats this merges into taller stands of salmon gum or morrell or gimlet.

North of the railway the vegetation of the sandplains changes to that of the northeastern wheatbelt with *Acacia neurophylla* thickets, mixed kwongan and *A. resinomarginea* thickets respectively on shallow, moderate and deep soil. The *E. transcontinentalis-E. flocktoniae* association is missing; mallee comprises *E. loxophleba* and *E. sheathiana*, and these species associate in tree form with salmon gum and gimlet. On red sandy soil there may be patches of mallee *(E. oleosa)* and spinifex *(Triodia scariosa)*.

Within these two zones outcrops of granite occur with their rock pavement vegetation.

In the major greenstone belt, the topography is controlled by the harder and more resistant strata, which form low ridges striking NNW-SSE, with alluvial flats between them.

The hardest strata are banded ironstones, which form the most prominent ridges, and are covered with thickets 2-3 m tall, in which *Acacia quadrimarginea* is dominant. Associated species include *Casuarina campestris, Acacia tetragonophylla, Eremophila oldfieldii, Scaevola spinescens, Dodonaea lobulata, Cassia nemophila, Ptilotus obovatus, Atriplex nummularia, Maireana sedifolia, Helipterum humboldtianum.*

Other ridges, which appear to be composed of

FIG 3.190

In granitic country a catena of soils and vegetation operates similar to that in the Wheatbelt. Next up the slope from the salt lakes come woodlands of salmon gum, gimlet, or in this case morrell *(Eucalyptus longicornis)*. These are fine tall trees having the general appearance of York gum but distinguished by their straight trunks. Morrell grows on calcareous soil and may have either a sclerophyll or saltbush understorey.

FIG 3.191

Further upslope the woodlands are less tall and of more mixed composition, generally *E. transcontinentalis* and *E. flocktoniae*. Undergrowth is normally sparse and consists of a few scattered shrubs. The large bush is boree, *Melaleuca pauperiflora*.

FIG 3.192

Higher ground in granitic country is capped with sandplains, frequently very extensive, and these have a rich kwongan flora. Towards the northeast they become much drier with the appearance of spinifex, *Plectrachne rigidissima* and *Triodia scariosa*. Also in this photograph are large coarse sedges, the desert blackboy *Xanthorrhoea thorntonii* and scattered shrubs of *Acacia* and *Casuarina*.

FIG 3.193
After fire or in disturbed ground such as along roadsides, two grevilleas act as pioneer species and frequently offer a conspicuous hedge effect along the Great Eastern Highway where it crosses sandplains. These are the golden-flowered G. eriostachya var. excelsior (formerly G. excelsior) and the white-flowered G. pterosperma. They are relatively short lived and will disappear in the absence of fire after a few years.

FIG 3.194
Inflorescence of Grevillea eriostachya var. excelsior.

more acid rocks, carry thickets of similar height of *Acacia acuminata* and *Eremophila oldfieldii*.

Ridges of a less abrupt nature, formed by meta-igneous rocks, carry a characteristic woodland described as the *Eucalyptus torquata-E. lesouefii* association. It occurs only within this area. *E. torquata* and *E. lesouefii* are co-dominant and abundant, and attain heights of 10-15 m. Associated trees are *E. campaspe*, *E. clelandii*, *Casuarina cristata* and *Grevillea nematophylla*. There is an open shrub understorey, largely of *Eremophila* spp., up to 1.8 m tall and of broombush habit, notably *E. scoparia*, *E. glabra*, *E. oldfieldii*, also *Dodonaea lobulata*, *Cassia cardiosperma* and *Acacia* species, interspersed with 1-2 m glaucous shrubs of the 'old man saltbush' *Atriplex nummularia*. Forbs include *Ptilotus exaltatus*. In the northeastern part the ridges carry *Casuarina cristata* instead of *Eucalyptus torquata*, and saltbush and bluebush join the understorey. A typical assemblage in this case would be *Casuarina cristata*, *Eucalyptus transcontinentalis*, *Grevillea nematophylla*, *Acacia oswaldii*, *Dodonaea lobulata*, *Eremophila alternifolia*, *E. oldfieldii*, *Atriplex hymenotheca*, *Maireana sedifolia*.

The woodlands of the middle slope in the catena are taller, 15-20 m, and of rich and varied composition, containing mainly *Eucalyptus lesouefii*, with *E. transcontinentalis*, *E. salmonophloia*, *E. oleosa*, *E. campaspe*, and as rarer species *E. clelandii*, *E. flocktoniae* and *E. gracilis*. *E. stricklandii* comes in on laterite breakaways. There are two types of understorey which may be characterised as the 'broombush' and 'saltbush' types. These interchange independently of the overwood and

FIG 3.195

Greenstone belts are generally hilly tracts with much variation in the nature of the soil, its depth and alkalinity. In general the hills are covered with woodlands of various kinds, reducing to shrublands along the summits, usually of *Acacia quadrimarginea*. Five different types of understorey can be recognised, i.e.:

a. Sclerophyll—mixed sclerophyll shrubs
b. Broombush—*Eremophila* spp. having a 'broombush' habit, see Fig 3.198
c. Bluebush—*Maireana sedifolia*
d. Saltbush—*Atriplex* spp.
e. Greybush—*Cratystylis conocephala*.

FIG 3.196

Characteristic of the deeper soils on greenstone is the goldfields blackbutt, *Eucalyptus lesouefii* (named after a Dr Le Souef who was director of the Perth Zoo), distinguished by the thick black bark on the butt. It associates with *E. clelandii* and *E. corrugata* which are both also blackbutts, *E. oleosa*, *E. transcontinentalis*, *E. campaspe* and *E. flocktoniae*. Variable understoreys are found as described under Fig 3.195.

FIG 3.197

A distinct association of *Eucalyptus torquata* with *E. lesouefii* occurs on rocky greenstone ridges north and south of Coolgardie. *E. torquata* is a relatively small tree with a rough bark and is extremely decorative in flower, the buds also being of a distinctive and intriguing shape, so that the tree is often planted as an ornamental. In the wild it associates with the other eucalypts listed under Fig 3.196. The understorey is of the broombush type. Taken at Bencubbin, in cultivation.

appear to be controlled by the alkalinity of the soil, the substratum beneath the saltbush being markedly calcareous. The broombush understorey appears on the less alkaline sites, usually on rising ground, and consists of tall shrubs, mainly of broombush habit, and reaching 1.8 m. *Eremophila scoparia* is typical — it is significant that its specific name means 'broomlike' — and other component species include *Eremophila alternifolia, E. decipiens, E.* sp. aff. *gilesii, E. glabra, E. ionantha, E. oldfieldii, E. pachyphylla, E. saligna, E. weldii, Acacia hemiteles, Atriplex nummularia, Cassia nemophila, Daviesia benthamii, Dodonaea adenophora, D. stenozyga, Grevillea oncogyne, Hibbertia pungens, Olearia muelleri, Prostanthera incurvata, Ptilotus obovatus, Scaevola spinescens, Thryptomene urceolaris, Westringia dampieri.*

The saltbush type of understorey appears usually on lower ground with high alkalinity. The actual saltbush, *Atriplex vesicaria,* is not always dominant and may be replaced by or mingle with *Cratystylis conocephala* (greybush) and *Maireana sedifolia* (bluebush), which are morphologically similar: sub-shrubs with glaucous semi-succulent leaves. Annuals may be found here in season, especially *Brachycome, Cephalipterum drummondii, Helipterum floribundum* and *Ptilotus exaltatus.*

Woodlands of the lower slopes and flats in the catena consist typically of *Eucalyptus salmonophloia* (salmon gum) often with *E. salubris* (gimlet), *E. lesouefii* (blackbutt) and *E. longicornis* (morrel). The soil is very calcareous with a pH of 8, except superficially where it may drop to 6, associated with a surface coating of moss and lichen. The woodlands are of less varied composition in tree species than those of the middle slope, usually with a tendency to single-dominant stands of either *E. salmonophloia* or *E. salubris*. They are also taller and reach heights of 20-25 m. There is a close association between tree species and soil type: *E. salmonophloia* is found on red loam soils overlying clay with calcium carbonate nodules in the subsoil, *E. salubris* on somewhat heavier clays, *E. longicornis* upon a calcareous hardpan, while *E. lesouefii* betrays the presence of basic rocks beneath or in the vicinity. The same two understorey types are found, but with increased salinity such salt-tolerant plants as *Sclerostegia arbuscula* and *Frankenia pauciflora* may appear. There is also a variant of the broombush understorey on heavy, periodically wet soils, the 'boree' community, where *Melaleuca pauperiflora* becomes dominant in the understorey or replaces all other species. This is a large broombush type of shrub growing to 2.5-3.5 m in height.

At the lowest points of the terrain there are a number of large salt flats incorporating playa lakes, notably Lakes Lefroy and Cowan which at 40 km and 90 km long, respectively, are among the largest playa lakes in Western Australia.

The country southeast of Lake Cowan is somewhat flat and featureless, so that the plant cover is in consequence correspondingly monotonous. There is a general expanse of woodland in the class of 15 m tall, with the rising ground bearing occasional granitic knolls with thickets on shallow, rocky soil, and rarely containing exposures of bare rock. On the lower ground the woodlands open to saltbush and bluebush communities around salt lakes. Small patches of marlocks or mallee and spinifex also relieve the monotony of the woodlands.

The predominant association is of *Eucalyptus oleosa* and *E. flocktoniae* (with *E. lesouefii* in general mixture) on a pink calcareous soil — containing calcium carbonate concretions — of a floury texture. *E. transcontinentalis* is occasional, also *Myoporum platycarpum* and *Grevillea pterosperma. Eucalyptus stricklandii* is seen locally. *E. celastroides* and *E. campaspe* are small trees which occur locally in a subordinate position.

Both *E. salmonophloia* and *E. salubris* may be present locally in mixture but more commonly form separate stands on a red, heavier soil, normally in the lowest situations. *E. dundasii* becomes common further south in the Norseman area.

The understorey throughout tends to consist of greybush *(Cratystylis conocephala)* with some mixture of saltbush *(Atriplex vesicaria),* and bluebush *(Maireana sedifolia, M. pyramidata)* and samphire *(Halosarcia).* Patches of both broombush and boree understorey also occur.

Between Norseman and Balladonia the highway crosses the Fraser Range which introduces an entirely different belt of country whose vegetation is to a large extent unique in Western Australia. The range consists of a chain of exposed bosses of granulite gneiss of Proterozoic age, disposed along an axis trending SW to NE. There are some outlying ridges parallel to the main chain, mainly in the south. The hills so formed are not high and probably rise no more than 100 m above the surrounding country. With the prevailing low rainfall they do not give rise to any rivers. The gneissic bosses are however somewhat bare due to their excessively rocky nature, and rise abruptly out of the surrounding woodlands and salt flats.

The rockiest, most bouldery of the hills carry rare

FIG 3.198

The broombush type of understorey is dominated by species of *Eremophila* which develop very numerous slender ascending branches so that the bush can readily be made into a bundle of twigs for a broom. Principal species are *Eremophila scoparia* (which means broom-like), *E. oldfieldii* and *E. glabra*. Smaller sclerophyll shrubs are also found, mainly *Dodonaea, Cassia* and *Acacia* spp., while the old man saltbush, *Atriplex nummularia*, is not uncommon. Annual herbs include *Ptilotus exaltatus*. The different types of understorey in the Interzone have not been mapped as they cannot be distinguished in aerial photography. It is a reasonable assumption that they represent different soil types, but this has not been studied.

FIG 3.199

It is probable that a sclerophyll shrub understorey occurs on soil which is acid to neutral, broombush where it is moderately alkaline, saltbush where it is highly alkaline and slightly saline, bluebush where it is highly calcareous. The above shows a saltbush understorey in salmon gum woodland which has been cut over for mining timber and not regenerated fully. The saltbushes comprise the smaller *Atriplex hymenotheca* and the larger *A. nummularia*.

FIG 3.200

Similar woodlands of gimlet *(Eucalyptus salubris)*, rather open and with a saltbush understorey, are often found in the more highly saline areas in the vicinity of salt lakes. Between Balladonia and Zanthus.

FIG 3.201

Extensive salt lakes occupy valleys leading towards the Nullarbor Plain and some flash-flood runoff may occasionally occur. Lake Raeside, above, is bordered by samphire *(Halosarcia)* with some *Atriplex* and *Frankenia*. Away from the lake this grades into saltbush–bluebush *(Atriplex hymenotheca–Maireana pyramidata)* and gradually shrubs and trees are added, typically *Acacia tetragonophylla, Casuarina cristata, Eremophila miniata, Grevillea sarissa, Myoporum platycarpum* and *Pittosporum phylliraeoides*.

FIG 3.202

Some lakes are bordered by red sand dunes on which the cypress pine *Callitris glaucophylla* appears. Associated plants are the trees and shrubs listed under Fig 3.201, together with saltbush and annuals. Lake Lefroy.

FIG 3.203

Dunes of whitish sand which are building up at the present day are colonised by mallee and spinifex—*Eucalyptus incrassata* and *Triodia scariosa*. This community occupies only small areas but is quite distinctive. East of Norseman.

FIG 3.204

The Fraser Range is a unique area, crossed by the main highway between Norseman and Balladonia. The range trends SW to NE and consists of low rounded hills of granulite gneiss. The dominant vegetation of the hills is a scrub 1–2 m tall of *Dodonaea microzyga* with a suite of less common shrubs and herbaceous plants, and scattered trees of *Casuarina huegeliana* and *Pittosporum phylliraeoides*. This is unique as a plant community. *Eucalyptus dundasii* and *E. lesouefii* form woodlands in the low ground between the hills.

FIG 3.205

In the Norseman area (Shire of Dundas) the Dundas blackbutt *E. dundasii* becomes very common in association with *E. lesouefii*, with *E. oleosa*, *E. flocktoniae*, *E. transcontinentalis* and *E. campaspe*. The understorey is formed predominantly by greybush *(Cratystylis conocephala)*, a small shrub of the Compositae with silvery leaves, but saltbush and broombush understoreys are also seen. Near Norseman.

scattered trees of *Casuarina huegeliana* and *Pittosporum phylliraeoides* some 6 m tall, between which is a scrub 1-2 m tall, with *Dodonaea microzyga* dominant.

The *Dodonaea* scrub, without trees, covers all the less rocky hilltops and extends to the country round them where this is of a sandy nature. On the latter sites it is joined by the spinifex *Triodia scariosa*, often also with mallee, the species being *Eucalyptus* sp. aff. *salubris, E. merrickiae* or *E. griffithsii*. The Dundas Hills near Norseman which are an outlier of the Fraser Range are covered mainly by mallee and spinifex *(E. griffithsii, Triodia scariosa)* with *Casuarina campestris, Calothamnus chrysantherus,* and *Melaleuca uncinata. Eucalyptus diversifolia* occurs on the foot-slopes.

A *Dodonaea* association has only been recorded elsewhere in Western Australia in the Billeranga Hills between Three Springs and Morawa, where the dominant species is *D. inaequifolia*.

Mostly the sandy soil is only present on the outcrop, or for a short distance from it. Otherwise the soil is red loam and carries woodland which normally fills the dips between the hills and sometimes runs up on to the ridges. Mallee (species as above) may form a fringe to the woodland. *Eucalyptus dundasii* and *E. lesouefii* are the principal species close to the outcrops, with *E. calycogona* and *E. conglobata*; further out, though, there is a gradual change to *E. oleosa, E. transcontinentalis* and *E. flocktoniae* on the typical pink, floury, calcareous soil, characteristic of this association, and there are also areas of *E. salmonophloia* and *E. salubris*, and of *E. eremophila,* which forms low woodland.

North and east of the Fraser Range the country

FIG 3.206

Further east towards the Nullarbor Plain the soil becomes more and more calcareous, leading to dominance locally both of *Casuarina cristata* (smaller trees in rear) and *Eucalyptus transcontinentalis* (foreground). The former is mostly seen on rocky ridges, while the latter is largely dominant east of the Fraser Range, between there and Balladonia, where the soil can be seen in road cuttings to contain large calcium carbonate nodules.
E. transcontinentalis is a very handome tree with its smooth grey bark and glaucous leaves. The yellow flowers are also moderately showy. Understoreys of saltbush and bluebush are usual. Hampton Hill.

FIG 3.207

Much of the woodland in the eastern part of the Interzone consists of *E. oleosa* varying in stature from large trees to mallee, growing in a pink soil of a floury texture. The understorey may be of bluebush or greybush. Karonie.

FIG 3.208

The more sandy soils in the east as the Great Victoria Desert is approached are covered with mallee and spinifex, intergrading with the *E. oleosa* woodlands. Chief mallee species are *E. oleosa* and *E. foecunda*, the spinifex species *Plectrachne rigidissima* and *Triodia scariosa*. Numerous small shrubs of *Acacia, Casuarina, Eremophila* etc. also occur.

becomes drier and the soil more calcareous with the approach to the Great Victoria Desert on the north side and to the Nullarbor Plain on the east. For the most part the vegetation is an intricate mosaic of woodland and mallee dominated by the same two species, *E. oleosa* and *E. transcontinentalis*, which may occur either in tree or mallee form with intermediates. As woodland they occupy a pink loam of floury texture over a layer of calcareous nodules and have a greybush understorey. As mallee they occupy a red sandy soil overlying a calcareous hardpan, and have a spinifex ground layer of *Triodia scariosa*. *E. foecunda* will be present as an additional mallee species. Woodlands of salmon gum, gimlet and goldfields blackbutt *(E. lesouefii)* occur locally in heavy loam soils.

Southeast of the Fraser Range the country is at first on granite, with occasional granite outcrops breaking through, then on the Nullarbor limestone where higher rainfall than on the plain itself enables the establishment of eucalypt woodland and mallee. The country is highly calcareous throughout. Over granite the surface soil is pink, of a floury texture and containing calcium carbonate concretions which become larger and more numerous with depth. The various plant communities are not always topographically related but in general there is a catenary sequence from ridge to flat. On ridges *Eucalyptus transcontinentalis* tends to form relatively tall dense woodland attaining 20 m. Woodlands of the middle slope are less tall, more open, and consist of *E. oleosa* and *E. flocktoniae*. On bottomlands *E. salubris* (gimlet) comes in and may be the sole tree species. Understoreys vary independently, mainly either sclerophyll with small *Eremophila* and *Acacia*, or greybush *(Cratystylis conocephala)*. Saltbush *(Atriplex)* replaces the latter in some gimlet stands and around salt lakes.

On the limestone the country is flat, with a soil of pink earth with nodules, overlying limestone at about 40 cm. Vegetation is mainly woodland of *E. oleosa* and *E. flocktoniae* with a sparse tall shrub layer of *Melaleuca* or *Eremophila* spp. and ground layer sporadically of saltbush, bluebush or greybush. Towards the coast of the Bight the woodland merges into mallee of *Eucalyptus socialis*.

BIBLIOGRAPHY

1 McArthur, W.M. and Clifton, A.L., 1975. Forestry and Agriculture in relation to soils in the Pemberton area of Western Australia. *Soils and Land Use Series 54*, CSIRO, Melbourne.

2 Smith, F.G., 1973. Vegetation map of Pemberton and Irwin Inlet. W.A. Dept of Agriculture, Perth.

3 Beard, J.S., 1979. The vegetation of the Albany and Mt Barker areas. Vegmap Publications, Applecross.

4 Smith, F.G., 1974. Vegetation map, Collie. W.A. Dept of Agriculture, Perth.

5 Havel, J.J., 1975. Site-vegetation mapping in the northern jarrah forest (Darling Range). I. Definition of site-vegetation types. II. Location and mapping of site-vegetation types. *Bull. For. Dept West. Aust.* Nos 86 and 87.

6 Wilde, S.A. and Low, G.H., 1978. Explanatory notes, Perth geological sheet, Geol. Surv. West. Aust. Perth.

7 McArthur, W.M. and Bettenay, E., 1960. The development and distribution of soils on the Swan Coastal Plain, Western Australia. *CSIRO Aust. Soil Pub. No.16.*

8 Smith, G.G., 1973. A guide to the coastal flora of South-Western Australia. *W.A. Naturalists Club Handb.* No. 10, Perth.

9 Beard, J.S., 1976. Vegetation Survey of Western Australia, 1:250 000 Series. a) Dongara, b) Geraldton, c) Ajana. Vegmap Publications, Perth.

10 Beard, J.S., 1972-80. Vegetation Survey of Western Australia, 1:250 000 Series. a) Southern Cross, b) Jackson (1972); c) Perenjori (1976); d) Bencubbin, e) Kellerberrin, f) Corrigin, g) Dumbleyung (1980).

11 Beard, J.S., 1972-3. Vegetation Survey of Western Australia, 1:250 000 Series. a) Newdegate and Bremer Bay, b) Hyden (1972); c) Ravensthorpe, d) Esperance and Malcolm (1973), e) Albany and Mt Barker (1979).

12 Beard, J.S., 1969-72. Vegetation Survey of Western Australia, 1:250 000 Series. a) Boorabbin and Lake Johnston (1969); b) Jackson, c) Kalgoorlie (1972).

4
THE EREMAEAN PROVINCE

The Eremaean Province, which is by far the largest of the three provinces occupying 70% of the State, forms part of the Arid Zone, the famous 'dead heart of Australia'. Technically it is all desert because rainfall is low and erratic without any assured growing season and the production of crops is impossible without irrigation. However it does not by any means conform to the conventional picture of a desert. Here we have no stony plains devoid of plant life, nor moving dunes of sand. There is always some plant cover, the vegetation is just as interesting as in better watered parts of the State and displays of wildflowers can, if there has been rain, be just as impressive or more so. Long droughts can occur when everything is dormant and no life is stirring, but the good years make up for this. On the average the Eremaea receives about 200 mm of rain each year and even the driest sector of all, the eastern Nullarbor Plain, receives an average of 150 mm. Some of the other of the world's deserts can be very much drier than this. In the centre of the Sahara for example the average is 45 mm. The interior of Western Australia therefore is arid, but it is not a howling desert.

A large proportion of the Eremaea is suitable for pastoral use with sheep and cattle grazing the natural plant cover, and in Western Australia this applies mainly to the Murchison, Gascoyne and Pilbara Regions (see Fig. 2.5) whereas the more easterly regions are not usable by pastoralists. This leads them to be popularly regarded as deserts — e.g. the Great Victoria, the Gibson, the Great and Little Sandy Deserts — whereas the others, being 'station country', are not. The difference derives basically from the geological structure since the granites and other ancient rocks of the Western Shield develop soils which grow palatable 'feed', whereas the sandstones of the sedimentary basins grow little but inedible spinifex. In the former case too stock water is more readily obtained from boreholes. The Nullarbor Plain is odd man out in this assessment. It represents a sedimentary basin but the rock is limestone and the vegetation eminently palatable to stock. Unfortunately water is only locally obtainable and then usually at great depth, so that there are relatively few pastoral stations. In human terms, Western Australia's 'deserts' are justly so regarded as they contain little or no permanent human habitation. The last Aborigines left the Great Sandy Desert in 1965 and the Great Victoria Desert in 1959. The Great Sandy contains the mining settlement of Telfer within its borders and the Great Victoria the Aboriginal community of Cosmo Newbery. Otherwise all these deserts are devoid of human habitation, something which is increasingly rare in the modern world.

While this general division of the Eremaean Province into east and west is imposed by geological structure, another division into north and south is imposed by climate. The northern half of the desert receives its rain in summer, which favours grassland,

FIG 4.1

The Western Australian Eremaea or Arid Zone is dry and contains much barren, rocky and sandy country but none the less receives sufficient rainfall to maintain a general plant cover. There is no part where plant life is scarce or absent. In this view in the Weld Range near Cue scattered shrubs of mulga *(Acacia aneura)* cover the hills; the white subshrub in the foreground is a mulla-mulla, *Ptilotus obovatus*. Mulga country is characteristic of the southern half of the Eremaea and is generally suitable for pastoral use.

FIG 4.2

The southern half of the Eremaea has a number of typical features. The country consists of plains and low ranges all covered with scattered trees and shrubs of mulga *(Acacia aneura)* and other species. In good seasons the ground layer develops a lush growth of annual grass and herbs, mainly of the daisy family (Compositae). Stock water is usually obtainable at no great depth, so that sheep can be run on an extensive scale, grazing the ground herbage. Carrying capacity is however low, requiring at least 12 hectares per sheep. Such station country is not popularly regarded as desert. Nambi station near Leonora; *Helipterum craspedioides* in flower.

Fig 4.3

The northern half of the Eremaea is quite differently vegetated, having a general cover of 'spinifex' forming hummock grasslands with scattered trees and shrubs. Spinifex, genera *Triodia* and *Plectrachne*, consists of species of spiny grass growing in dense domed hummocks. Except for one species, *Triodia pungens*, it is inedible, so that most spinifex country is unsuitable for pastoral use and is popularly regarded as desert. Spinifex country covers almost the whole of the northern half of the Eremaea and extends to the southern half also on sandplains, mainly in the Gibson and Great Victoria Deserts. In this view on the upper Oakover in the Pilbara the trees are *Eucalyptus leucophloia*, the spinifex *Triodia basedowii*.

Fig 4.4

In the northern Eremaea displays of colour from wildflowers in good seasons may be seen between the spinifex or on loamy plains where mulga persists. However the Compositae give way to other species, the most famous being Sturt's desert pea *Clianthus formosus* seen here in profusion. Also present are a purple pea-flowered *Swainsona* at left centre and a white *Ptilotus obovatus* in the background. Melrose station.

so that the plant cover consists almost entirely of spinifex grasslands. In the southern half the different rainfall regime favours woody vegetation so that the typical plant cover is mulga (*Acacia aneura* low woodland). However there is a complication. Mulga does not grow readily in sand, therefore in the southern half sandplains and sand dune areas tend to be covered with spinifex as in the north; but mulga is always there in less sandy soil, unlike in the northern half. The Nullarbor Plain is again a law unto itself. Where sand dunes have overridden the plain at the margins they have a spinifex vegetation. Mulga and other acacias cover a large part of the plain where there is some depth of soil but are eliminated by shallow, rocky soil mainly in the central and eastern parts, forming the well-known treeless plains.

4.1 *Nullarbor Region*
EUCLA BOTANICAL DISTRICT

Bluebush steppe of perennial *Maireana sedifolia* (formerly *Kochia sedifolia*) with annual grass and forbs; treeless in the centre, peripherally with low trees of *Acacia papyrocarpa, A. aneura, Casuarina cristata*.

Climate: Arid non-seasonal (rain in any month); annual precipitation 150-200 mm.
Geology: Sedimentary basin, exposing Eocene to Miocene sediments, mainly limestone.
Topography and soils: Gently undulating, featureless plain, with shallow calcareous loam.
Area: 148 764 km².
Boundaries: Edge of bluebush plains on limestone. The boundary coincides with the limestone boundary except in the southwest where higher rainfall brings *Eucalyptus* woodland onto the limestone.

The region more or less corresponds to the Nullarbor Plain which extends over the border far into South Australia and is one of the largest limestone plains in the world, about 200 000 km² in extent. It is renowned for its flatness, emptiness and monotony, and was named from the Latin *nulla arbor* (no tree) by E. A. Delisser who was exploring for the government of South Australia in 1867. Its character is shown graphically by the fact that the Eyre Highway on one section runs absolutely straight for 150 km, and the railway line for 500 km! Actually it is only the centre of the plain which is treeless — the periphery is wooded to various degrees.

Movements in the crust of the earth associated with the breaking apart of Australia and Antarctica

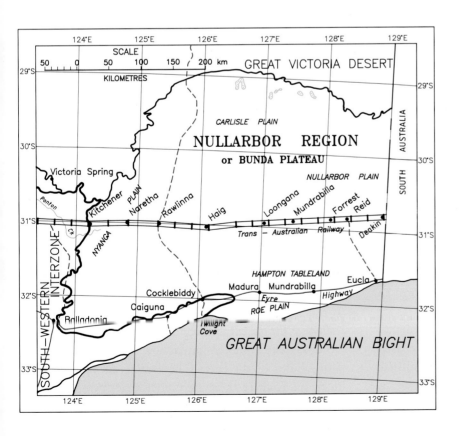

FIG 4.5

Southernmost of the regions of the Eremaea is the Nullarbor, a vast limestone plain, flat, empty and monotonous, named from the Latin for 'No Tree' by E.A. Delisser in 1867. The centre of the plain is in fact totally treeless, and in a poor season when there is no herbaceous growth presents an extremely desolate appearance, with no plant life except for the perennial bluebushes *(Maireana sedifolia)*. Chunks of rock show that the limestone is close to the surface. On the road from Reid to Eucla.

caused the Eucla Basin to begin to subside in the lower Cretaceous. During the Eocene a considerable thickness of limestone was deposited, and after an interval there was further deposition during the Lower Miocene. The sea finally withdrew in the Middle to Upper Miocene stage, about 12 million years ago. The Miocene limestones are much thinner than their precursors, but now form the surface of the plain. The Eocene limestones are exposed in the cliffs along the edge of the Bight. These cliffs have been formed by marine erosion since the Miocene emergence and stand at heights of 60 to 100 m above the sea. As there are no rivers, creeks or valleys on the plain, the cliffs form an unbroken line, unscalable and undercut by the sea, for 150 km from Point Culver to Twilight Cove. Further east to the State boundary at Wilsons Bluff, a coastal plain has formed at the foot of the cliffs which continue on the inland side as a well-marked feature known as the Hampton Tableland. The highway runs at the foot of this and on the coastal plain from near Cocklebiddy to Eucla.

The Plain slopes very gently up to the north from the cliffs to reach about 250 m above sea level at its northern edge where the sand dunes of the Great Victoria Desert are encountered. The Plain undulates very gently, but shows almost no trace of river action or co-ordinated drainage. Its extreme flatness calls for some explanation and many authorities have taken it to be an exposed sea floor. However the geological survey of the Plain[2] concluded that up to 100 m of limestone have been removed from the surface, progressively more towards the south, so that the Plain is remarkable not so much for the preservation of the Lower Miocene seafloor as for the extreme regularity of weathering. This is due to the combination of low rainfall and high permeability of the surface limestone. As is usual in limestone country there are numerous caverns formed by solution.

Fig 4.6

Fortunately in a good season the Nullarbor Plain presents quite a different aspect. The bluebush is more luxuriant and is filled in with a lush growth of the annual grass *Stipa nitida*, flowers such as *Podolepis* and *Helipterum*, and numerous other herbs. This appearance deceived two surveyors, Mason and Yonge, who had to walk out 260 km in 1896 after Aborigines had driven off their camels, and reported that they had discovered millions of acres of the finest agricultural and pastoral land in the world! It is obvious that they cannot have seen it in the conditions of Fig 4.5. Near Rawlinna, 1968.

Fig 4.7

This photograph, also taken near Rawlinna but two years earlier than Fig 4.6, emphasises the contrast between good and poor seasons on the Nullarbor Plain. A mutilated *Pittosporum* tree, the dried remains of weeds brought in by pastoral stocking, all else has disappeared leaving total desolation until the next rains should fall.

The soil is typically a pinkish-brown silt having a floury texture and containing lime nodules. Limestone is encountered at shallow depth, mostly at 30-40 cm, while lumps and slabs of limestone often appear at the surface. Soil is deeper beneath the wooded parts of the Plain, and is very shallow and rocky under the treeless area of the centre. Much of the soil is believed to have been blown away by wind during arid phases of climate such as the last Ice Age ±15 000 years ago.

The vegetation of the Plain is simple and monotonous. The character-plant is the bluebush *Maireana sedifolia*, belonging to the Chenopod family which includes also salt bushes and samphires. Bluebush is a subshrub about 45 cm tall, with silver-tomentose succulent leaves. The bluebushes are variably scattered, perhaps about 2 m apart in well-stocked areas, so that the cover is only about 5%. Other groundlayer plants occupying space between the bluebush are almost entirely annuals whose presence and density vary with the season. In drought the bluebush plants stand isolated in expanses of bare pink soil studded with rocks. In an ample season they almost disappear in a sea of grass and herbs. An annual grass, *Stipa nitida*, becomes exceedingly abundant at these times. Perennial grasses such as *Danthonia caespitosa* and *Stipa eremophila* tend to be confined to depressions. Associates next in rank of abundance appear to be *Zygophyllum ovatum*, *Lepidium oxytrichum*, *L. rotundum* and *Helipterum floribundum*. Other annuals recorded by the writer were: *Angianthus conocephalus*, *Sclerolaena obliquicuspis*, *S. patenticuspis*,

FIG 4.8

Before myxomatosis reduced their numbers, much damage was done by the rabbit plague on the Nullarbor. Rabbits existed in such numbers that they ate everything during times of drought, even the perennial blue-bushes. At this site north of Forrest on the plain, the bluebush had entirely disappeared except for the dead stumps which could be seen. Plant growth then consists only of the annuals, which may or may not sprout well according to the season. When this photograph was taken in October 1966 it was a poor year, and the annuals rather short, already drying off.

FIG 4.9

Depressions in the plain naturally tend to accumulate more moisture and show better growth. In some cases the bluebush is interspersed with perennial grasses such as *Danthonia caespitosa* and *Stipa eremophila* which provide a more permanent herbaceous cover. North of Caiguna.

FIG 4.10

One of the most attractive flowering plants of the plain, found in good years in depressions, is *Erodiophyllum elderi*, worthy of a place in anybody's garden. It was named after Sir Thomas Elder of Adelaide who promoted a number of exploring expeditions in the 19th century.

Fig 4.11

Only the central part of the Nullarbor Plain is completely treeless. The railway line crosses the centre and provides a good view of it, but the highway passing nearer the coast only offers glimpses. A small tree layer is provided on the outer parts of the plain typically by the myall, *Acacia papyrocarpa* seen here with the other typical components, bluebush *Maireana sedifolia* and the grass *Stipa nitida*.

Fig 4.12

In the southwest, towards Balladonia a tree component is provided by *Myoporum platycarpum* which attains a larger size than the myall. Somewhat more rarely trees of *Eucalyptus salmonophloia* and *E. salubris* (salmon gum and gimlet) may be seen in this area. The ground layer remains unaffected.

Fig 4.13

In the most southerly part of the plain, south of the highway, trees and shrubs occur more densely and are interspersed with glades of grassland composed of the perennial grasses *Danthonia caespitosa* and *Stipa eremophila* without bluebush. The soil beneath these glades appears to be of a heavy clay, whereas under the adjoining woodland it is a loam of floury texture. Caiguna.

FIG 4.14
The outermost parts of the plain in the north and west are relatively thickly wooded with such trees as *Casuarina cristata* (centre), *Acacia aneura*, *Myoporum platycarpum* and *Eucalyptus oleosa*. Undershrubs here are *Cassia nemophila* (yellow) and *Ptilotus obovatus* (white) while *Atriplex nummularia* (old man saltbush) shows as a grey shrub in the background.

Eriochiton sclerolaenoides, Brachycome ciliaris, Calotis breviradiata, Clianthus formosus, Erodium cicutarium, Euphorbia drummondii, Gnephosis skirrophora, Goodenia aff. *pinnatifida, Helipterum haigii, Maireana* aff. *georgei, Nicotiana goodspeedii, Lawrencia squamata, Podolepis* cf. *rugata, Sida calyxhymenia, Swainsona campestris*. Introduced weeds have found their way into the assemblage with grazing.

The saltbushes *Atriplex vesicaria, A. acutibractea* subsp. *whyallensis* and *A. cryptocarpa* in the east join the bluebush locally, or in some limited parts of the Plain replace it. In the well-wooded parts the similar soft silver-tomentose shrubs *Ptilotus obovatus* and *Cratystylis conocephala* also appear. Before myxomatosis much damage was done to the vegetation by rabbits and there are stretches where they eliminated the bluebush. In such cases the stumps of the latter could still be seen on fieldwork in 1966 but the plant cover had been reduced to annuals only. As observed north of Forrest these were the species of *Sclerolaena, Zygophyllum* and *Angianthus* listed above.

Larger shrubs are exceedingly sparse but include *Acacia burkittii, A. oswaldii, A. tetragonophylla, Atriplex nummularia, Eremophila alternifolia, E. longifolia, E. maculata, Geijera linearifolia, Lycium australe*.

The tree component is provided mainly by the myall *Acacia papyrocarpa* in the areas described as 'lightly wooded'. This is a small flat-topped tree, generally 3-5 m in height. Bark is coarse and stringy, the phyllodes greyish-green and long but delicate, giving a soft appearance to the crown. Spacing is irregular and increases in density towards the perimeter of the Plain.

Other trees which may somewhat rarely occur with the myall are *Myoporum platycarpum, Pittosporum phylliraeoides, Heterodendrum oleaefolium, Acacia aneura* and *Casuarina cristata*. They attain no greater size.

The presence or absence of trees appears to make little difference to the composition of the ground layer, in which bluebush remains conspicuous. Towards the northern and western margins of the Plain, where the country can be described as 'thickly wooded', *Acacia papyrocarpa* is almost entirely replaced by *A. aneura*, *Casuarina cristata* and *Myoporum platycarpum*, sometimes with a few *Eucalyptus oleosa*. The proportions vary. *Acacia aneura* appears to be dominant in the northern part as a small tree of 4-5 m with scattered *Casuarina* and *Eucalyptus* up to 9 m and *Myoporum* of 6 m tending to form groups. In the west *A. aneura* seems to be rare and replaced by the others, principally *Casuarina*. In the far southwest, at and beyond Balladonia, due to higher rainfall the limestone plain becomes covered with eucalypt woodlands and is included in the Southwestern Interzone. *E. oleosa* is the commonest species, with *E. flocktoniae* and *E.* sp. aff. *anceps*. Saltbush, bluebush and greybush all form part of the understorey. Along the highway this woodland can be seen gradually merging into the more open plain.

A feature of the Nullarbor Plain is the frequent occurrence of shallow saucer shaped depressions, formed by solution and typical of limestone country. These are known locally as 'dongas' or 'dongers'. Whether this is a misapplication of the South African word *donga* which means an erosion gully, is not clear. In the northern part of the Plain they are mostly shallow, gentle

Fig 4.15

The Nullarbor Plain terminates along its southern edge in a spectacular cliff eaten away by the sea. From Twilight Cove to Eucla there is a coastal plain at the foot of the cliff along which the highway runs, the cliff forming a scarp known as the Hampton Tableland. However west of Twilight Cove the cliff fronts the sea, 100 m high in places, and unbroken for 240 km. This is one of the most remarkable cliffed coastlines in the world, but is not approached by the highway and must be reached by tracks from Caiguna or Cocklebiddy. Vegetation on the cliff top is reduced by exposure to mat and cushion plants. *Westringia dampieri* is in flower foreground. *Correa reflexa* occurs in this habitat, its sole locality in Western Australia.

Fig 4.16

Inland of the clifftop, vegetation grows taller, and there is a belt of mallee in which *Eucalyptus socialis* is dominant, with *E. oleosa* and *E. conglobata*. It is interspersed with the grassy glades of Fig 4.13, and merges inland into the more open parts of the Nullarbor Plain. Locally on the clifftop there are sand dunes with an interesting flora allied to that of the Esperance Plains and forming the last easterly extension of it. South of Caiguna.

FIG 4.17
The coastal mallee may be mixed with *Melaleuca lanceolata* and *M. quadrifaria* and have an understorey which at first sight resembles that of the rest of the Nullarbor but may turn out to be not bluebush but greybush *(Cratystylis conocephala)*. The latter does not belong to the Chenopod family as do bluebush and saltbush but is a composite related rather to the sagebrush *(Artemisia)* of steppe country in the northern hemisphere. Eucla.

depressions vegetated with perennial grasses. In the north centre there is an area of deeper depressions with a dense tree vegetation of *Acacia aneura, Myoporum, Heterodendrum* and *Pittosporum*, south of which there is again a grassy belt, while on the south side of the Plain depressions are relatively large and have a saltbush vegetation.

The Hampton Tableland is topped with a belt of mallee about 15 km wide between the cliffs at the southern edge and the open plain to the north. *Eucalyptus socialis* is dominant with *E. oleosa* and *E. conglobata, Melaleuca lanceolata* and *M. quadrifaria*. Greybush seems to be the principal undershrub but both saltbush and bluebush of several species also occur. There are numerous grassy glades with the perennial grasses *Danthonia caespitosa* and *Stipa eremophila*.

The coastal plain at the foot of the Hampton Tableland — the Roe Plain — is varied, containing in part poorly vegetated sand dunes, or stony plains covered with mallee. As seen along the highway the inland portion has open saltbush plains in part, or lightly wooded bluebush steppe.

More detail on the region is to be found in (1d).

4.2 Great Victoria Desert
HELMS BOTANICAL DISTRICT
Mulga low woodland on hardpan soils between dunes; otherwise tree steppe of *Eucalyptus gongylocarpa, E. youngiana, Triodia basedowii*.

Climate: Arid with summer and winter rain; annual precipitation 200 mm.
Geology: Quaternary sandplain overlying Permian and Mesozoic rocks which are occasionally exposed.
Topography and soils: Undulating, somewhat featureless, mostly with longitudinal dunes. Shallow earthy loams overlying red-brown hardpan frequently occur between the dunes; otherwise red earthy sands, with red-brown sands in the dunes.
Area: 209 206 km².
Boundaries: Edge of desert sandplains and dunefields all round. N—transition to laterite plains country of Gibson Desert and Precambrian country of Warburton Region. S—edge of Nullarbor Plain, bluebush plains on limestone. W—edge of mulga country on Yilgarn Block.

The Great Victoria Desert was named in honour of the Queen by Ernest Giles in 1875 after a crossing with camels from the east. His party was saved after travelling 325 miles in 17 days without finding water when they came upon a small claypan full. Giles promptly named it Queen Victoria Spring, and while camped there recorded in his journal: 'The great desert...which will most probably extend to the west as far as it does to the east, I have also honoured with Her Majesty's mighty name, calling it the Great Victoria Desert.' Giles went on to apologise for having discovered nothing better than a desert to name after her! The desert was crossed again from the northeast by the Elder Exploring Expedition of 1891 which made a dash for Queen Victoria Spring after getting into difficulties from lack of water, only to find it dry. They were able to save

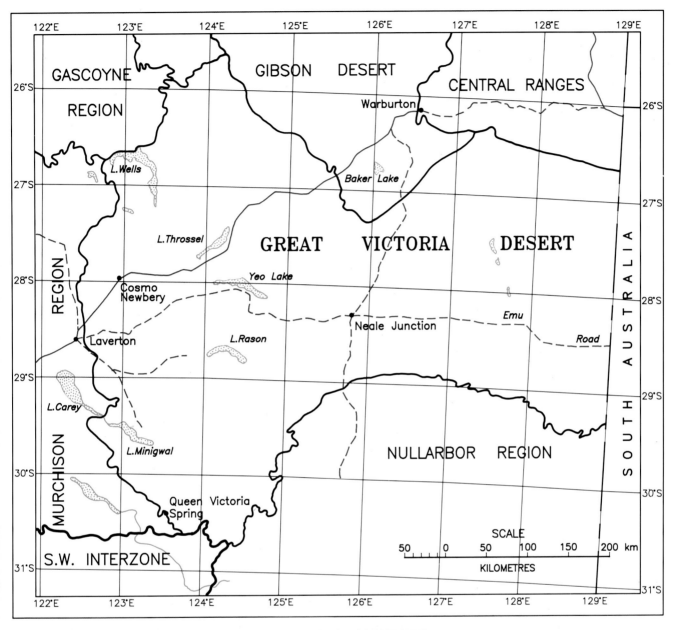

FIG 4.18

The Great Victoria Desert was named by Ernest Giles in 1875 in honour of Queen Victoria, while he was encamped at a claypan on its western edge which he also named after her as Queen Victoria Spring. It is not in fact a spring but a claypan which is, it seems, usually dry, as it was at the time of the writer's visit in 1966. Local Aborigines had dug a waterhole (centre) to utilise the last of the water.

FIG 4.19

The Great Victoria Desert is underlain by Mesozoic rocks, mainly sandstones, which occasionally crop out in breakaways, but for the most part have broken down to form extensive sandplains and dunefields. Breakaway country is typically covered with mulga (Acacia aneura) in tree or shrub form, and mulga may also occur in sandhill country on the flats between the dunes if the soil is lateritic or loamy. Mulga does not normally grow in sand and is absent from sandplains and the dunes themselves. East of Neale Junction.

themselves by pressing on to the Fraser Range. This expedition was accompanied by a botanical collector named Helms after whom the botanical district is named. Professional botanists did not enter the area until 1958, after which several parties made visits. The writer made the first north-south crossing of the desert with Mr Alex George in 1966. The area remains very inaccessible, with few tracks, except for the road from Laverton to the Warburton Range which has become a tourist route.

The Great Victoria Desert is situated entirely on sandstone country of the Officer Basin, which gives it a unity of character. Weathering of the sandstones has produced predominantly sandy soils so that apart from relatively small areas of mesaform hills and ranges, and some breakaway country, the entire region consists of sandplains, almost always with linear dunes which trend east-west. The Great Victoria is therefore essentially a waterless, sandy desert. It begins along the northern edge of the Nullarbor Plain and extends northwards between 150 and 250 km rising as it does so from about 275 m above sea level to about 425 m. It extends as well far to the east into South Australia and westward almost to Laverton and to Lake Wells.

The sandy country of the Great Victoria is continuously mantled with 'hummock grassland' formed by the spinifex *Triodia basedowii*, in which there are scattered trees, mallees, sclerophyll tall shrubs and small ericoid shrubs. The characteristic tree is the marble gum *Eucalyptus gongylocarpa*, a very handsome tree resembling the wandoo of the southwest, with glaucous leaves and a whitish bark mottled with small peeling flakes. It grows to heights of 9-12 m, with a stout trunk, and is irregularly distributed, congregating especially on rises in the uneven sandplain, probably where the sand is deeper. On the other hand it does not grow on the sand dunes unless also present locally between the dunes. Many groves of these trees contain few other

Fig 4.20

Sandplains in the Great Victoria Desert have a spinifex cover, and frequently, where the sand is deep, have groves of the beautiful marble gum, *Eucalyptus gongylocarpa*, which may also occur less commonly on the sandhills. This tree is the most important character plant of the Great Victoria, occurring nowhere else but within the confines of the desert in Western Australia and the adjoining part of South Australia. The spinifex here is *Triodia basedowii*, looking rather dried up in this photograph, under poor seasonal conditions. West of Neale Junction.

Fig 4.21

Eucalyptus gongylocarpa is not universal in the Great Victoria, and is frequently replaced by the mallee *E. youngiana*. In places the spinifex may be mixed with ericoid shrubs, an attenuated form of the flora of the kwongan of southwestern sandplains. These include a species of *Baeckea* and as seen here *Wehlia thryptomenoides*, in flower. Between Laverton and Cosmo Newbery.

FIG 4.22

Eucalyptus youngiana is one of the Western Australian mallees which have large red showy flowers 5 cm across. In a sandplain on Nambi station.

FIG 4.23

Sandhills in the Great Victoria are vegetated differently from the intervening sandplains. An ericoid shrub *Thryptomene maisonneuvii* is normally dominant on the flanks of the dunes, ousting the spinifex or reducing it to a few scattered plants. The Rottnest pine *Callitris preissii* may be locally abundant and sometimes *Eucalyptus gongylocarpa*. Taller shrubs include *Grevillea stenobotrya, Acacia ligulata, Gyrostemon ramulosus* and *Crotalaria cunninghamii* which are universal desert dune plants throughout the Eremaea. South of Giles.

FIG 4.24

In a different view from Fig 4.23, this photograph is taken on a sandhill showing the dense growth of *Thryptomene maisonneuvii* in contrast to the sandplain below which is covered with spinifex and scattered mallee and other shrubs. 65 km east of Lake Violet station.

Fig 4.25
Flowers and foliage of *Thryptomene maisonneuvii*.

Fig 4.26
Another ericoid shrub found less commonly on the dunes is the yellow-flowered *Micromyrtus flaviflora*. A third is the pink-flowered *Calytrix carinata*. All of these are related to the kwongan plants of the southwest.

Fig 4.27
Numerous other plants of this sandy country are related to and recall kwongan. This spectacular *Grevillea* is *G. juncifolia*. Mt Fanny.

Fig 4.28
The desert blackboy *Xanthorrhoea thorntonii* occurs occasionally, usually in scattered colonies. It may be found in this way in sandplains throughout the southern half of the Eremaea. 20 km east of Cosmo Newbery.

Fig 4.29
The quandong, *Santalum acuminatum*, occurs freely in the desert but is not confined to it, being seen just as frequently in the southwest. The brilliant red ripe fruits attract attention. The rind is edible and was used by early settlers to make jam. Quandong is related to the sandalwood *Santalum spicatum* whose similar spherical fruits are smaller and remain green.

FIG 4.30

One of the most remarkable plants of the desert sandplains is the subshrub *Leptosema chambersii*, which occurs here and there among the spinifex clumps. A leafless plant with a mass of rigid, prickly green stems, it produces a mass of showy flowers at ground level around the base of the plant. This presumably indicates that the flowers are visited and pollinated by some ground-dwelling creatures, perhaps mice. North of Bandya station.

FIG 4.31

One of the strange things about Australian desert flora is the almost total lack of succulent plants. In other deserts of the world, succulents are normally expected to be a conspicuous feature, notably in the family Cactaceae in the Americas, and in the Euphorbiaceae and Liliaceae in Africa. The latter families occur in Australia and have in fact produced a few succulents, but they are never common or conspicuous. This photograph shows *Sarcostemma australe*, a member of the Euphorbiaceae which has leafless green succulent stems and small yellow flowers. It is occasionally seen throughout the desert. It is possible that the infertility of Australian desert soils is responsible for this effect, favouring sclerophylly in plants.

plants besides marble gum and spinifex but others are mixed with mallees and other large shrubs, and these become generally dominant wherever marble gum is not present. All sandy areas are a mosaic of tree and shrub communities grading into one another.

The commonest and most beautiful mallee is *Eucalyptus youngiana* which has large red flowers. Other mallees occurring locally are *E. rigidula*, *E. leptopoda*, *E. concinna*, *E. oleosa*. Larger sclerophyll shrubs include *Acacia ligulata*, *A. helmsiana*, *A. murrayana*, *Alyogyne pinoniana*, *Eremophila leucophylla*, *Grevillea juncifolia*, *G. pterosperma*, *Hakea multilineata*, *H. suberea*, *Melaleuca leiocarpa*. Smaller ericoid shrubs are *Baeckea cryptandroides* and *Wehlia thryptomenoides*. Small subshrubs: *Dicrastylis exsuccosa*, *Leptosema chambersii*, *Newcastelia cephalantha*, *Ptilotus obovatus*. Creeper: *Kennedia prorepens*. Annual herbs: *Brunonia australis*, *Helichrysum apiculatum*, *Helipterum stipitatum*, *Podolepis* sp., *Ptilotus polystachyus*, *P. exaltatus*, *Waitzia acuminata*. The desert blackboy *Xanthorrhoea thorntonii* occurs in groups at widely scattered intervals.

The small ericoid shrubs in this assemblage, *Baeckea* and *Wehlia*, become very abundant in some patches, and with the larger acacias, grevilleas and hakeas impress a physiognomy closely recalling the scrub-heath of southwestern sandplains. The same is true of the vegetation of the dunes, here as elsewhere unique and differing from the interdune. In this southerly latitude

FIG 4.32

Within the Great Victoria Desert, plains where the soil is less sandy have a cover of mulga *(Acacia aneura)* as shown here, but attempts to establish sheep stations have been generally unsuccessful due to lack of ground herbage. In this example near Cosmo Newbery the ground layer consists partly of inedible spinifex. There is some growth of yellow-flowering *Waitzia*, but by no means an abundant cover.

FIG 4.33

Another adverse factor in the Great Victoria is recurrent drought. At the time of the writer's first visit, much of the country was dying or dead, and rainfall records showed that there had recently been a period of 40 months—more than 3 years—during which only one effective fall of rain had been received. This area evidently was formerly a stand of mulga like that in Fig 4.32, but almost all of it had died of the drought. This of course would be completely disastrous for any pastoral occupation. On the Gunbarrel Highway, 1966.

FIG 4.34

The drought had broken in most places by October 1966, and plant growth was recovering. In spinifex plains, much of the spinifex had dried off or died back severely and regrowth was occurring, accompanied by a flush of growth by shorter-lived adventitious plants, such as appears after spinifex has been burnt. Conspicuous in this case are *Dicrastylis exsuccosa* (yellow flowers) and *Halgania* sp. (blue).

Thryptomene maisonneuvii assumes complete dominance over the spinifex, *Triodia* and *Plectrachne* which are relegated to occasional plants. In general the flanks and lower slopes of the dunes are densely covered with *Thryptomene* with minor numbers of the similar ericoid shrubs *Micromyrtus flaviflora* and *Calythrix carinata*. On the summits it is rather more sparse and one finds the universal *Grevillea stenobotrya*, *Acacia ligulata*, *Gyrostemon ramulosus* and *Crotalaria cunninghamii*. Locally, small trees of *Callitris preissii* may become abundant.

In the south on the boundary of the Nullarbor Plain *Eucalyptus oleosa* becomes the dominant mallee, and *Triodia scariosa* replaces *T. basedowii*, a change that may be related to the presence of a calcareous hardpan.

Many areas of mulga occur throughout the region, on hills and breakaways, on plains and even to a considerable extent between sandhills, where the soil is fine-textured. In the latter case there is usually a sandy strip along the foot of the dune up to a quarter of a mile in width, which carries *E. youngiana* shrub steppe or a community with *Grevillea juncifolia*, *G. eriostachya* and *Hakea suberea* in *Triodia basedowii*. This merges into mulga and spinifex *(A. aneura + T. basedowii)* and this into pure mulga. Mulga here reaches 3-4.5 m in height, moderately dense, mainly *A. aneura* with some *A. pruinocarpa* and *A. linophylla*. Understorey shrubs include *Eremophila latrobei*, *Cassia* spp. and *A. aciphylla*. Herbs and grasses are generally sparse even in good season so that attempts to set up pastoral stations utilising the mulga areas have not been successful. *Schoenia cassiniana*, *Podolepis canescens*, *Waitzia acuminata*, *Velleia rosea* and *Eragrostis eriopoda* were recorded.

On breakaways *A. aneura* is usually joined by *A. tetragonophylla* at the foot, and by *A. quadrimarginea* on the top. Height on the summit is reduced to 1.5–1.8 m and small eremophilas *(E. latrobei, E. fraseri)* are conspicuous. Halophytes appear on the footslope — *Atriplex*, *Maireana*, *Sclerolaena* and *Rhagodia*.

In the valleys which are ancient drainage lines, and around salt lakes, there has been much deposition of calcium carbonate, reflected by corresponding changes in the vegetation. In sand over kunkar and travertine mallee becomes dominant, either pure or with mulga. Near Lake Throssel *Eucalyptus comitaevallis* is the species concerned, forming stands near the lake and occupying an ancient drainage line linking Lake Throssel with Lake Wells. There is some *Triodia basedowii* as a ground layer. In the Neale Junction area the mallee is *E. oleosa*, and *T. scariosa* joins *T. basedowii*. Further to the east in the sandy desert aerial photography shows mallee between sandhills in depressions, but its identity has not been verified.

The presence of a well-developed calcareous hardpan is thought to be responsible for the appearance of *Casuarina cristata* which forms low woodlands especially at Lake Throssel and Lake Rason. Trees may reach 12 m in height, generally 7.5-9 m, and occur patchily mixed with *A. aneura*, *Pittosporum phylliraeoides*, *Eremophila miniata*, *Ptilotus obovatus*, grasses and forbs. Smaller trees of only 6 m cover the kopi dunes in Lake Throssel, appearing stunted and miserable with a few *Eremophila miniata*, *Acacia* sp. and *Senecio lautus*. The soil is quite bare. Samphire covers the bed of Lake Throssel and around it in the casuarina woodlands and mulga there are open saltbush flats of *Atriplex*, *Maireana carnosa*, *Frankenia* and *Zygophyllum*.

4.3 *Murchison Region*

AUSTIN BOTANICAL DISTRICT

Predominantly mulga low woodland *(Acacia aneura)* on plains, reduced to scrub on hills. Tree steppe of *Eucalyptus* spp. and *Triodia basedowii* on sand plains.

Climate: Arid with summer and winter rain; annual precipitation 200 mm.

Geology: Archaean granite with infolded volcanics and meta-sediments (greenstones) of like age, forming the Yilgarn Block.

Topography and soils: Undulating, with occasional ranges of low hills, and extensive sandplains in the eastern half. The principal soil type is shallow earthy loam overlying red-brown hardpan; shallow stony loams on hills and red earthy sands on sand plains.

Area: 316 239 km².

Boundaries: N, E & W—vegetational expression of geological boundaries of the Yilgarn Block. S—climatically determined 'Eucalyptus-Acacia line' between *Acacia* low woodland and *Eucalyptus* medium-height woodland on lower-slope soils, parallel changes in sandplains.

The Murchison Region takes its name from the Murchison River with whose basin it largely coincides. The river was named by George Grey in 1839 after Sir Roderick Murchison, President of the Royal Geographical Society and promoter of exploration, a gentleman who has left his name in many parts of the world including for example the Murchison Falls on the Nile in Uganda.

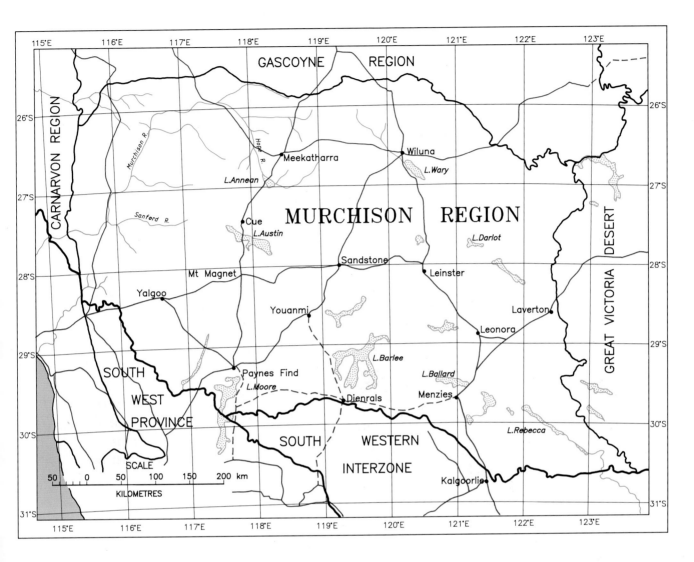

The region is not only the largest of the Eremaean regions, but is the most accessible and best known to the general public due to its widespread pastoral use. It adjoins the northern boundary of the Southwest Province and Interzone, so that it is normally the first Eremaean region entered by travellers. It is confined to the granite country of the Western Shield so that it does not extend onto the sedimentary rocks of the Carnarvon Basin on the west which underlie the Carnarvon Region, nor those of the Officer Basin on the east which underlie the Great Victoria Desert. In the north the Murchison Region ends where ranges of Proterozoic rocks overlying the granite basement are encountered. This boundary follows in the west the watershed between the Murchison and Gascoyne catchments, passing further east to the south of the Robinson Range and Lake Nabberu. The western half of the region more or less coincides with the basin of the Murchison River, the eastern half embraces the drainage of former rivers, now dry, draining towards the Eucla Basin. The western half is dissected, consisting of broad valleys and wide plains, with ranges of hills formed of harder, more resistant rocks, and few sandplains. The eastern half is less dissected and preserves a landscape more similar to that of the Wheatbelt Region in the Southwest Province with catenas comprising sandplains on the higher ground, loam soils on the slopes and plains, and salt lakes in the valley bottoms. In some cases there are 'low level sandplains' in the valleys formed of sand transported from the upper parts of the landscape. The soils formed differ however from those of the wheatbelt, and the most important types may be described as:

1 Shallow stony earthy loams on hills and ranges.
2 Red earthy sands on upland sandplains, with red sands on occasional dunes.
3 Earthy loams overlying red-brown hardpan, often with a surface stone layer, on undulating terrain.
4 Shallow acid or neutral red earths in intimate mosaics with (3) on extensive flat and gently sloping plains.
5 Saline soils associated with salt lakes.

FIG 4.35

The Murchison Region is characterised above all by mulga country. Mulga *(Acacia aneura)* forms the dominant vegetation on both hills and plains almost throughout the region. Viewed from any line of hills, the prospect is of an almost flat plain stretching to the horizon, with red soil, covered with little grey trees; and on the horizon another line of low hills. On reaching these, the prospect is again precisely the same, and so on for hundreds of kilometres.

The red-brown hardpan is a very characteristic feature. It lies usually between 20 and 100 cm depth, but is sometimes exposed at the surface. Plains with exposed hardpan become more common towards the north. Sandplain soils are red here. The change in colour from yellow to red occurs patchily along the boundary of the Southwest Province, and is one of the distinguishing features of this transition.

The natural vegetation of the Murchison is palatable to stock and as water is obtainable in permanent pools of the river and its tributaries, sheep were brought into the area in large numbers from the 1870s. Paddocking was introduced later with provision of water from wells and boreholes, and the number of stations gradually spread until today virtually the entire region is under pastoral use. The stations hold the land on leasehold from the crown. The Murchison has become the principal sheep ranching region of the State. There were 2.5 million sheep in the pastoral areas in 1912 out of a State total of 4.5 million, rising to 5.5 million in 1934 out of a State total of 11 million. A prolonged drought in 1936–41 reduced the sheep in the pastoral areas to about 3 million, at which figure numbers have been approximately stabilised ever since, whereas numbers have continued to rise in the agricultural areas—where they are run on artificial pasture—to exceed 16 million in recent times.

The Murchison is essentially the mulga region of Western Australia. *Acacia aneura*—mulga—is dominant or a significant component in the most extensive communities. Mulga does not extend further to the west since climate and soils in the Carnarvon Region do not favour it. It extends to the north into the Gascoyne Region, where it is widely developed, but geology, topography and soils differ. It extends also to the east onto different geology, topography and soils in the Great Victoria Desert but its occurrence is limited by the unfavourable substrata. The climate throughout that area is favourable to it, and if the Yilgarn Block had extended further in that direction, there is no doubt that it would have been covered with mulga. Conditions within the Murchison Region favour mulga more generally than in any other part of Western Australia.

Acacia aneura grows in the form of a tree with a single erect trunk on the more favourable soils and then forms low woodland. On less favourable sites it takes the form of a shrub, and such areas have been mapped as scrub. The low woodland may be continuous or interrupted by bare patches. *A. aneura* is at its best on plains with a red loam soil overlying a siliceous hardpan. In its shrub form it extends to hills with other soil types. It tends to be absent or only sparingly present on sandplains and on heavy alkaline and saline soils. Height and density of the mulga are both greater in the south than in the north, and throughout mortality is a very noticeable feature; dead trees are everywhere.

Structure consists of an open low tree or tall shrub layer of more than 3 m, a sparse low shrub layer of 1-2 m and a ground layer of ephemeral herbs which may be a closed one in a favourable season. In an unfavourable season this layer may not be present at all. There are also sparse perennial and annual grasses.

The more important component species in the mulga low woodland formation are as follows:

Small trees or large shrubs > 3 m *Acacia aneura* (very abundant: all other spp. rare or localised), *A. coriacea*, *A. grasbyi*, *A. kempeana*, *A. ligulata*, *A. linophylla*, *A. pruinocarpa*, *A. ramulosa*, *A. sclerosperma*, *A. tetragonophylla*, *A. victoriae*, *Bursaria occidentalis*, *Canthium latifolium*, *Hakea suberea*. Scattered *Eucalyptus kingsmillii*, *E. lucasii* and *E. oleosa* in a tall mallee form may be present in some areas.

Undershrubs 1-2 m *Acacia craspedocarpa*, *Cassia desolata*, *C. leurssenii*, *C. sturtii*, *Dodonaea microzyga*, *Eremophila clarkei*, *E. duttonii*, *E. exilifolia*, *E. foliosissima*, *E. fraseri*, *E. georgei*, *E. gilesii*, *E. granitica*, *E. latrobei*, *E. leucophylla*, *E. mackinlayi*, *E. macmillaniana*, *E. oppositifolia*, *E. platycalyx*, *E. punicea*, *E. spathulata*, *E. spectabilis*, *E. viscida*, *Grevillea deflexa*, *Halgania preissiana*, *Ptilotus obovatus*, *P. rotundifolius*, *Sida calyxhymenia*, *Solanum lasiophyllum*.

Perennial herb *Ptilotus drummondii*.

Perennial grasses *Monachather paradoxa*, *Eragrostis lanipes*, *E. eriopoda*, *Eriachne helmsii*, *E. mucronata*.

Annual grasses *Artistida contorta*, *Eragrostis dielsii*, *Eriachne pulchella*.

Ephemeral herbs *Brachycome ciliocarpa*, *Brunonia australis*, *Calotis multicaulis*, *Cephalipterum drummondii*, *Clianthus formosus*, *Erodium cygnorum*, *Gnephosis burkittii*, *Goodenia concinna*, *G. hirsuta*, *Haloragis odontocarpa*, *Helipterum charsleyae*, *H. craspedioides*, *H. floribundum*, *H. splendidum*, *H. venustum*, *Myriocephalus guerinae*, *Peplidium muelleri*, *Podolepis auriculata*, *P. kendallii*, *Ptilotus aervoides*, *P. polystachyus*, *P. exaltatus*, *P. helipteroides*, *P. macrocephalus*, *Schoenia cassiniana*, *Swainsona beasleyana*, *S. villosa*, *Symphiobasis* sp., *Velleia rosea*, *Waitzia aurea*, *W. citrina*.

Acacia pruinocarpa is the largest tree, reaching about 8 m in height: it is absent from some areas. *A. aneura*, *Bursaria* and *Hakea* reach 5-6 m, other species 3-5 m. *Acacia coriacea* is mainly on drainage, *A. grasbyi* in creek beds or on granite, *A. ramulosa* on sand, *A. sclerosperma* on calcrete and *A. victoriae* on salty flats.

Acacia craspedocarpa is a common undershrub but mainly in the Wiluna area. The commonest undershrubs are the *Cassia* species, *Eremophila fraseri* and *E. foliosissima*, which are normally present in good stands of mulga. *E. fraseri* seems to indicate stony ground, and *E. foliosissima* deep loam.

The perennial grasses are normally confined to patches of sandy soil, known as 'wanderrie country', in which the sand tends to occur in low, raised banks. *Acacia aneura* associates with *A. linophylla* on these, and the grasses are frequently so dense as almost to form a closed stratum. The normal mulga association occurs between the banks. Annual grasses are favoured by summer rains and forbs by winter rains. The display of colour from the flowers in this country in a favourable season is very striking. Given plentiful winter rains, by August the ground is covered with a carpet of flowers which may be continuous for hundreds of kilometres. White and yellow colours predominate with lesser pink, mauve and blue. Red and orange are absent except in *Clianthus formosus*. The commonest herbs are *Cephalipterum* (which has two colour forms, white and yellow), *Erodium*, *Velleia rosea*, *Haloragis*, *Helipterum* spp., *Myriocephalus*, *Podolepis auriculata*, *Ptilotus polystachyus*, *P. helipteroides* and *Schoenia*.

Optimum conditions for mulga are found on plains which have a deep red loam soil overlying siliceous hardpan. Here mulga adopts a tree form in relatively

FIG 4.36

The plains in mulga country are underlain by a siliceous hardpan developed within the soil by natural processes. This must not be confused with the laterite which generally underlies the topsoil on the higher ground. The cementing agents in laterite are oxides of iron and aluminium, and were deposited there millions of years ago under different climatic conditions from those of today. The silica-cemented hardpan or 'silcrete' is much more recent and may be forming at the present time. The trees here are *Acacia grasbyi* (minnieritchie), not mulga. Mount Gould station.

FIG 4.37

The mulga tree exists in several different forms, all classified taxonomically as *Acacia aneura*, some co-existing and some occupying different habitats. The form shown here, variety *latifolia*, has relatively broad, silvery leaves and occupies certain types of plains country. The herbaceous plants are the mulla mulla *Ptilotus exaltatus*. West of Sandstone.

FIG 4.38

The commonest form of mulga, var. *aneura*, is that shown here, where the leaves are linear, that is, very narrow and parallel-sided, grey green, and held erect, pointing towards the sky, thus reducing insolation. In the broad-leaved form of Fig 4.37 the leaves are held horizontally and on edge, with the blade perpendicular to the ground, which also reduces insolation. The linear-leaved form is found generally on stony plains, pediments and hills. Thundelarra station.

FIG 4.39

A distinct form of mulga with a horizontal branching habit is restricted to calcrete platforms, where limestone has been deposited locally in the evaporation of ground water. Byro station.

FIG 4.40

On the rough, stony hills of the Murchison Region mulga is normally of the narrow-leaved form, much reduced in size and often only a shrub. In the north spinifex may form part of the ground layer. The white-flowered subshrub here is Ptilotus obovatus, a perennial mulla mulla. In the Teano Range.

FIG 4.41

Along the boundary of the mulga in the south there is a broad belt containing a mixture of cypress pines (Callitris glaucophylla) with the mulga, as well as occasional eucalypts. The cypress pines appear where the soil is slightly more sandy and may reach large sizes, overtopping the mulga. Along the rabbit-proof fence north of Bonnie Rock.

FIG 4.42

While *Acacia aneura* is dominant in mulga country it is not the sole species, and some half dozen other *Acacia* species associate with it in lesser numbers. These include the 8 m tall *A. pruinocarpa* (gidgee) found in some areas, *A. coriacea, A. grasbyi, A. ramulosa, A. sclerosperma* and *A. victoriae*. *A. grasbyi* (minnieritchie) is shown here; it occurs mainly in creek beds or on granite. Both bark and foliage are distinctive. Wanna station.

FIG 4.43

The bark of *Acacia grasbyi* is an immediate distinguishing feature, a striking red-brown in colour, covered with small peeling flakes. Several other minor species share this bark characteristic.

tall, dense stands which have tended to remain longest in good condition. On other soils both further up slope and further down slope towards the rivers, mulga was originally less tall and dense and has deteriorated more since the introduction of grazing. On stony plains and stony pediments in many parts of the region nearly all the mulga is dead, and in some cases the death took place so long ago that the trees have rotted and disappeared. Under these conditions the understorey species of *Cassia* and *Eremophila* become larger and more numerous, also secondary species of *Acacia* such as *A. victoriae* and *A. tetragonophylla*.

The death and regeneration of mulga is thought to take place in regular cycles. Opening of the mulga in this way is different from the condition of irregular and patchy growth. This is found on plains where there has been strong development of the red-brown hardpan, which is later brought to the surface by erosion, resulting in the exposure of indurated pavements. These are treeless, so that the woodland is interrupted by frequent open patches, or in extreme cases is reduced to groves following sand-filled depressions and channels in the hardpan. The indurated pavements are frequently bare except for rare shrubs, grasses and ephemerals in season, many of which adopt a prostrate or ground hugging habit.

Hills on granite and gneiss are normally covered with mulga in a shrub form, height 2–3 m. *Acacia aneura* is associated with *A. quadrimarginea, A. ramulosa* and *A. grasbyi* and has understoreys as before of *Cassia* and

FIG 4.44

While mulga and its associates form the tree and large shrub layers in this formation, smaller shrubs of 1–2 m in height are also present, varying from sparse where the mulga is dense, to more abundant where it thins out. These consist mainly of species of *Cassia* and *Eremophila* with some smaller acacias such as *A. craspedocarpa* and *A. tetragonophylla*. The cassias and eremophilas are both showy in flower. Both are seen here with mulga in the background. Yalgoo.

FIG 4.45

The eremophilas are known by the general name of 'poverty bushes' because they will survive in eaten-out country where mulga has died. However prominent species have their own names. *Eremophila fraseri* is known as turpentine bush as the leaves have a sticky varnish-like coating. The flowers are decorative, with a blend of blue and red colours. Yalgoo.

Eremophila shrubs, and ephemerals. Granite outcrops and abrupt rocky ranges of resistant Archaean rocks tend to be similar to the granite hills.

Sandplains cover remnants of the earlier Tertiary land surface throughout the region but more especially in the eastern sector. In the west there are only few, small sandplains and these have a cover of *Acacia* scrub formed mainly of *A. linophylla* and *A. ramulosa* with some mulga, *Thryptomene decussata* and typical mulga undershrubs. In the eastern half however there are numerous much larger sandplains which have a spinifex cover of *Triodia basedowii*, and these share the flora of the Great Victoria Desert described in the previous section. *Eucalyptus gongylocarpa* is to be seen locally.

Frequently the sandplains are bounded by laterite scarps or breakaways, which generally carry shrubby *Acacia aneura*, *A. grasbyi* and *A. quadrimarginea*, but sometimes also trees of *Callitris glaucophylla* and *Eucalyptus carnei*. It is common to find an open area at the foot of the breakaway formed of white gritty clay leached from the pallid zone on which there are small *Frankenia* or *Halosarcia* and other halophytes.

Along the southwestern boundary of the region sandplains are transitional and reflect a mingling of elements of the hummock grasslands of the Eremaea and the southwestern heaths. The spinifex changes to *Plectrachne rigidissima* and *Triodia irritans*. Small ericoid shrubs are dominant in patches and include *Baeckea floribunda*, *Daviesia grahamii*, *Eriostemon tomentellus*, *Halgania viscosa*, *Olearia propinqua* and *Thryptomene*

FIG 4.46

Eremophila belongs to the Australian family Myoporaceae. Flowers are always decorative and of very varied colours. This one is *E. margarethae* which boasts a mauve corolla. In many species, especially *E. fraseri*, there is also a coloured calyx which persists for a long time on the bush after the corolla has fallen, and prolongs the show of colour while the fruit is ripening. Leaves of *E. margarethae* are greyish and tomentose.

FIG 4.47

The contrasting corolla and calyx are well illustrated in *Eremophila platycalyx* which has a short lived cream coloured corolla and a persistent pink calyx.

FIG 4.48

The *Cassia* species are shrubs of the same size, 1–2 m tall. As with the eremophilas a number of species is represented. All have yellow flowers, most have green leaves but some are glaucous like this *Cassia helmsii*. Helms was the collector for Baron von Mueller in the Great Victoria Desert, on the Elder Exploring Expedition of 1891. Yalgoo.

FIG 4.49

Flowers and foliage of *Cassia helmsii*. In cultivation.

Fig 4.50

The ground layer in mulga, apart from a few perennial grasses, consists of annuals and therefore varies with the season. Rainfall is erratic. If it is received in summer a crop of annual grasses is likely to respond. If it is received in winter the result is a crop of annual herbs, typically 'everlastings' of the daisy family Compositae (now called Asteraceae). These are at their peak in August when they can form continuous carpets of colour for hundreds of kilometres. In this scene the pink colour is provided by *Schoenia cassiniana*, the white by *Cephalipterum drummondii* and *Helipterum* spp. Paynes Find.

Fig 4.51

A feature of the Paynes Find area, not observed so markedly elsewhere, is that pink and white colours of everlastings are found on flats and depressions, and yellow colours on rising ground. *Cephalipterum drummondii* has two colour forms, white and yellow, otherwise identical, which segregate in this way, the yellow form associating with other yellow-flowered species such as *Helipterum craspedioides* in this case. Near Paynes Find.

urceolaris. Larger shrubs, which are more general, include *Casuarina acutivalvis, C. campestris* and *Melaleuca uncinata* as well as *Acacia* and *Eremophila* spp. A number of mallee eucalypts occur, principally *E. kingsmillii* which is conspicuous for its large flowers.

The mulga association also becomes transitional along its southern boundary. Mulga trees are taller and denser, and are mixed with cypress pines *(Callitris glaucophylla)* and *Eucalyptus oleosa.*

In the vicinity of salt lakes, which are found principally in the eastern sector down slope from the plains of optimum mulga growth, the landscape units become more alkaline, often floored with calcrete deposit (a form of limestone), or saline if there is not active drainage. In the first stage, which may be characterised as mulga with salty openings, the mulga low woodland is interrupted by patches of scrub of *Acacia sclerosperma, A. tetragonophylla, A. victoriae* and *Hakea preissii.* This

FIG 4.52

Flowers of *Cephalipterum drummondii*, yellow form. The ray florets in this species form a spherical head, unlike the conventional daisy flower where they are spread out in an outer ring. Mt Singleton.

FIG 4.53

Elsewhere all colours may appear together, but are still confined to the range pink, white and yellow. Red and orange colours are not found in composites in this habitat in Western Australia, unlike South Africa where they are common, even dominant. Other herbaceous species further north produce red colours and even blue, but not the Asteraceae. Near Paynes Find.

FIG 4.54

The sward of everlastings is a feature of the southern part of the mulga country. Further north they are replaced by herbaceous species belonging to other families which in good seasons can make as brave a showing. The most brilliant of these is undoubtedly Sturt's desert pea *(Clianthus formosus)* with its deep red petals and black centre. It is localised and not always seen. The mauve flowers in this photograph are of another pea, a *Swainsona.* White is provided by mulla mullas, *Ptilotus* spp., while the small shrub at left is a *Solanum.* Melrose station.

FIG 4.55

The blue-flowered *Brunonia australis* can be common in rougher country of poor hilly mulga, with some of the yellow *Podolepis canescens*. *Brunonia* used to be placed in a family of its own, the Brunoniaceae, but has recently been transferred to the Goodeniaceae, the family of *Dampiera* and *Lechenaultia*. Melrose.

FIG 4.56

Flowers and foliage of *Brunonia australis*. On the road from Laverton to the Warburton Range.

FIG 4.57

This unusual and bizarre combination was photographed one day near Wiluna. Sturt's pea creeping over the ground is overtopped by the white-flowered *Helipterum sterilescens*. The flowers cover open ground on a calcrete platform where limestone has been deposited by evaporating ground water coming from a creek on the left which is marked by a line of river red gums, *Eucalyptus camaldulensis*.

FIG 4.58

The glorious displays of colour illustrated above must not mislead us into thinking that the country is always like that. The writer's pictures are a selection of the best seen over many years, and the displays shown were the result of good seasons. In poor seasons the ground can be completely bare, the sheep starving and the pastoralist in despair. Diemals Find.

FIG 4.59
Fauna in mulga country is not abundant but can include the brush turkey, *Ardeotis australis*.

FIG 4.60
The frogmouth owl too is a denizen of the acacia country, using his protective colouring and appearance to pretend to be a dead branch!

FIG 4.61
Drainage in the eastern parts of the Eremaea is disorganised, meaning that there are no defined active rivers. Drainage flows into salt lakes and evaporates. The lake itself has a bed of bare salt, and the margins are salt-affected with zonations of more and less salt-tolerant plants. At the edge of the lake the dark coloured samphires congregate (*Halosarcia halocnemoides* and spp.) adjoined by a saltbush zone of *Atriplex*, *Maireana* and *Frankenia*. The mulga is held back to higher ground a little away from the lake. Nambi station.

FIG 4.62
In some lake beds gypsum predominates over common salt and gives rise to a suite of plants tolerant of it. This is particularly the case at Lake Austin near Cue, where the most striking and conspicuous gypsum plant is *Lawrencia helmsii*. The stems are covered with a dense mass of tightly curled succulent leaves making the plant look at first sight like a leafless stem-succulent cactus or euphorbia. Actually it belongs to the hibiscus family, the Malvaceae, and its seedlings in early life have tiny petiolate hibiscus-like leaves.

Fig 4.63

In the Murchison Region sandplains occur frequently on higher ground, and sometimes in the valleys if sand has been transported to a lower level. These have a spinifex vegetation of *Triodia basedowii* much as in the Great Victoria Desert, and even *Eucalyptus gongylocarpa* may occasionally be seen in deep sand though mallee of *E. oleosa* is more common as the tree component. Other species of spinifex replace *T. basedowii* in the south while in the western part of the region spinifex is replaced by scrub of *Acacia ramulosa*. On the rabbit-proof fence west of Diemals.

Fig 4.64

An extended period of pastoral use has taken its toll of vegetation in some areas especially in the lower Murchison catchment where stations were taken up early and initial stocking rates were excessive. Removal of ground vegetation by sheep exposes the soil to sun and weather, the surface may become sealed by the beating of rain drops so that water no longer penetrates freely, the mulga trees dry out the soil and eventually die. This process appears to have taken place in this landscape where *A. aneura* has largely disappeared. Other species of acacia with the cassias and eremophilas have survived but the tree cover has been much reduced. 25 km north of Nookawarra.

FIG 4.65

The mortality which is commonly seen in mulga, while it may be accelerated by pastoral use and weather conditions, does not altogether have such simple explanations. There is evidence that there is a cyclic regeneration process in mulga where stands of trees may be much the same age and on reaching maturity may start to die off, to be replaced gradually by a succeeding more or less even-aged crop. The mature trees here are beginning to thin out, while regeneration can be seen beginning to come up in the foreground. Youinmery station.

FIG 4.66

At a later stage the mulga has died out, and the regeneration is well established but still small.

FIG 4.67

Some years later the young mulga is beginning to cover the ground. Remains of a single dead tree can be seen poking up in the background. Lake Way station.

FIG 4.68

Eventually the young mulga begins to approach maturity, and in its turn may approach senility and die so that the cycle is repeated. The ecosystem is however fragile and can be ruptured if regeneration is eaten off by stock. Yakabindi station.

scrub in turn is interrupted by openings bare of trees and large shrubs, carrying scattered *Maireana pyramidata* and *M. triptera* surrounded peripherally by *Eremophila pterocarpa* or *Melaleuca uncinata*. With increasing salinity samphire *(Halosarcia)* replaces the *Maireana*. All such country tends to be badly eaten out and reduced to patches of mulga in open bare areas containing nothing but *Eremophila pterocarpa* and prostrate ephemerals.

Calcrete has been widely deposited along drainage lines past and present, and in many cases valleys are floored with sheets of it. Later erosion may convert the material into raised benches. Scrub of *Acacia sclerosperma* with *Pittosporum phylliraeoides* and *Grevillea nematophylla* is characteristic and there is a form of *Acacia aneura* with horizontal branching habit which occurs in some areas. The ephemerals *Helipterum sterilescens* and *H. humboldtianum* are also characteristic. Both are silver-tomentose with white and yellow flowers respectively. Other species include the grass *Enneapogon caerulescens* and the succulent *Sclerolaena*, *Calandrinia* and *Zygophyllum* spp.

Towards the rivers, where there is active drainage, scattered trees of *Eucalyptus camaldulensis* and *Casuarina obesa* appear in the mulga, and they also line the drainage channels. Locally *Eucalyptus camaldulensis* may form woodland on flood plains, in claypans and on calcrete areas.

Where drainage is disorganised, salinity of the lower plains increases further, with the formation of extensive salt flats vegetated with small halophytes such as *Atriplex*, *Maireana* and *Frankenia*. Sometimes scattered acacias persist. In the most extreme cases playa lakes have formed with beds of bare mud and salt crystals or with samphire colonies. Some of these lakes have accumulated gypsum which carries a particular group of plant species.

The abundance of dead trees in mulga areas has been remarked on numerous occasions all over Australia. It is usually coupled with observations of degradation and erosion due to pastoral misuse as in the CSIRO Survey of 1963[3], where there was a photograph illustrating the process and a caption reading:

> *Uncontrolled grazing has resulted in severe degradation, particularly the loss of more palatable grasses and shrubs. Increased wind and water erosion have caused loss of topsoil in some areas and widespread sealing of soils elsewhere, resulting in widespread death of mulga on shallow soils, and removal of shrub and ground layers.*

The country is unfortunately very vulnerable to such effects. The cover of the mulga trees is fairly sparse, and the shrub understorey is even more scattered. In good seasons however a dense and continuous ground layer of annuals springs up, and dries off after a few months. Under undisturbed conditions this plant material remains on the site protecting the surface soil against insolation, wind erosion and compaction by raindrops. If on the other hand the dried herbage is consumed by sheep, the soil is left bare and exposed to destructive forces. Seeds from which the next crop should regenerate are swept away by the wind, and the surface soil becomes compacted and impermeable. Rain no longer penetrates as it did, and the trees gradually exhaust soil moisture and die. This process is particularly noticeable in the western sector, attributed to excessively high rates of stocking in the early days of settlement. There has been widespread death and disappearance of mulga, leading to more open country in which it has been succeeded by secondary acacias and the former understorey species of *Cassia* and *Eremophila*. In the eastern sector such deterioration of rangeland is less in evidence.

At the same time there seems to be good evidence that mulga trees are not long lived and that whole communities may be more or less even aged, reaching the end of their life to a degree simultaneously, dying off and being replaced by regeneration. Severe droughts can also trigger this mortality but fire is not a normal factor. Mulga stands of all ages can be seen, and the process of cyclic regeneration can continue indefinitely so long as young plants are permitted to establish successfully. If eaten out by sheep, for example, the cycle will be disrupted.

4.4 *Carnarvon Region*

CARNARVON BOTANICAL DISTRICT

Mainly *Acacia* scrub and low woodland becoming tree and shrub steppe in the north, and with halophytes along the lower river courses.

Climate: Semi-arid bixeric at the coast becoming arid with summer and winter rain further inland; annual precipitation 200 to 250 mm.

Geology: Sedimentary basin with locally exposed rocks of Permian to Recent age; most of the surface covered by alluvium and colluvium.

Topography and soils: Gently undulating plain with mesa shaped remnants in the east and fields of longitudinal dunes. Hard alkaline red soils predominate in the plains with red sands in the dunefields.

Area: 91 046 km².

Boundaries: The eastern boundary is drawn according to the vegetational expression of the geological boundary of the Carnarvon Basin.

The Carnarvon Region coincides broadly with the geological Carnarvon Basin and is thus situated upon sedimentary rocks to the west of the Western Shield. The topography includes low plateaux, coastal plains, occasional tabular hills and some low folded ranges. The vegetation is very varied, and is generally *Acacia*-dominated in the south, changing to *Triodia*-dominated in the north. It includes a number of unique plant communities.

From the south the Carnarvon Region is entered on crossing the boundary of the Southwest Province which runs diagonally from SE to NW across the Toolonga Plateau which is situated between the Murchison River and Shark Bay. It is a vast, flat, monotonous and featureless sandplain rising to 300 m above sea level in the centre, sloping down to north and south. On the eastern part the plant cover is a dense thicket overwhelmingly of *Acacia ramulosa*, a bushy, spreading shrub reaching heights of 3-4.5 m. Other members of this shrub layer are *Thryptomene decussata*, *Grevillea eriostachya*, *Hakea* sp., *Acacia murrayana*, *A. sclerosperma*, *A. rhodophloia* and *A. tetragonophylla*. Occasional low trees rise above the canopy level, principally *Callitris glaucophylla* which is locally quite common and reaches heights of 10 m. Other trees are *Eucalyptus oleosa*, *E. eudesmioides* and *E. oldfieldii*. The bush is too dense on this part of the plateau to permit pastoral use. On the western part, with limestone closer to the surface, the shrub layer is more open and there is a number of pastoral stations. Several different *Acacia* associations occur. *A. sclerosperma* and *A. ramulosa* are dominant on relatively deep sand, *A. victoriae* and *A. eremaea* on patches of clay, *A. eremaea* and *A. subtessarogona* with *Acacia* sp. aff. *coriacea* on limestone with little soil cover. There are other associated shrubs, numerous undershrubs including *Cassia* and *Eremophila* spp., and a rich growth of annuals in a good season. The sides of the North-west Coastal Highway along this stretch are one of the best places in the State to view spring annuals. The pink-flowered *Schoenia cassiniana* is the principal species.

It will be noticed that mulga, *Acacia aneura*, does not appear among the acacias listed. Mulga does not normally occur in the Carnarvon Region, presumably as it is intolerant of the soil conditions.

The peninsulas and islands of Shark Bay form part of this region — the inner Peron Peninsula, the outer peninsula known as Edel Land of which Dirk Hartog Island is a detached northern portion, and the Bernier and Dorre Islands. Most of the geographical names are French, having been bestowed by the Baudin expedition in 1803 and the Freycinet expedition in 1818. Other names such as Shark Bay itself are translations from the French. The Peron Peninsula consists of sandhills often with saltpans in the swales while the outer sequence is built up of hills and platforms of coastal limestone overridden by calcareous dune sands.

The impression conveyed by the vegetation is one

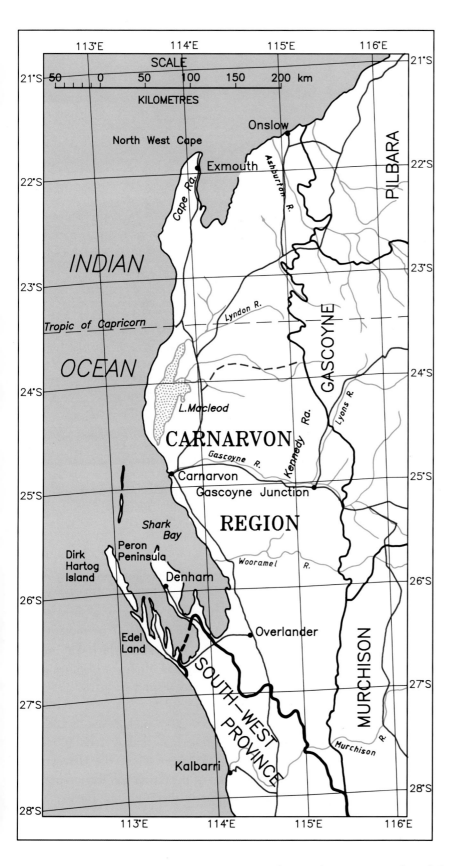

of arid and wind-swept desolation. *Acacia ramulosa* scrub covers the east side of the Peron Peninsula and Faure Island, but is replaced by a low thicket of *Acacia ligulata* and *Lamarchea hakeiifolia* in the northwest. The remainder of the peninsula consists of spinifex country with *Triodia plurinervata*, a local endemic species peculiar to Shark Bay. It is mixed with small wind-pruned shrubs of *Acacia ligulata* and others. This community also occurs

FIG 4.69

The Carnarvon Region, lying along the west coast from north of the Murchison River to the Northwest Cape, is typically *Acacia*-vegetated, but the mulga, *Acacia aneura*, is replaced by other species. The principal dominant is *A. ramulosa* (bowgada) a slightly smaller tree or large shrub of similar appearance as shown here. Understoreys of smaller shrubs and annuals are much the same.

in Edel Land where it may additionally (also on Dirk Hartog Island) form a mixed association with the ericoid shrubs *Melaleuca cardiophylla* and *Thryptomene baeckeacea* and another local spinifex *Plectrachne* sp. The plant cover is reduced to an open growth of cushion plants on cliff tops and similar exposed situations, but becomes thicker and taller on lower more sheltered ground, with larger shrubs of *Acacia ligulata* and *Diplolaena dampieri*. Saltbush communities occur also.

On the mainland the Toolonga Plateau merges into the Yabalgo Plain, a flat plain mainly covered by linear sand ridges. These support a unique community, a low woodland with trees 6 m tall of an unnamed species of *Acacia*. This is an erect tree with a straight trunk, and frequently forms pure stands virtually without associated species. Between the sandhills other *Acacia* associations hold sway as on the Toolonga Plateau.

Further north as the lower course of the Gascoyne River is approached, the country flattens still more and becomes marshy. The sandhills are reduced to low sandy rises at first with *Acacia ramulosa* dominant, then *A. sclerosperma*. In the flats, samphires *(Halosarcia)* are dominant in the wettest claypans, otherwise there are saltbush *(Atriplex)* and bluebush *(Maireana)* communities. A subshrub *Ptilotus polakii*, of a genus which does not normally include halophytes, is often conspicuous in distinct communities in and around claypans.

Along the floodplains of the Gascoyne River, *Eucalyptus camaldulensis* up to 20 m tall lines the channels, together with *Acacia citrinoviridis* and *A. coriacea*.

To the north of the point where the Gascoyne flows out onto the plains of the Carnarvon Region from the interior plateau stands the Kennedy Range, an extensive tableland 80 km long standing about 90 m higher than the surrounding plains and bordered by abrupt escarpments. Its surface is thickly covered with sand piled into linear dunes. The vegetation is a mixture of spinifex with small ericoid or 'heath' shrubs and scattered larger shrubs similar to that found on Edel Land and Dirk Hartog Island, except that the species

Fig 4.70

An associated shrub which for once is not an *Acacia* and produces some floral interest, is *Thryptomene decussata* (centre). It is common on the Toolonga Plateau.

Fig 4.71

The peninsulas between Shark Bay and the coast can adopt a desolate appearance, affected by both wind and salt spray. The silvery shrub here is a saltbush, *Atriplex bunburyana*, mixed with low wind-pruned shrubs of *Acacia ligulata* and the ericoid shrubs *Melaleuca cardiophylla* and *Thryptomene baeckeacea*. Hills of the west coast appear in the rear, and a samphire flat in the middle ground. Carrarang station, Edel Land.

Fig 4.72

On the Peron Peninsula on the way up to Denham the main road crosses spinifex country of *Triodia plurinervata* mixed with domed shrubs of *Acacia ligulata*. The latter owe their shape to exposure to wind. This association occurs also in Edel Land but is confined to these two peninsulas. Near Denham.

FIG 4.73

On top of the cliffs next to the ocean the vegetation is further reduced to sprawling and cushion plants, for example *Scaevola crassifolia* (dark green), *Acanthocarpus preissii* (pale green) and *Atriplex paludosa* (grey). The plant cover is battered continually by wind and spray on an already dry substrate of calcareous sand. Edel Land.

FIG 4.74

The Zuytdorp cliffs (Fig 3.85) continue northward as far as Edel Land and then, rather lower and less spectacular, along the coast of Dirk Hartog Island. In the north the cliffs are vertical and undercut by the sea but further south on Tamala they slope down and are vegetated with halophytes such as small *Atriplex*, *Frankenia*, *Senecio lautus* and *Swainsona* sp.

FIG 4.75

North of Carnarvon along the coast towards Quobba there are the famous blowholes where the sea spurts up from caverns in the coastal rock pavement.

Fig 4.76

The *Acacia ramulosa* scrub of the Toolonga Plateau which is traversed by the Northwest Coastal Highway is another of the State's most spectacular localities for spring annuals—in good seasons. Plants in flower here, from left to right, include *Ptilotus macrocephalus, Solanum orbiculatum, Swainsona* sp. and *Cephalipterum drummondii* (yellow form). The display continues for perhaps 100 km on either side of Overlander and the Denham turnoff.

Fig 4.77

Schoenia cassiniana is usually the most abundant of the annuals along this stretch and extends from the roadside in a mass here with yellow *Cephalipterum* in the background. Both species are 'everlastings' and belong to the Asteraceae.

Fig 4.78

After a break along the highway in marshy country, the floral display is resumed on sandhills south and east of Carnarvon. In flower here are mainly *Schoenia cassiniana* with the orange-flowered *Tephrosia flammea*, a *Brachycome* sp. (ray-flowered daisy) and bushes of *Eremophila leucophylla*.

Fig 4.79

A spectacular plant of low sandy ridges south of Carnarvon is the climber *Brachycome latisquamea*, which scrambles over other plants. The grey-leaved shrubs in the rear are *Eremophila pterocarpa* which is pretending to be a saltbush and grows around claypans and in salty depressions.

represented are different. Here the spinifex is *Triodia basedowii* and *T. pungens*, heath shrubs are mainly *Pileanthus peduncularis* (coppercups) with *Baeckea* sp. and *Calytrix brevifolia*. Larger shrubs include *Acacia ramulosa*, *Banksia ashbyi*, *Calothamnus chrysantherus*, *Grevillea gordoniana*, *G. stenobotrya*, several spp. of *Cassia* and *Eremophila*, and a *Eucalyptus* sp. in mallee form.

On the plains north of the lower Gascoyne there is a broad sand belt with north-south trending sand ridges. With the higher rainfall nearer the coast this is *Acacia* country but scattered spinifex plants are still present. On the dunes *Acacia ramulosa* is dominant as a small tree to 5 m with *Grevillea stenobotrya*, *Eremophila* sp., *Hakea stenophylla* and *Santalum acuminatum*. The smaller ericoid shrub species of the Kennedy Range are present also, plus *Verticordia forrestii*. Between the dunes the flats are dominated by *Acacia subtessarogona*, a tree to 8 m which is a local endemic, with smaller more shrubby *A. tetragonophylla*, and *A. xiphophylla* (snakewood), *Cassia*, *Eremophila*, *Scaevola* and *Solanum* spp.

North of the Minilya River there is an alternation of sandy plains with dunes and gravel plains, the latter having a surface layer of stones overlying clay. Spinifex is now dominant on sand, and Acacia on the heavier soils. On the sand hills *Plectrachne schinzii* (feathertop spinifex) provides a general cover with scattered shrubs as listed in the preceding paragraph. On sandy flats both *Triodia basedowii* and *Plectrachne schinzii* provide the general cover with scattered *Acacia pyrifolia* as the typical shrub. The typical species which inhabits gravel plains is the snakewood, *Acacia xiphophylla*, and it is widespread also in the Gascoyne and Pilbara Regions. It is a large spreading shrub of 2-2.5 m in height and a twisted or gnarled form. It may be sparingly associated with *Acacia sclerosperma*, *A. victoriae* and *A. aneura*. The most typical undershrub is *Eremophila cuneifolia* with other *Eremophila* and *Cassia* spp., *Acacia bivenosa* and *A. tetragonophylla*. There is little ground vegetation other than annuals in season when they can make a colourful show. At other times such country normally presents a desolate appearance.

Between the Exmouth Gulf and Lake Macleod there is a low range of limestone hills. Along the crest there is a thin cover of *Triodia* with sparse *Acacia bivenosa*. Surrounding pediplains with more soil are better vegetated with *Acacia* scrub, principally *A. sclerosperma*, *A. victoriae*, *A. coriacea* and/or *A. xiphophylla*. Locally this has a saltbush understorey, mainly *Atriplex amnicola*.

Along the ocean coast there is a belt of sand ridges covered in general with spinifex — *Triodia pungens*, *T. basedowii*, *T.* sp. aff. *angusta* — but with very numerous shrubs, chiefly of *Acacia*, *Calothamnus*, *Labichea* and *Thryptomene*. On the sandhills below the southern end of the Cape Range we again find an association of spinifex and small ericoid 'heath' shrubs. The spinifex is *Plectrachne schinzii*; larger shrubs include *Acacia spathulifolia*, *Banksia ashbyi*, *Grevillea eriostachya*, *G. stenobotrya*, *Hakea stenophylla*, *Mirbelia ramulosa*; small shrubs *Calytrix brevifolia*, *Hibbertia spicata*, *Scaevola globulifera*, *Thryptomene baeckeacea*, *Verticordia etheliana*.

The Cape Range which forms the peninsula running out to the North West Cape, with the smaller Rough Range, are rugged limestone ranges, locally with steep cliffs. The area is dominated by bare limestone with pockets of soil. Clumps of spinifex establish themselves here and there among the rocks, *Triodia pungens* and an undescribed species. *Eucalyptus prominens* is common, as a tree to 6 m or more often as a mallee. *Brachychiton obtusilobus* and *Grevillea* sp. are other small trees, *Eucalyptus oleosa* and *E. prominens* as mallees, *Ficus platypoda* in dense clumps. Very scattered small shrubs are also found.

An extensive coastal plain borders the opposite (east) side of the Exmouth Gulf and forms the northernmost portion of the Carnarvon Region. The coast is lined with mangroves, behind these salt flats, claypans and coastal sand dunes, behind these again extensive plains, frequently sandy and with linear dunes, and occasional rocky hills. Four species of mangrove are found here — *Avicennia marina*, *Rhizophora stylosa*, *Ceriops tagal* and *Aegialitis annulata*.[4] Behind the mangroves as on the Pilbara coast there is a belt of bare hypersaline mud. This is flooded only at highest tides and develops a salinity so extreme that nothing can grow there. In this area the belt attains a width of 10 km. On somewhat higher country there are extensive plains with a very patchy vegetation of snakewood scrub, grassland, claypans and bare patches of gravel and shingle. The snakewood *(Acacia xiphophylla)* is normally rather stunted (1.5-2.5 m) and spreading, mixed with *A. victoriae* and *A. tetragonophylla*. *Triodia basedowii* sometimes forms a ground layer, sometimes annual herbs such as *Ptilotus* and *Swainsona* spp., sometimes there is none. Besides these gravel plains there are extensive sandy plains mostly covered with linear dunes of a north-south trend. The flora here is strongly reminiscent of the Great Sandy Desert but there are also ericoid shrubs of southern affinity, particularly *Verticordia forrestii* which makes a conspicuous show in spring in the dune belt immediately

FIG 4.80

Marshy country at the mouth of the Wooramel River, south of Carnarvon, is a maze of claypans and low sandy rises. Scattered succulent shrubs of *Maireana polypterygia* dot the claypans, while scrub of *Acacia eremaea*, *A. ramulosa*, *A. sclerosperma*, *A. tetragonophylla* and *A. victoriae* covers the rises.

FIG 4.81

Much of the country along the lower Gascoyne and Wooramel Rivers consists of sandy salt flats where the vegetation appears to be of typical chenopodiaceous halophytes, but surprise, these belong to other families and are only pretending to be saltbushes. The larger shrub here with glaucous leaves is *Eremophila pterocarpa*, and the smaller one in flower is *Ptilotus polakii*, neither of them expected to be salt-tolerant plants. A few acacias are also present.

FIG 4.82

Flowers and foliage of *Ptilotus polakii*. 150 km east of Carnarvon.

Fig 4.83

Towering above the marshy country of the lower Gascoyne is the 90 m escarpment of the Kennedy Range, an extensive mesa 80 km long. It is formed of Permian sediments duricrusted and covered with sand piled into linear dunes. The range is notable for the discovery of a fossil banksia cone from the early Tertiary, but its modern vegetation is also interesting. Merlinleigh station.

Fig 4.84

The sandplain with dunes atop the Kennedy Range has a general spinifex cover—*Triodia basedowii* and *T. pungens*—which shares space with larger shrubs such as *Grevillea eriostachya* (centre) and a number of ericoid shrubs related to southern kwongan. These include *Baeckea* sp., *Calytrix brevifolia* and *Pileanthus peduncularis* (coppercups).

Fig 4.85

In the northern part of the Carnarvon Region, spinifex becomes more conspicuous and may associate with the acacias or replace them. Here in the High Range, Mangaroon, it forms the ground layer in acacia scrub.

Fig 4.86

Salty country, wherever it occurs, has its own specialised vegetation. *Acacia victoriae* is salt tolerant and can reach very large sizes, though such trees are rare. The ground layer is of *Eremophila pterocarpa*. Merlinleigh station.

Fig 4.87

A belt of sandhill country traversed by the highway running north from Carnarvon has a similar flora to the Kennedy Range but without spinifex. *Acacia ramulosa* is dominant on the dunes as a small tree to 5 m with the ericoid species listed on the Kennedy Range and larger shrubs such as *Hakea stenophylla*, *Eremophila* sp. and *Santalum lanceolatum*. On the interdune flats a local endemic species *Acacia subtessarogona* forms a mulga-like low woodland 8 m tall, with a mulga-like understorey and ground layer. 120 km north of Carnarvon.

Fig 4.88

An unusually good season in 1963 brought these shrubs into flower on sandhills alongside the highway 120 km north of Carnarvon. The orange flowers belong to *Pileanthus peduncularis* (forma) and the pink ones to an undescribed species of *Calytrix*. Although they were right along the highway, the writer never succeeded in finding them again. Such myrtaceous ericoid shrubs are relatives of the kwongan plants of the southwest, and continue to occur on sandy country all the way up the coast as far as the Northwest Cape.

FIG 4.89

As the Ashburton River is approached sandhill country contains this magnificent *Verticordia forrestii*, the largest of all the morrisons. It may occur in a pale pink form as well as red.

FIG 4.90

The peninsula running out to the Northwest Cape has a limestone spine, dissected into some canyons which are quite scenic. The vegetation is sparse, consisting mainly of spinifex, with scattered trees and shrubs.

FIG 4.91

The limestone of the Cape Range has a general cover of the spinifex *Triodia wiseana* which is normally expected on limestone and dolomite. The scattered trees are *Eucalyptus prominens*. *Ficus platypoda* is also seen, and two mallees, *Eucalyptus oleosa* and *E. prominens*. Rare shrubs include *Acacia pyrifolia* and *A. bivenosa*, other acacias, sundry *Cassia*, *Eremophila* and *Grevillea*.

FIG 4.92

Sandhill country at the southern foot of the Cape Range reveals the same mixture of spinifex and kwongan elements which is encountered in the Carnarvon Region all the way up from the south. The spinifex here is the feather-top, *Plectrachne schinzii*, amongst which the photograph shows *Calytrix brevifolia* (pink) and *Banksia ashbyi* (at rear). Other kwongan elements include *Acacia spathulifolia*, *Grevillea eriostachya*, *Hakea stenophylla*, *Hibbertia spicata*, *Mirbelia ramulosa*, *Thryptomene baeckeacea* and *Verticordia etheliana*.

west of Barradale Crossing. The spinifex *Plectrachne schinzii* provides a general cover on the dunes, *Triodia basedowii* and *T. pungens* on the flats between them. Shrubs on the dunes include *Acacia victoriae*, *A. coriacea*, *A. translucens*, *Calytrix brevifolia* and *Grevillea stenobotrya*. The desert walnut *Owenia reticulata* occurs near the coast but inland tends to be replaced by *Grevillea gordoniana* which is particularly conspicuous west of Barradale Crossing as noted above.

Between the dunes *Owenia reticulata* is conspicuous as a scattered tree near the coast and is replaced further inland by *Eucalyptus dichromophloia*, *E. setosa*, *Hakea suberea* and *Grevillea pyramidalis*. *Acacia pyrifolia* is the most characteristic shrub, with various other acacias, cassias, eremophilas and others, but very sparse. The general impression is one of dreary, sparsely vegetated country.

4.5 Gascoyne Region

ASHBURTON BOTANICAL DISTRICT

Almost entirely mulga *(Acacia aneura)* often with snakewood *(A. xiphophylla)* and other *Acacia* spp. as scrub on the hills and as low woodland on the plains. Some areas of dwarf scrub of *Eremophila* and *Cassia*.

Climate: Arid with summer and winter rain; annual precipitation 200-250 mm.

Geology: Middle Proterozoic rocks, mainly sandstone, and some granite.

Topography and soils: Mountainous, with low ranges divided by broad, flat valleys. Chiefly shallow earthy loams overlying red-brown hardpan on the plains, with shallow stony soils on the ranges.

Area: 181 453 km².

Boundaries: N & E—boundary between *Triodia*-dominated steppe and *Acacia*-dominated scrub. S & W—geological boundaries of the Yilgarn block and Carnarvon Basin respectively, in their vegetational expression.

The Gascoyne Region comprises the middle and upper valleys of the Gascoyne and Ashburton Rivers together with a projection to the southeast beyond the Gascoyne catchment, reaching as far as Lake Carnegie. This portion has been termed the Carnegie Salient[1C]. The region is essentially mulga-covered, distinguishing it from the regions to the east, north and west. It is distinguished from the Murchison Region to the south by its different geological structure which produces different soils and a different character in the landscape. The Murchison Region lies upon the granitic Yilgarn Plateau whereas the Gascoyne Region is mainly upon Middle Proterozoic rocks consisting of sandstone and conglomerate with shale and siltstone, and some Archaean or Lower Proterozoic metamorphic rocks and granite. These form hills and ranges which strike ESE-WNW, with plains between them. The northern boundary of the region is determined by a major change in vegetation from shrubland to hummock grassland, a transition which must basically be climatically determined with the spinifex on hotter, drier country than the mainly *Acacia* shrubland. The change-over occurs however at a different threshold on each rock type, so that in the Ashburton Valley with the rocks striking to the WNW the boundary zigzags from one

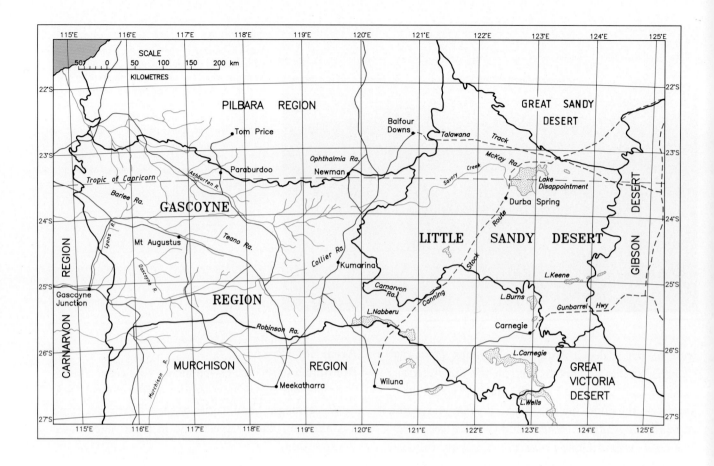

outcropping formation to the next. This is best observed in the area of Wyloo Station.

The geological structure also imposes poorer conditions for tree and shrub growth than in the Murchison. Gravel plains are a major component of the middle valleys of both rivers and have proved particularly sensitive to pastoral practices, arousing some official concern[5]. The Ashburton Valley has besides a particularly low rainfall. In consequence some of the poorest and most degraded country in the State is to be seen here.

On the Lyons River as far up as Minnie Creek and on the Gascoyne itself as far up as Mt Pickford there is a particularly poor stretch of country underlain by shale, greywacke and granite. Almost all of this is extremely poor and stony with a thin soil cover. It is poorly vegetated with signs of degradation due to stocking, the small trees present being frequently dead or in bad condition and the ground layer appearing denuded. There is a general but sparse cover of large shrubs of *Acacia aneura*, *A. xiphophylla*, *A. victoriae*, *A. tetragonophylla*, *A. sclerosperma* and *A. ligulata*. *Hakea preissii* appears where the soil is salty, and *H. suberea* on granite. Smaller shrubs of *Cassia*, *Eremophila* and *Solanum* occur as elsewhere and numerous species of ephemerals form a continuous cover in a good season.

Much the same species are found on the plains and ranges further upstream but taller, denser, and generally in better conditions. The hills and ranges which divide the Ashburton from the Lyons River tributary of the Gascoyne are almost entirely covered with *Acacia* scrub and low woodland, the scrub on the numerous rocky ranges and the low woodland on the pediments and valley plains. In the former case mulga tends to be the sole shrub with *Cassia* and *Eremophila* as smaller components. On stony plains and pediments mulga associates generally with snakewood, or the latter may become the sole species in extreme cases on plains surfaced with large stones, which are perhaps best described as shingle plains rather than gravel plains. In the latter case ground plants tend to be halophytes, e.g. *Halosarcia*, *Sclerolaena*, *Cassia pilocarina*, *Frankenia magnifica*, *Atriplex lindleyi*, *Ptilotus murrayi*.

On the loam flats between the ranges typical mulga low woodland comes into its own, mulga as a small tree in association with gidgee (*Acacia pruinocarpa*) and *Canthium latifolium*. Gidgee is usually a larger tree than

FIG 4.93
The Gascoyne Region comprises the catchment of the Gascoyne River and its tributaries such as the Lyons River, inland of the coastal plain. It has also an eastward prolongation called the Carnegie Salient. The region consists of poor stony plains, hills and ranges; its hilly nature distinguishes it from the Murchison Region to the south accompanied by a geological change. The vegetation is still essentially mulga, but often more open and mixed with other species. Typical elements are shown here: the mountain ridge, the stony plain, the open mulga and smaller shrubs of *Eremophila cuneifolia*. Coodardoo Gap, Wanna station.

mulga but is not so common. A large range of *Cassia* and *Eremophila* spp. provides the typical undershrubs and there is the usual ground layer of annuals in season. In this region the annuals do not belong to the Compositae as they do further south, but to various families, e.g. *Clianthus formosus, Goodenia maideniana, Ptilotus exaltatus, P. helipteroides, P. macrocephalus, Swainsona canescens, Trachymene glaucifolia*. The loam flats are underlain by the usual mulga hardpan at variable depth, and where it surfaces or is thinly covered the vegetation also thins out or becomes absent. The low woodland of the valley bottoms is therefore typically patchy and interrupted by bare hard-surfaced plains. In the east as the Little Sandy Desert is approached the soil generally becomes more sandy, and the mulga may have locally a spinifex understorey of *Triodia basedowii*. There are even a few sandplains. The highway crosses one 20 km north of Kumarina; the largest is south of the Kunderong Range. *Triodia basedowii* provides the main cover with the mallee *Eucalytpus gamophylla* as the most conspicuous shrub, however where the sand overlies calcrete (deposited limestone) the components are *T. longiceps* and *E. oleosa*.

The vegetation of the Carnegie Salient is essentially similar. It contains the Frere Range on which there is scrub of mulga with *Acacia quadrimarginea* and *A. grasbyi*. On either side of the range are the salt flats and playas of the Lake Nabberu System, with wide plains sloping gently towards them, covered with mulga low woodland including *A. pruinocarpa*. Further to the east are the Lee Steere Range which is spinifex covered and has some spinifex plains to the north of it, the Princess Ranges which carry mulga, and more wide expanses of salt flats connecting eventually to Lake Carnegie. Samphire communities occur in the lowest and wettest sections (or surfaces of bare salt in the lake itself), a *Frankenia-*

FIG 4.94

Mount Augustus, a large isolated massif, is now becoming a tourist attraction. It is formed of coarse-grained sandstone, from which a sandy plain around its foot is derived. The mountain itself is very rocky and sparsely vegetated with *Acacia linophylla* (the narrow-leaved form of bowgada) with some *A. aneura* and *Cassia* spp. There is spinifex on the summit. Around the foot there is a low woodland of mulga and bowgada (*A. ramulosa*) with other species and the usual understoreys of *Cassia–Eremophila*, and annuals in season.

FIG 4.95

Mulga low woodland on stony plain with understorey of *Eremophila abietina* and *Cassia* spp., and ground layer of *Goodenia forrestii*. The flat stones littering the surface, forming a *hamada* or desert pavement, are a constant feature in this region. Wanna station.

Fig 4.96

Although the Gascoyne and Lyons Rivers only flow after rain and shrink to chains of pools between rain episodes, the flow when it occurs is strong enough to shape a river channel of impressive size, bordered by trees of river red gum *(Eucalyptus camaldulensis)* and paperbark *(Melaleuca leucodendra)* which draw adequate moisture from the river bed to support them throughout the year. Lyons River.

Fig 4.97

Degradation due to pastoral use is perhaps more severe in this region than any other. In this case dead acacias have been replaced by inedible shrubs of *Solanum lasiophyllum* and *Cassia* spp., nature's own built-in protection at work. Mt Sandiman station.

Fig 4.98

The small succulent plants called 'parakeelya'—*Calandrinia* spp.—may sometimes form a sward of roadside colour after rain. Milgun station.

Fig 4.99 and 100

The commonest species of *Eremophila* on the Gascoyne is perhaps *E. cuneifolia*, a low shrub but very floriferous. It is one of those with a persistent calyx. The corolla is blue, the calyx deep pink to magenta and remains on the bush after the corolla has fallen. Lyndon station.

Fig 4.101

River plains are frequently salt-affected. Salt-tolerant plants here comprise *Acacia victoriae* surrounding an open area in a depression vegetated with small ericoid shrubs of *Frankenia magnifica* and ground cover of small billybuttons, *Gnephosis* sp. Mt Phillips station.

Fig 4.102 and 103

One of the most glorious of the mulla mullas is *Ptilotus rotundifolius*, a perennial shrub with silver-tomentose leaves. This would clearly make a good garden subject, for the right climatic conditions. Yinnietharra station.

Fig 4.104

An important alternative species to mulga in the Gascoyne is the snakewood, *Acacia xiphophylla*. It associates with mulga on many of the hills and ranges, and becomes dominant on its own account on certain stony plains which have a clay base and are covered with particularly large stones. The writer has termed these 'shingle plains'. Snakewood is a large spreading shrub rather than a tree, with contorted branches perhaps suggesting the name. The ground plants here are *Sclerolaena cuneata* and *Cassia pilocarina*. Wanna station.

 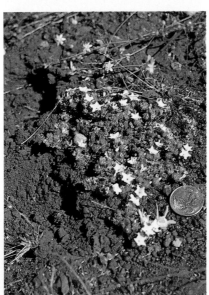

Fig 4.105

A feature of mulga-covered plains in the Gascoyne is that the siliceous hardpan in the soil may become shallowed or exposed by erosion so that the mulga itself is confined to strips of deeper soil forming glades with bare patches between. These bare sheets of silcrete may reveal their own flora after rain, of tiny plants emerging from the stones. In this case *Chthonocephalus* sp. (Asteraceae). Cunyu station.

Fig 4.106

Heliotropium heteranthum (Boraginaceae).

Fig 4.107

Trianthema oxycalyptra (Aizoaceae).

FIG 4.108
The upper Ashburton valley is also included in the Gascoyne Region. Some of it can be very desolate but at the same time scenic, an arresting landscape. The Red Hill which gives its name to Red Hill station is capped by a band of iron ore and rises from a desolate spinifex plain.

FIG 4.109
Further south on Wyloo station an iron ore band, not horizontal but tilted this time, again controls a line of hills. This area is at the transition from *Acacia*-dominated to *Triodia*-dominated vegetation. Both appear together in the foreground, while in the rear spinifex can be seen to cover the drier hill slopes while snakewood covers the flats. Wyloo station.

Fig 4.110
The Ashburton valley above Wyloo consists of wide plains with low hills of shale and in bad times can give an exceedingly desolate impression. The alluvial flat in the foreground consists of an extensive bare scald grading into *Acacia* scrub, while the hills appear red in colour and at a distance seem not to carry a trace of vegetation. Actually in closer view (foreground) they have a scattered cover of shrubs.

Fig 4.111
The shale hills are covered with an open growth of low shrubs 60–100 cm tall, principally species of *Cassia* and *Eremophila*, e.g. *C. nemophila, C. luerssenii, C. oligophylla, E. abietina, E. cuneifolia*. It was a good season when this photograph was taken, witness the *Ptilotus rotundifolius* in flower. Even so the plant cover is sparse enough.

Fig 4.112
This area has been a plain covered with snakewood but has degenerated to a bare scald with dead bushes. Regeneration of such scalds requires special measures which have been developed in recent years, as it will not occur naturally. Wyloo station.

Fig 4.113

Ashburton country in good condition and in a good year gives a very different impression. The larger shrubs are *Acacia wanyu* (feathery foliage, left) and *A. victoriae*. *A. wanyu*—wanyu is its local name—is a local endemic species. Shrubs of *Cassia* (grey) and *Eremophila* (green) form the lower layer. In the right foreground are plants of the tall herb *Trichodesma zeylanicum*. Wyloo station.

Fig 4.114

Mininer Hill and a denuded shingle plain on the upper Ashburton. Some *Acacia wanyu* persist along a draw at rear.

Fig 4.115

Rarely, under good conditions, denuded country will cover up and produce a crop of flowers such as these billybuttons, *Gnephosis brevifolia*. Snakewood in the rear. Red Hill station.

Fig 4.116

The giant pea *Swainsona maccullochiana* comes into prominence on the Ashburton in good seasons. It is an annual, but is capable of growing to 2 m tall. Flowers are generally reddish purple, but may be white or bronze.

Atriplex community on relatively higher surfaces and *Acacia* scrub on sandhills surrounding the lakes.

Looking now at the Ashburton Valley, the valley lies in a geological trough formed by the erosion of rocks less resistant than those to north and south of it. On the north side stand the Hamersley Ranges, on the south the Capricorn Range and others, the former spinifex covered and the latter *Acacia* covered. The valley itself contains substantial areas of wide riverain plains interspersed with hills of a shale dipping at a very high angle. These have little soil, a very thin vegetative cover, and appear barren and of a red-brown colour at a distance. The valley plains tend to be covered with water-worn stones derived from the adjoining ranges, creating both gravel and shingle plains.

The main river channels as along the Lyons and the Gascoyne are lined by *Eucalyptus camaldulensis* up to 20 m in height with smaller *Melaleuca leucadendra*. Alluvial flats away from the river vary according to soil. The loam flats are best vegetated and have a general cover of *Acacia victoriae, A. sclerosperma* and *A. wanyu* with smaller shrubs, grasses and annuals. *Swainsona maccullochiana*, the Ashburton pea, is a conspicuous feature of this community in a favourable season. Although an annual herb, it grows to some 2.5 m in height. Masses of flowers are produced of which the predominant form combines pink and mauve colours, but there are also white and bronze colour forms. Clay flats on the other hand are comparatively poorly vegetated and and covered with gravel and shingle. Snakewood (*Acacia xiphophylla*) is the characteristic species, about 2 m tall, rather scattered, sometimes with mulga or wanyu (*A. wanyu*, a local endemic species — wanyu is the local name). There is a general ground cover in season of *Lepidium platypetalum* and the small shrub *Eremophila cuneifolia*; the succulent subshrubs *Sclerolaena cuneata* and *Atriplex lindleyi* may also be present. The bush thickens up considerably on washes where it contains *Acacia aneura, A. citrinoviridis, A. grasbyi, A. wanyu* and *A. bivenosa*.

The bare-looking shale hills do actually carry some plant growth, scattered low shrubs 60-100 cm tall, principally species of *Cassia* and *Eremophila*. Either *Cassia oligophylla* or *C. nemophila* appears locally to be dominant, associated with *C. luerssenii, Eremophila cuneifolia, E. abietina, Acacia tetragonophylla, Corchorus walcottii, Ptilotus drummondii, P. obovatus* and *Solanum lasiophyllum*, also annuals in season. Occasional small stunted *Acacia* may be present. This is a very unusual dwarf shrub community, which is a feature of the Ashburton Valley. It may be regarded as a very attenuated form of mulga where the adverse conditions have reduced the acacias to sparse stunted specimens, leaving the smaller shrubs, which normally form a lower layer, to assume dominance.

4.6 *Little Sandy Desert Region*

KEARTLAND BOTANICAL DISTRICT

Shrub steppe of *Acacia* and *Grevillea, Triodia* spp. on and between dunes, patches of desert oak and mulga.

Climate: Arid tropical with summer rain, annual precipitation 200-250 mm.

Geology: Quaternary sandplain with longitudinal dunes developed over locally exposed Proterozoic siliceous rocks.

Topography and soils: Sandplain with numerous low hills and small ranges, with mainly bare rock and shallow stony soils on these; red earthy sands in the plains.

Area: 110 314 km².

Boundaries: N—change in underlying rocks from Precambrian to Mesozoic. E—edge of laterite plains country of Gibson Desert. S & W—edge of desert sandplains and dunefields.

The name Little Sandy Desert was suggested for this region by the writer in 1969 and has been officially accepted. It is of the same general character as the Great Sandy Desert, consisting predominantly of sandy plains with linear dunes, but the two are almost completely separated by the Throssell, Broadhurst and McKay Ranges and there is an important difference in the underlying geology. The Great Sandy Desert is underlain by horizontally bedded sandstones which emerge here and there as tabular hills bounded by scarps. The Little Sandy Desert is underlain by quartzitic Middle Proterozoic rocks which form hills and mountains of rounded outline, in the shape known as 'whalebacks', rising from a sea of sand. Some of these may be 15 km long and 8 wide, and many attain notable heights. Mt Essendon rises to 910 m. Physiographically the entire region is a basin with Lake Disappointment at its lowest point. The Little Sandy Desert remains one of the most inaccessible parts of the desert and is uninhabited. The Talawana Track is the only graded track which crosses it, and that along the northern edge. The Canning Stock Route crosses it diagonally and can be followed by four-wheel drive vehicles through Durba Springs. The

Fig 4.117
The Little Sandy Desert Region is bounded by the Gascoyne Region, the Gibson Desert and the Great Sandy Desert. It is a sandhill region like the latter but is of different underlying geological structure and is cut off from it by the Throssel, Broadhurst and McKay Ranges. The principal plant cover is spinifex of *Plectrachne schinzii* and *Triodia basedowii* with scattered shrubs, mainly *Acacia* and *Grevillea* with some mallee and smaller plants. The region is traversed by the Canning Stock Route. View in the McKay Range.

Fig 4.118
Near the southern end of the Canning Stock Route lies Windich Spring, a beautiful permanent pool forming a veritable oasis. It is named after an Aboriginal guide employed on John Forrest's expedition of 1874.

Fig 4.119

On 2 June 1874, John Forrest and his party discovered a plentiful source of water which they named Weld Spring (after the governor) and camped there for 17 days. Forrest recorded that the spring ran down the gully for 20 chains, was as clear and fresh as possible, and the supply unlimited. In 1905 Well 9 of the Canning Stock Route was sunk at the site, and in recent years a windmill has been erected over it. The surrounding area has been eaten out and the spring has ceased to flow. While in camp here Forrest's party was attacked by Aborigines, and built a stone breastwork to defend themselves, which still exists.

Fig 4.120

Well 24 of the Stock Route is where the east-west Talawana Track crosses. It is in a little valley with spinifex, and mulga on the ridge overlooking. Each of the stock route wells originally had a timbered shaft, a windlass, a pulley, an iron stock-watering trough and iron covers to keep animals out of the well—all these carted in on camel back.

Fig 4.121

In good seasons when the pans are full of water, budgerigars quickly breed up into large numbers circling in clouds and feeding on the seeds of the spinifex which also responds to the rain by flowering. Alas, when the waters begin to dry up the little creatures must try to survive from drying waterholes and a desperate situation may develop. Glen Ayle station.

FIG 4.122

Not a gidgee tree in this case but a budgie tree!

FIG 4.123

The galah, *Cacatua roseicapella*, ranges all over the State and is to be seen in the desert just as he is in Perth. These were at Weld Spring.

vegetation is generally known but has not been studied in detail.

The hills are generally undulating, but with some steep cliffs and gullies. There is a high proportion of bare rock, so that the vegetation must be described as very sparse. The tops of the hills are generally vegetated with *Acacia aneura* as scattered trees of 4-5 m, or in groups, with *Grevillea* sp. inedit., *Hakea* sp. and *Thryptomene* sp., but there are also areas of hummock grassland consisting of both *Triodia basedowii* and *Plectrachne melvillei*. Scattered larger trees of *Eucalyptus camaldulensis* occur here and there. The gullies have *E. microtheca* and *Callitris glaucophylla*. The latter and *Ficus platypoda* also occur on steep slopes and screes as do *Eucalyptus setosa*, *Pittosporum phylliraeoides* and *Melaleuca nervosa*. This occurrence of *Callitris* on rocky ranges, where it is protected from fire, has been reported elsewhere in the desert.

Mulga low woodland of *Acacia aneura* and *A. pruinocarpa* covers stony pediments of the ranges. It has the usual understorey of *Eremophila* spp. and ground cover of annuals with occasional *Eragrostis eriopoda*.

The sandplains carry spinifex country of *Triodia basedowii* with numerous tall shrubs, principally of *Hakea suberea* and *Acacia* spp. with *Hakea rhombale*. *Xanthorrhoea thorntonii* is also conspicuous. South of Mount Methwin an area was noted in which *Thryptomene maisonneuvii* was codominant with *Triodia*.

This species is normally numerous only on the flanks of dunes, where indeed it occurs here sharing dominance with *T. basedowii*. The desert bloodwood, *Eucalyptus chippendalei*, occurs as scattered trees along the dune crests, with an *Acacia*, probably *A. ligulata*; *Grevillea stenobotrya* is presumably also present. Groves of desert oak, *Casuarina decaisneana*, occur scattered in some areas in depressions between the sandhills.

4.7 Gibson Desert

CARNEGIE BOTANICAL DISTRICT

The cover of the laterite plains is a mosaic of mulga (*Acacia aneura* low woodland) and shrub steppe (*Hakea*, *Acacia*, *Triodia basedowii*). Shrub steppe covers the sandy areas.

Climate: Arid with summer rain; annual precipitation 200 mm.

Geology: Flat-lying Cretaceous and Jurassic sandstones.

Topography and soils: Monotonous, gently undulating plain with a few sandstone mesas. Stripped laterite surfaces are general on the uplands with sands in the valleys, often with longitudinal dunes.

Area: 149 784 km².

Boundaries: N, S & W—edge of laterite plains country. E—edge of sedimentary basin.

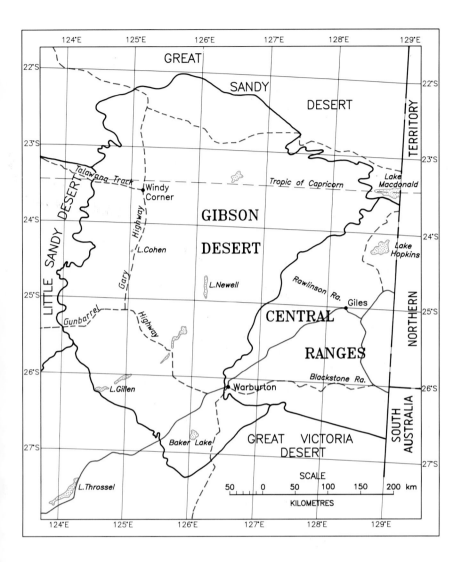

The Gibson Desert is essentially a region of laterite plains from which emerge occasional tabular sandstone hills and ranges, gently undulating with shallow valleys trending south-southwest. Peripheral to the main area the valleys are relatively deep and are filled with sand, frequently with dunes, and with small salt lakes or evaporite deposits in their lowest portions. On the higher part of the plains the valleys are shallow, less sandy, and show no evidence of salt accumulation.

The sedimentary rocks of the Officer Basin appear to have been uplifted at the close of the Cretaceous period. Uplift was greatest in the centre of the marine trough which ran from the Indian to the Southern Ocean, so that a high plain was formed at about 420 to 500 m above present sea level, sloping gently down towards the sea both to the south and the northwest. It is this high plain which constitutes the Gibson Desert, gently undulating, monotonous, relieved only by some low tabular hills of sandstone, and with a surface of laterite. During the Tertiary era after uplift the climate appears to have been very humid so that the surface rocks weathered deeply and a crust of laterite was formed. Laterite is usually covered by a layer of sandy soil, however the high plain has apparently been largely swept bare by the wind which has deposited the sand in the deeper valleys, often with linear dunes. The high plain itself consists of bare laterite exposed at the surface and now subject to gradual weathering and decay. This therefore represents an inhospitable environment, whose vegetation is correspondingly desolate and desertic. The region defeated the explorer Giles who endeavoured to cross it from the east in 1874 and led to the death of his follower Gibson whose name has been given to it. Giles wrote, 'I called this terrible region that lies between the Rawlinson Range and the next permanent water that may eventually be found to the west, Gibson's Desert, after this first white victim to its horrors'.

Hilly ground is covered with low mulga, a scrub of *Acacia aneura* not usually exceeding 3 m in height and of fair density. *A. grasbyi* may be found with *A.*

FIG 4.124

The Gibson Desert lies to the east of the Little Sandy and is somewhat central to the desert areas of the State. It consists mainly of high plains forming a watershed, with a laterite surface. The Gibson is therefore flat, monotonous, desolate and poorly vegetated. It is crossed by the famous Gunbarrel Highway seen here, a track formed from Alice Springs to Carnegie station in 1956 and so-called from the long straight stretches. Seen here in a poor season, even the spinifex has died and is beginning to recover.

FIG 4.125

The laterite plains of the Gibson Desert have been swept bare by the wind of any fine sandy and earthy material, leaving a surface of laterite pebbles overlying more massive laterite beneath. The laterite is believed to have been formed in the remote past, and is decaying at the present day. Spinifex has largely died of drought here. Some clumps are recovering and some seedlings are to be seen.

FIG 4.126

This signpost was set up at a junction on the Gunbarrel Highway by Len Beadell, surveyor in charge of road construction in 1956. The three different directions show Carnegie, Giles and Warburton Mission. As the road from Warburton to Giles has been improved and rerouted, this marker may no longer exist.

Fig 4.127
Seen in a better season when the spinifex has been flowering, the Gibson takes on a more genial appearance. The track here is the Gary Highway which joins the Gunbarrel and Windy Corner by a north-south traverse. The occurrence of mulga either as scattered trees or denser groves in depressions has led this vegetation type to be termed the 'mulga parkland' as it is shown as unit 11 on the frontispiece map to this book.

Fig 4.128
A few hills appear here and there in the Gibson, in spite of its monotony. This one is called Charles' Knob. While the general cover is of spinifex, mulga develops wherever the soil is deeper and more penetrable. Climatically, this should be mulga country. It only reaches the limit of its range at the north end of the Gibson Desert. Poor, dry conditions however favour spinifex south of its general latitude.

FIG 4.129

The Canning Stock Route traverses the northwest corner of the Gibson Desert with a line of wells dug along a depression where calcrete has been deposited and water is available at shallow depth below the hardpan. Such habitats are marked by thickets of *Melaleuca lasiandra* and *M. glomerata* (in background). The well here is no. 33 and has the usual equipment. One of the iron well covers has been thrown aside in the foreground.

FIG 4.130

Shallow drainage lines traverse the Gibson Desert but are little more than strings of claypans. However these, when they hold water, are fresh, not saline. The largest of them is called Lake Cohen, most often a bed of dried mud with a few coolabah trees *(Eucalyptus microtheca)*.

FIG. 4.131

In a good year when everything fills with water, one cannot help feeling that it seems rather incongruous! This is when the budgerigars breed furiously (Fig 4.121) and all the plants grow and flower, but the water usually only lasts a few months. Near Mt Webb on the Sandy Blight Junction track, 1967.

Fig 4.132
As shown in Fig 4.34, death of spinifex from drought may be followed by pioneer regeneration of other subclimax species. Such plants here include *Ptilotus clementii*, *P. polystachyus*, *P. schwartzii* and *Goodenia scaevolina*. Windy Corner.

Fig 4.133
Good seasons bring out pioneer plants which cannot otherwise tolerate the competition of the spinifex. *Newcastelia cephalantha* (Chloanthaceae) has responded by germinating in disturbed soil alongside the Gary Highway.

Fig 4.134
Flowers and foliage of *Newcastelia cephalantha*.

Fig 4.135
Among other plants of the Gibson Desert is the shrub *Clerodendrum tomentosum*. The flowers are white, but it is more conspicuous when in fruit with a dark blue seed borne on a red persistent calyx. Gary Junction.

FIG 4.136

A herbaceous plant with the typically formed blue flowers of its genus, is *Goodenia scaevolina* subsp. *stobbsiana*. Leaves and stems have a sticky coating. Near Well 24.

FIG 4.137

A shrub of the *Cassia* family (Caesalpiniaceae) is *Petalostylis cassioides*, well named from the resemblance. The flowers have a conspicuous red mark at the centre. On road to the Warburton Range.

FIG 4.138

A small animal which seems to epitomise the harshness of desert life is the lizard *Moloch horridus*, mountain devil. While quite harmless, he is well protected like an ancient dinosaur. In the American deserts a similar aspect is assumed by the so-called horned toad or gila monster, another lizard of this kind.

aneura, mainly on top of breakaways. Smaller shrubs include *Eremophila latrobei* and spp., *Dodonaea rigida*, *Calythrix carinata*.

Characteristic of the laterite plains is a composite formation known as the mulga parkland. This is essentially a mosaic of mulga scrub and shrub steppe. The laterite plains undulate gently; in the draws and depressions there is some loamy soil, but on the rises there is only hard ironstone. Mulga grows quite normally on the lower ground, in groves with scanty ground vegetation. Rising uphill the mulga opens out and a ground cover of *Triodia basedowii* becomes established, a stage that may be described as 'mulga and spinifex'. On the hilltops, growth is reduced to a few scattered *Hakea suberea*, *Acacia pruinocarpa* and *A. aneura* in the *Triodia* cover, a stage which could be classified as *Hakea-Acacia* shrub steppe. On a large scale these communities could be mapped separately. Associated species were recorded as follows:

Shrubs *Acacia dictyophleba, A. grasbyi, A. helmsiana, Eremophila leucophylla, Eucalyptus kingsmillii* (mallee), *Grevillea juncifolia*.

Subshrubs *Burtonia polyzyga, Calythrix carinata, Cassia notabilis, Dicrastylis exsuccosa, Halgania solanacea, Petalostylis cassioides*.

Herbs *Dampiera* sp., *Goodenia azurea, Helipterum stipitatum, Ptilotus polystachyus, P. drummondii, P. macrocephalus, Trachymene glaucifolia*.

Grass *Amphipogon caricinus*.

At the time of the writer's fieldwork the subshrubs and forbs were mainly in evidence on areas which were recovering from drought and where much of the spinifex had died. It is apparent that this was a successional stage.

The valleys and depressions on the higher part of the plain develop something resembling a tree savanna formation, owing to the increased moisture from run-on. There is no evidence of salinity, saltbushes and samphire being absent. Both *Eucalyptus dichromophloia* and *E. microtheca* occur, more rarely *E. aspera*, always somewhat scattered, often with *A. aneura*. *Plectrachne schinzii* comes in on the ground and tends to replace *Triodia basedowii*. Soft grasses appear also and replace spinifex altogether on clay soils. There are numerous claypans, of which Lake Cohen is the largest. These fill with water after rain and have beds of dried mud at other times.

The deeper valleys peripheral to the high plain are filled with sand, frequently with sand ridges having an east-west trend. The sandy interdune features a mixed shrub steppe comprising the following:

Eucalyptus gamophylla (mallee), *E. kingsmillii* (mallee), *Acacia helmsiana, A. linophylla, A. grasbyi, A. pachyacra, Dicrastylis exsuccosa, Eremophila leucophylla, Grevillea juncifolia, G. eriostachya, Newcastelia cephalantha, Thryptomene maisonneuvii, Triodia basedowii, Plectrachne schinzii*.

Mulga replaces this assemblage if the soil of the interdune changes to loam. In the valley bottoms where travertine and other evaporites accumulate, there is a low open scrub composed of *Acacia aneura, A. ligulata, Melaleuca* spp., *Santalum acuminatum, Lamarchea* sp., *Templetonia* sp., *Alyogyne pinoniana, Plectrachne schinzii, Triodia basedowii, Atriplex hymenotheca*. Open *Atriplex* flats also occur, patches of *Melaleuca* and occasional desert oak trees *(Casuarina decaisneana)*.

The sandhills have as ever a completely distinctive vegetation which differs from that of the interdune. Within the Gibson Desert area it is transitional in character, changing from north to south from the tree steppe of the Great Sandy Desert to the scrub-heath of the Great Victoria. In the north scattered trees of the desert bloodwood *(Eucalyptus chippendalei)* are characteristic and the dune flanks are thickly covered with *Plectrachne schinzii*. In a southerly direction the desert bloodwood gradually becomes scarcer, until it vanishes by about the 27th parallel. Spinifex *(Plectrachne)* is gradually replaced by low ericoid shrubs among which *Thryptomene maisonneuvii* is the most abundant. These first assume dominance on the south slopes of the dunes, gradually invading the north also with more southerly latitude. The taller shrub components of this community tend to be more consistent and to range right through. These are *Grevillea stenobotrya, Acacia ligulata, Gyrostemon ramulosus* and *Crotalaria cunninghamii*, also the herb *Trichodesma zeylanicum*. Noted locally were *Acacia linophylla, Eremophila willsii* and *Anthrotroche pannosa*. The dune flora is not a rich one.

4.8 Central Ranges Region

GILES BOTANICAL DISTRICT

Acacia scrub on quartzitic ranges, mulga on volcanics and the heavier plain soils; otherwise tree and shrub steppe. Stands of desert oak are common in depressions.

Climate: Arid with both summer and winter rain; annual precipitation 200 mm.

Geology: Ranges and hills of Proterozoic rocks including both volcanics and quartzites, interspersed in Quaternary sandplains with some Permian exposures.

Topography and soils: Shallow rocky loams on the ranges; red earthy sands and red earths in the plains.

Area: 60 788 km².

Boundaries: Throughout, the limit of underlying Precambrian rocks. In the south and northwest arbitrary lines have had to be drawn across sandplains as these rocks have no surface outcrop.

The Central Ranges Region occupies an area along the State boundary in the east from the Blyth Range in latitude 27° to Lake Macdonald in latitude 24°, extending westward for a maximum distance, at the Warburton Range, of 260 km. It represents an extension into Western Australia of the highland area of the Centre, bringing with it a number of plant species not otherwise known in the west. The boundaries of this region are geological and follow the division between pre-Cambrian and post-Cambrian rocks.

Within this region we have the Warburton Range Mission and Giles Weather Station, both permanent settlements where meteorological data have been recorded for some years. Over the 25 years of records kept at the Warburton Range, annual rainfall has varied from 35 to 680 mm with an average of 216 mm. Giles' figures, kept for a shorter period, are very similar. Nil rainfall may be recorded in any month of the year and heavy falls of rain may also be experienced in any month except, generally speaking, in spring. September and October are consistently dry months. Mean temperatures range from 29°C in January to 13°C in July. Both frost in winter and great heat in summer can be experienced.

The physiography of the region consists of a low plateau standing about 480 m above sea level from which rise numerous isolated and rather abrupt small mountains and ranges. These fall into two groups, a northern acid group and a southern basic group. In the north stratified massive quartzites attributed to the middle Proterozoic form a number of relatively long continuous ranges. The Petermann Ranges in the Northern Territory are continued in the Schwerin Mural Crescent, an impressive south-facing escarpment, and in the Rawlinson Range. Numerous smaller isolated ranges occur further north. The southern area consists of Proterozoic granites and metamorphics, largely basic in character, with large intrusions of basalt. The plains surrounding these mountain remnants are known in places to carry Permian sediments but are mostly overlain with more recent material derived from the weathering of the Proterozoic outcrops. In the north therefore sandhills and shrub steppe predominate, and in the south, plains with mulga.

The quartzite ranges are exceedingly rocky with very little actual soil. The lower slopes carry mostly a thin cover of low shrubs such as *A. orthocarpa, Eremophila elderi, E. gilesii, Prostanthera* sp. inedit., among tufts of *Triodia basedowii* and *Plectrachne melvillei*. Small eucalypt trees, *Eucalyptus aspera* and *E. grandifolia* occur, mainly in drainage. The upper slopes, however, are more thickly covered with larger shrubs 1.7-2 m tall, among which *Acacia ? cyperophylla* is dominant. Others are *A. pruinocarpa, A. aneura, A. grasbyi, Eucalyptus oxymitra* and *Hakea rhombale*.

Callitris glaucophylla becomes dominant locally in very rocky and inaccessible places. The footslopes of the main ranges usually carry a belt of mulga, often rather open and mixed with *Eucalyptus gamophylla, E. oxymitra, Hakea lorea* and *Grevillea wickhamii*, with patches of *Casuarina decaisneana*.

Sandhill country between the ranges is as described for the Gibson Desert. Groves of desert oak (*C. decaisneana*) are a common feature here, however, and are thought to survive on underground water resulting from run off from the rocky ranges absorbed by the surrounding sand of the plains. On major drainage lines running from the ranges large trees are often present, *Eucalyptus papuana, E. grandifolia* and *E. dichromophloia*.

The bed of Lake Hopkins consists of samphire and saltbush communities as described for the preceding region, with gypsum dunes. *Casuarina cristata* has been reported on these from aerial photography but was not seen by the writer. At the point examined the dunes had a fairly dense cover of large *Melaleuca lasiandra* with smaller *Acacia victoriae, A. ligulata, A. tetragonophylla* and *Santalum acuminatum*.

FIG 4.139
On entering the Central Ranges Region east of the Gibson Desert the country changes, becomes hilly and mountainous, and much more pleasant and scenic. We are now in a part of that greater region of Central Australia which contains Ayers Rock and the Macdonnell Ranges, and protrudes over the Western Australian border for 250 km. It is a region of low rocky ranges separated by sandy plains, often with sand ridges. The view here is to the southeast from the Walter James Range towards the Petermann Ranges. Sand ridges can be seen snaking across the plain below.

FIG 4.140
One has the feeling here of an exceedingly ancient landscape where the mountains have been worn down to the very root, and the products of their erosion spread out in flat plains below them. Even here spring flowers make a brave show, particularly the spinifex daisy *Helipterum stipitatum*. Various shrubs are in flower among the spinifex, such as *Grevillea juncifolia* in the middle ground. On the track to Giles from Sandy Blight Junction.

FIG 4.141
There are two principal ranges in the region, in the north the Rawlinson Range named after Sir Henry Rawlinson, president of the Royal Geographical Society, which is quartzitic; and the southern Warburton Range named after Colonel Warburton the explorer, which is basaltic. This view east from the Rawlinson Range near Giles shows the rocky nature of the range and the sparse vegetation. Below in the valley conditions are more favourable so that mulga woodland covers much of the plain.

FIG 4.142
In places the plains carry very good mulga, perhaps some of the tallest mulga in the State. The ground layer here consists mainly of the annual *Ptilotus helipteroides* (purple), a mulla mulla, with some of the everlastings, *Helipterum floribundum* and *H. fitzgibbonii* (white). North of Warburton.

FIG 4.143
A closer view of the understorey plants from the previous photograph, *Ptilotus helipteroides* and *Helipterum floribundum*. At far right *Ptilotus obovatus*, at left *Brunonia australis*. Some plants in the foreground, probably *Lepidium* sp., have already dried off.

Fig 4.144

In the Rawlinson Range plant growth is kept sparse by the rocky ground. A tree of *Eucalyptus grandifolia* at left, acacias in the ravine, but the rocky slopes have only odd tufts of spinifex—both *Triodia basedowii* and *Plectrachne melvillei*—with small shrubs such as *Acacia orthocarpa*.

Fig 4.145

This beautiful *Prostanthera* was discovered in the Rawlinson Range by Mr A.S. George and the writer on their expedition in 1966. It has still not been given a name.

Fig 4.146

Part of the Rawlinson Range was grandiosely named the Schwerin Mural Crescent by the explorer Giles. The bluff at centre is Gill's Pinnacle.

Fig 4.147

In the south, east of Warburton, the Blackstone Range is named from the outcrops of bare basaltic boulders which appear black at a distance. Another similar range was named the Bell Rock Range by Giles because the boulders gave a metallic noise when walked on. Mulga trees at the foot and a plain covered with annuals, now dry.

FIG 4.148
In the background Mt Aloysius, named after the governor, Sir Aloysius Weld. The plain is covered with spinifex and scattered mulga, with spinifex daisies (*Helipterum stipitatum*) providing a riot of colour.

In the southern sector the rocky hills are most frequently vegetated with mulga and spinifex on the lower slopes, that is with open *Acacia aneura* over *Triodia basedowii* and *Plectrachne melvillei*. Mulga thins out higher up and the upper slopes carry little but spinifex with rare stunted mulga and *Eucalyptus dichromophloia*. On and around Mt Aloysius where the rock weathers into large rounded boulders, *Callitris glaucophylla* replaces mulga, growing in rock crevices. It appears to have been eliminated by fire on areas capable of supporting a ground vegetation. On the other hand basic dykes in the Blackstone Range weather to large masses of naked boulders but *Callitris* is not found here, the only species being a rare *Ficus platypoda*.

The basalt Warburton Range is covered with open mulga with rich lower layers of shrubs (*Eremophila* and *Cassia* spp.) subshrubs, forbs and grasses. Spinifex is absent except for *Plectrachne melvillei* in the rockiest places. The plains at the foot of the hills in this region are typically covered with mulga of this nature, some of it up to 8 m tall, being the best quality mulga in any part of the desert. The tree layer comprises *Acacia aneura* (in both normal and weeping forms), *A. pruinocarpa*, *A. tetragonophylla*, *Canthium latifolium*, *Heterodendrum oleaefolium*; shrubs *Eremophila latrobei* and spp., *Cassia desolata*, *C. sturtii*; subshrubs *Ptilotus obovatus*; forbs *P. polystachyus*, *P. exaltatus*, *P. clementii*, *P. helipteroides*, *Waitzia acuminata*, *Helipterum stipitatum*, *H. floribundum*, *H. fitzgibbonii*; grasses *Monachather paradoxa*, *Eragrostis eriopoda*, *Enneapogon caerulescens*, *Eucalyptus dichromophloia* occurs on drainage. At the foot of the Blackstone Range there are some small black-soil plains with *Astrebla* sp. (Mitchell grass) but these are of insignificant extent.

FIG 4.149
Many of the sandy plains between the ranges carry groves of the desert oak, *Casuarina decaisneana*, with a spinifex floor. Such large trees must have access to underground water which is probably available from storage if the sand is deep and may be contributed by run-off from the ranges. There has been some mortality from drought at this site. In sandhills west of the Rawlinson Range.

FIG 4.150
Desert oaks regenerate freely in good seasons, probably after fires. When young the tree has very short branches and has a bizarre appearance, suggesting a fairy-tale scene of hairy old men. Curtin Springs, N.T.

FIG 4.151
These desert oaks had all run out of moisture and died of drought. Fire does not have this effect as they have very thick protective bark. Recovery from drought is shown by the ground layer. Near Lake Hopkins.

FIG 4.152
Unlike desert oaks, cypress pines are tender to fire and must depend for their survival on some natural protection against it. They are found throughout the Eremaea but only among rocks or on cliffs where little other herbage grows and cannot carry a severe fire. The cypress pines *Callitris glaucophylla* have taken root here among the rocks of this very bouldery hill where little grass or spinifex comes near them. Mt Aloysius in the distance; dense mulga covering the plain.

FIG 4.153
Calcrete is deposited along drainage lines and as elsewhere carries a specialised vegetation principally of mallee—*Eucalyptus oleosa* and another species—and spinifex *Triodia pungens* and *T. scariosa*. Van der Linden Lakes.

FIG 4.154
We say goodbye to the Central Ranges Region with a sunset scene in the Rawlinson Range.

4.9 Pilbara Region

FORTESCUE BOTANICAL DISTRICT

Essentially tree- and shrub-steppe communities with *Eucalyptus* trees, *Acacia* shrubs, *Triodia pungens* and *T. wiseana*. Some mulga occurs in valleys and there are short-grass plains on alluvia.

Climate: Arid tropical with summer rain; annual precipitation 250–300 mm.

Geology: A basement of Archaean granite and volcanics, overlain by massive deposits of Proterozoic sediments (including jaspilite and dolomite) and volcanics.

Topography and soils: A mountainous region rising to 1250 mm. Chiefly hard alkaline red soils on plains and pediments, shallow and skeletal soils on the ranges.

Area: 178 017 km².

Boundaries: On east and west, according to vegetation expression of the margins of the Carnarvon and Canning Basins. On the south the boundary is that between *Triodia*-dominated steppe and *Acacia*-dominated scrub, climatically controlled with local geological modification.

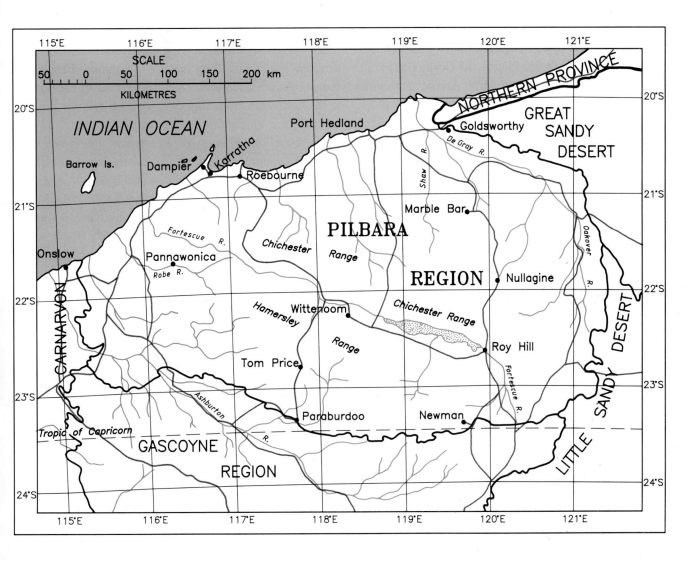

The Pilbara Region of the northwest has become much better known since the exploitation of iron ore began in the 1960s and brought more population and activity into the area. Visitors today can even find themselves assisted by a neat handbook on the Pilbara coastal flora[6] with photographic illustrations. The Pilbara Region derives its name from the small Pilbara Creek, a tributary of the Yule River, where gold was discovered in 1888, leading to the proclamation of the Pilbara goldfield and application of the name to the whole district. The Pilbara Creek was not very productive but more consistent finds were made at Marble Bar and Nullagine which were important enough to justify a railway from Port Hedland to Marble Bar, in operation from 1912 to 1951. Tin, copper, manganese and asbestos have also been worked but it is the iron ore which has become the major source of activity since the 1960s. New port facilities were constructed at Dampier and Port Hedland, standard gauge railways from Dampier to Mt Tom Price and Paraburdoo, from Cape Lambert to Pannawonica and from Port Hedland to Mt Goldsworthy and Mt Newman. New mining townships have been created, and population and economic activity drawn into a region which previously was but sparsely occupied by a few sheep and cattle stations.

The Pilbara Region occupies the northernmost portion of the ancient Western Shield and is bordered to east and west by the Canning and Carnarvon Basins with their sedimentary rocks. The southern edge of the region is determined by a major biogeographic boundary, the *Acacia-Triodia* line, to the north of which spinifex vegetation is predominant and to the south woody *Acacia*-dominated vegetation. *Acacia* woodlands and scrub are not entirely absent from the Pilbara, but spinifex country is the characteristic landscape element. The Pilbara actually receives a slightly higher average rainfall than most of the Eremaean Province, up to 250-300 mm annually due to the prevalence of cyclones on the coast, but this is not enough to modify the essentially desertic appearance of the plant cover.

The northern part of the region consists of a partly granitic, partly alluvial coastal plain, the Abydos Plain, broken by rough and abrupt ranges, the Gorge Ranges, formed of harder and more resistant Archaean rocks folded into the granite; and to the east of these the Oakover Valley, dissected country descending from the uplands to the south. The whole of this complex is dominated on the south by the escarpment of the Chichester Range trending ESE to WNW and rising 150-200 m to a summit plateau. The range consists of volcanics — tuff and basalt — of Lower Proterozoic age capped by the hard resistant Marra Mamba Iron Formation which forms the plateau surface. The latter slopes down very gently to the SSW with the dip of the rocks until after 65 km there is a second and more abrupt escarpment rising about 250 m to form the Hamersley Range, exposing hard, fine-grained rocks, jaspilite, shale and dolomite. This escarpment is also controlled by a still harder capping, the Brockman Iron Formation. The frontal escarpment of the Hamersley Range is an impressive feature stretching for over 400 km. At its foot, in the trough between it and the Chichester Range, lies the curious course of the Fortescue River. This, an intermittent stream running only after heavy rains, rises at the extreme eastern end of the range in the area of the Mt Newman mine. Its numerous headwaters take a northerly to northwesterly course and unite above Roy Hill Station, only to lose themselves in an enormous salt marsh 100 km long and up to 16 km wide, where the waters evaporate and flow no further. This marsh is not one of the usual playa lakes of the interior as it is well vegetated with halophytic plants and there are no expanses of bare mud or salt. Further west along the foot of the Hamersley escarpment a possible former extension of the river's course is marked by a chain of grassy flats but these are normally dry. Short rivers rise in the Hamersley Range and flow out to this valley but their courses are quickly lost in the plain. Some of them have cut the impressive gorges into the escarpment such as the Wittenoom and Dales Gorges which are popular with visitors. Further west, two of the largest of these streams, known once more as the Fortescue, are active enough to maintain a flood channel along the floor of the valley to the vicinity of Millstream Station, where powerful springs rise in and near the channel bed, dramatically creating a permanent river in the desert. These springs are said to flow steadily at about 50 000 kgl daily at a temperature of 25°C and have now been exploited to provide water for coastal towns. They create a perennial stream flowing quietly for some 15 km through several still deep reaches and minor channels overgrown with cadjuput trees, until it leaves the area to the northwest and plunges into a young valley leading through the mountains to the sea.

Above the Hamersley escarpment one finds a different topography of high but gently rounded hills reaching 900 m above sea level dissected by valleys. Some 40 km further south the rocks become folded and form a broken country of impressive ranges with flats between them. This part of the range includes Mt Meharry, 1235 m, the highest eminence in Western Australia.

The north coast of the Pilbara is for the most part low lying and muddy. Only between Capes Preston and Lambert is it largely rocky. The low coast is fronted by mangroves with wide bare flats of hypersaline mud behind them. Five species of mangroves occur here. *Avicennia marina* typically forms the deep-water fringe with *Rhizophora stylosa* behind, and *Ceriops tagal* coming in on the landward side. Near to mean sea level gently sloping sandy areas are often colonised by the club mangrove *Aegialitis annulata* or along creeks by the river mangrove *Aegiceras corniculatum*[6]. Saltmarsh communities generally inhabit the landward fringe of mangroves and may consist of the succulent samphire *Halosarcia halocnemoides* with *Suaeda arbusculoides* on clay soil, or of a mixed herb and grass community on more sandy ground comprising *Muellerolimon salicorniaceum* and *Sporobolus virginicus* or *Halosarcia indica*, *Frankenia ambita* and *Hemichroa diandra*[6]. Some of the unvegetated bare mud flats at Dampier and Port Hedland are now utilised for the production of salt from sea water by evaporation.

The extensive alluvial plains behind the coast are for the most part sandy, but where material is derived from basaltic rocks, mainly in the west around Roebourne and towards Onslow, the soil may consist of 'cracking clay', a dark earth which cracks deeply in the dry season. The latter typically carries short bunch-grassland either pure or mixed with spinifex. Mitchell grass (*Astrebla pectinata*) may once have been common and survives in pockets. The principal grass today appears to be *Eragrostis setifolia*. Spinifex species are *Triodia pungens* and *T. wiseana*. Scattered trees are often present, notably coolabah (*Eucalyptus microtheca*) ghost gum (*E. papuana*), bauhinia (*Lysiphyllum cunninghamii*) and waterwood (*Atalaya hemiglauca*). It was these grass plains which brought the first pastoralists to the Pilbara in the 1860s, and led Diels who visited Roebourne by sea in 1900

Fig 4.155
The Pilbara Region of the northwest is noted for its iron ore mining but is also interesting scenically and botanically. The region is very varied, rising from a coastal plain by stages to the heights of the Hamersley Range. The coastal plain is perhaps the least interesting part, a dreary expanse covered with spinifex—*Triodia pungens*—and scattered shrubs of *Acacia pyrifolia*, relieved only by a few low hills and dykes of black rock.

Fig 4.156
Further inland where the hills and ranges begin to rise, the country at once has scenic value. In this northerly latitude mulga survives only on valley plains and all else is covered with spinifex, generally with scattered eucalypts. While generally widespread these trees congregate more thickly on drainage lines. Around Marble Bar.

FIG 4.157
Sturt's desert pea, the glory of the north, is frequently present to enliven the spinifex with colour.

to create a long lasting false impression that the Pilbara is covered by 'savannas'.

Sandy coastal plains are inhospitable, covered by *Triodia pungens* interspersed with numerous very low spreading shrubs of *Acacia translucens*. Further inland on granitic soil *T. pungens* is joined by *T. lanigera, T. longiceps* and other species, and there are taller shrubs reaching 3 m, mainly of *Acacia pyrifolia, Grevillea pyramidalis* and *Hakea suberea*, fire-resistant shrubs which form an association throughout the Pilbara. Others which are killed by fire but regenerate from seed include *Acacia pachycarpa, Carissa lanceolata* and *Grevillea wickhamii*. Rare trees of *Eucalyptus dichromophloia* and *E. papuana* are seen, usually on drainage lines. *Triodia pungens*, known as 'soft spinifex', unlike other species is palatable to stock when young. Pastoralists running sheep on this country therefore endeavour to maintain the spinifex in a young stage by frequent burning, which has the added advantage that after fire a considerable number (over 40 have been listed) of annual and short-lived perennial species, many of which are also edible, regenerate and flourish until the spinifex gradually reestablishes its cover, when in the absence of further fire they would die out.

The granite plain is broken by numerous stony rises, low hills, basic dykes, large granite outcrops and ranges of hard, resistant rocks. These are mainly characterised by replacement of *T. pungens* by *T. wiseana, T. longiceps* or *T. angusta* and thinning of the associated shrub flora. Basic dykes consist often of bare boulders appearing black at a distance, stretching in long lines like ramparts across the country. Spear Hill at Pilga on the Shaw River is a granite outcrop covered with trees to 8 m tall of an undescribed *Acacia* which has long branches and was presumably sought after at one time by Aborigines for making spears. The bark is of the red-brown peeling 'minnieritchie' type.

The high, steep parts of the Gorge Ranges have only a thin cover of spinifex — *Triodia brizoides* and *T. wiseana* with *T. pungens* and scattered small trees of *Eucalyptus leucophloia*. These same communities carry through into the Oakover valley, the latter on the steeper, more rocky and stony country and the shrub association of *Acacia pyrifolia, Grevillea pyramidalis* and *Hakea suberea* on more even ground where soil cover is present. *Triodia pungens* tends to be limited to deeper soils.

The very rugged dissected country along the frontal escarpment of the Chichester Range repeats the landscape of the Gorge Ranges, with *Eucalyptus leucophloia* and *Triodia wiseana* on the steepest and stony ground, *Acacia pyrifolia* and associates plus *T. pungens* on the gentler slopes. The latter association continues to provide the general cover on the summit plateau, but on the flatter sites and in depressions there are treeless grass plains of the bunch grasses *Aristida latifolia, A. contorta* and *Iseilema vaginiflorum* with a range of herbs such as *Ptilotus carinatus, Crotalaria dissitiflora* and *Streptoglossa odora*. The ptilotus may at times become dominant and extremely colourful when in flower. Descending into the Fortescue Valley, mulga *(Acacia aneura)* appears in the valleys.

At the eastern end of the Fortescue Valley there are extensive sandplains, along the foot of the Hamersley

Fig 4.158
The mining settlement of Marble Bar is named from a dyke of banded jasper which impounds a pool, Chinaman's Pool, on the local creek.

Fig 4.159
The 'marble bar' itself is an outcrop of banded red and white jasper, which may be seen at Chinaman's Pool.

Fig 4.160
Cassia venusta, one of the most beautiful cassias and worthy to become a garden subject, enlivens a granite outcrop in this otherwise dreary plain near Marble Bar, looking towards Mt Edgar, the pimple on the horizon.

escarpment gravelly outwash plains, while the central axis of the valley contains valley plains and flood-out zones with deeper, finer soil. On sandplains *Triodia basedowii* provides the general cover, only occasionally with *T. pungens*. The corkwood *Hakea suberea* is the most conspicuous woody species, usually also with much *Eucalyptus gamophylla*, a mallee with glaucous foliage. As the specific name implies ('wedded leaf') the leaves are opposite and united in pairs at the base. Occasional trees include *Eucalyptus dichromophloia* and *E. setosa;* there are numerous other shrubs such as *Acacia pachycarpa, A. pyrifolia, Cassia* spp. and *Grevillea juncifolia*, as well as small woody plants and ephemerals. Mulga is present along creeks. This community is also seen on the

Fig 4.161

The Chichester Range rises in a series of scarps to a summit plateau at 150–200 m above sea level and is formed of volcanics capped with an iron formation. The general cover is of the spinifex *Triodia wiseana* on the rockiest sites and *T. pungens* on better soil. The white-barked tree *Eucalyptus leucophloia* generally associates with *T. wiseana*. Between Roy Hill and Nullagine.

Fig 4.162

In the less mountainous sections *Triodia pungens* is dominant and forms a characteristic association with the shrubs *Acacia pyrifolia* (seen here), *Hakea suberea* and *Grevillea pyramidalis*. Bonnie Downs station.

outwash plains but tends to be replaced by the *Acacia pyrifolia* association in the western end of the valley.

Valley plains are typically covered with mulga (*Acacia aneura*) forming low woodland. This is further north than mulga normally occurs and its presence is presumably due to favourable bottomland soils with added moisture from run-on. In the east there are 'buckshot plains' where the soil surface is covered with fine black gravel. Coarser gravels and shingle are seen elsewhere, the latter bringing in the snakewood *Acacia xiphophylla*. The mulga has the usual understorey shrubs of *Eremophila* and *Cassia*, and annuals, chiefly *Ptilotus exaltatus*. Flood-out zones of the valley plains are somewhat different. These begin along the branches of the eastern Fortescue as high as Ethel Creek and widen out into an extensive plain above Roy Hill covered with tree savanna of scattered trees of coolabah (*Eucalyptus microtheca*) with a thick grass and herb layer. Below Roy Hill the channels discharge into a salt marsh 100 km long and up to 16 km wide which is well vegetated either with samphire (*Pachycornia triandra*, *Halosarcia halocnemoides*) or saltbush (*Atriplex cinerea*, *Enchylaena tomentosa*) communities. Other halophytes are *Atriplex eardleyae*, *Sclerolaena* and *Frankenia* spp., *Muellerolimon salicorniaceum*, *Swainsona kingii*. The paperbarks *Melaleuca lasiandra* and *M. pauperiflora* appear in fringing communities, and long lines of *Eucalyptus camaldulensis* extend out into the upper end of the marsh along channels. Downstream the line of the marsh is continued along the valley in grassy flats of *Aristida contorta*,

Fig 4.163

The plateau of the Chichester Range slopes gently to the south into the valley of the Fortescue River, beyond which rises the escarpment of the Hamersley Range. Numerous grassy flats on the plateau and in the valley may in certain seasons be swamped by a mass of annuals, in this case *Ptilotus carinatus* and *Gomphrena cunninghamii*.

Fig 4.164

Such photographs as Fig 4.163 are evidence that floral displays can occur, but as elsewhere we must remember that they are dependent upon good seasons. In a bad season only the abomination of desolation may be visible. This is a buckshot plain where the surface is covered with tiny spherical pebbles, unlike the gravel and shingle plains (Figs 4.95, 4.104, 4.114). Marillana station.

A. latifolia, *Eragrostis setifolia*, *E. xerophila* and *Panicum whitei*.

At the Millstream Station at the western end there is literally an oasis with long permanent pools of water lined by *Eucalyptus camaldulensis* and *Melaleuca leucadendra*, and some introduced date palms have become established. Below the Deep Reach in permanent flowing water there is a small forest of *M. leucadendra* with an understorey of *Cyperus vaginatus*. The flats beside the rivers and creeks carry an irregular woodland including the local endemic palm *Livistona alfredii*.

The Hamersley Plateau and its escarpment has predominantly a spinifex vegetation, but mulga can occur in valleys and on basaltic rocks, and grass savanna on valley plains. The spinifex here is *Triodia wiseana*, a smallish inedible species. The landscape is given most of its character however by the scattered trees of *Eucalyptus leucophloia*, known as mygum, which has glaucous foliage and white bark. This was formerly confused with *E. brevifolia* (snappy gum), a northern species which it resembles, but has been recognised as distinct since 1976. There are few large shrubs, which include *Cassia desolata*, *C. pleurocarpa*, *Dodonaea viscosa*, *Grevillea wickhamii* and *Hakea suberea*. However there is a rich flora of small shrubs and herbaceous species which can produce a rich show of colour in spring in a good season. Many of these are related to southern 'heath' plants, such as *Burtonia polyzyga*, *Calytrix exstipulata* and *Gastrolobium grandiflorum*, but others are typically Eremaean, e.g. *Petalostylis labicheoides*, *Ptilotus*

Fig 4.165

The escarpment of the Hamersley Range rises to a general height of 900 m with a maximum of 1235 m at Mt Meharry on the plateau behind. The frontal escarpment is dissected by impressive gorges such as Dales and Wittenoom Gorges. The horizontally bedded rocks are extremely hard and consist of jaspilite and dolomite beds between which there are veins of blue asbestos. The view of Wittenoom Gorge here shows the asbestos mine which formerly operated here before the danger to health of this mineral was fully realised. *Triodia wiseana* on the slopes, river red gums, *Eucalyptus camaldulensis*, in the valley bottom.

Fig 4.166

Some of the gorges terminate in narrow canyons with waterfalls. This is Joffre Falls, originally Bismarck Falls and renamed after the French marshal during the first world war. Vegetation *Triodia wiseana* and *Eucalyptus leucophloia*.

Arthur Fairall
Arthur Fairall

Fig 4.167

Bee Gorge in the Hamersley Range. Paperbarks and reeds contrast with the bare cliffs. Pieces of the rock are so hard that they are like iron plate.

Arthur Fairall

FIG 4.168

Eucalyptus leucophloia does well to obtain a foothold!

FIG 4.169

Fortescue Falls in Dales Gorge are a favourite haunt of the visitor. It is interesting that the cliffs here are another locality for the cypress pine, *Callitris glaucophylla*, which has survived due to natural fire protection (Fig 4.152).

FIG 4.170

Above the gorges the Hamersley Range presents a landscape of rounded summits, rocky and with little soil. The general cover is of *Triodia wiseana* with scattered trees of *Eucalyptus leucophloia*. This eucalypt is known as mygum and was formerly referred to *E. brevifolia*, but has been named as a distinct species since 1976.

Fig 4.171
The general aspect of this vegetation is shown here. In dry periods only the conspicuous perennial species, chiefly the trees and spinifex, appear to the eye, but there is a rich associated flora which appears after rain.

Fig 4.172
Spring flowering in the Hamersley Range can be as eye catching as anywhere. At this point *Burtonia polyzyga*, a yellow-flowered pea, and *Goodenia scaevolina* (blue, see Fig 4.136) provide the main interest. Above Wittenoom Gorge.

Fig 4.173
The Fortescue River, rising in the east near Newman, is really two distinct streams linked only by a dry valley. The eastern one collects all its tributaries above Roy Hill where it first traverses extensive savanna-like flats of grassland scattered with coolabah trees *(Eucalyptus microtheca)*. Chief grasses are *Eragrostis setifolia* and *Panicum decompositum* but most of the ground cover consists of *Ptilotus gomphrenoides*, *P. clementii*, *P. helipteroides* and *Calotis multicaulis*.

Fig 4.174
Below Roy Hill homestead the eastern Fortescue discharges into an enormous salt marsh 100 km long and 10 km wide. It is not a salt lake as the whole area is vegetated and salt does not appear on the surface. At the Roy Hill end there is a saltbush community with *Atriplex cinerea*, *Eremophila* aff. *pterocarpa* and *Enchylaena tomentosa*, but seen here halfway along the marsh it is a samphire community with a red samphire *Pachycornia triandra* dominant and a yellowish species *Halosarcia halocnemoides* with it. Other species present include *Atriplex eardleyae*, *Eremophila* sp., *Maireana luehmanii*. A number of annuals may appear in a good season, which it was not when the photograph was taken. Marillana station.

FIG 4.175
Northwest of the salt marsh, after the Fortescue valley has been dry for some distance, streams flowing out of the Hamersley Range renew the flow of the river. At Millstream this is augmented by powerful underground springs which permanently fill an extensive 'deep reach' of the river. This then forms a permanent oasis, set about with coolabah, river gum and paperbark trees. Millstream.

FIG 4.176
The Deep Reach at Millstream is a tranquil spot, very popular for camping. This water supply has now been harnessed for coastal towns.

FIG 4.177
The Millstream oasis is home to an endemic palm, *Livistona alfredii*, found nowhere else but related to *L. mariae* of the Macdonnell Ranges in the Centre. Both species no doubt are derived from a population widespread in more pluvial times.

FIG 4.178
Hibiscus panduriformis, the commonest and most attractive of the northern species, is occasionally seen throughout the Pilbara and further north. Plants of this kind appear to be weeds, appearing casually in any piece of empty or disturbed ground, rather than members of established plant associations.

rotundifolius and spp., *Clianthus formosus* and *Goodenia maideniana*. The cliffs of the gorges are a special habitat notable for local or endemic plants. Best known of these is *Astrotricha hamptonii*, a shrub otherwise occurring in Queensland, and a most beautiful eremophila, *E. 'magnifica'* nov.sp.inedit., discovered clinging to the rock in Wittenoom Gorge by the writer in 1963.

Mulga in the valleys of the plateau is as described elsewhere but on basaltic country mainly in the centre around Mt Tom Price it forms a mosaic with the *Acacia pyrifolia-Triodia pungens* association, the latter on the more stony sites and the mulga on the better soils. An intermediate phase is common, however, where the spinifex forms an understorey to the mulga.

Grass plains are on cracking clays known as 'crabhole plains'. The Mitchell grass *Astrebla pectinata* seems to be generally dominant, or sometimes kangaroo grass *Themeda australis*. Other associated grasses are *Aristida latifolia*, *Chrysopogon fallax* and *Eragrostis setifolia*.

4.10 Great Sandy Desert

CANNING BOTANICAL DISTRICT
Tree-steppe grading to shrub-steppe in the southeast, comprising open hummock grassland of *Triodia pungens* and *Plectrachne schinzii* with scattered trees of *Owenia reticulata*, *Eucalyptus* spp. and shrubs of *Acacia* and *Grevillea*.
Climate: Arid tropical with summer rain; precipitation 200-300 mm per annum.
Geology: Quaternary sandplain overlying Cretaceous and Jurassic sandstones which are exposed locally.
Topography and soils: Gently undulating plain dominated by longitudinal dunes of varying frequency tending mainly WNW-ESE. Chief soils are red earthy sands and red siliceous sands, with exposures of ironstone gravels locally.
Area: 291 017 km².
Boundaries: Drawn all round at the limit of desert sandplains and dunefields on Mesozoic rocks.

The Great Sandy Desert was discovered by Augustus Gregory in 1856 when his party followed the Sturt Creek from the north to its termination and found impenetrable sandhills ahead. His brother Frank Gregory came to the opposite, southwestern side from the Oakover River in 1861 and endeavoured to penetrate further but was driven into a hasty retreat by the sand, lack of water and feed for the horses. To this day a dump of guns, saddles and other gear must be lying out there waiting to be rediscovered. The Great Sandy Desert was first crossed — and given its name in the process — by Col. P.E. Warburton's expedition in 1873, suffering appalling hardships. The sandhills make the going on foot very difficult, there is very little drinkable surface water, and ants tormented them. In the course of the journey one member of the party lost his reason, Warburton himself became partly blind and the whole number all but perished. Since that time the four wheel drive motor vehicle and the aeroplane have fortunately taken the horrors out of interior exploration and the forbidding desert can be traversed in virtual comfort.

The Canning Stock Route between Halls Creek and Wiluna, the nearest southern railhead, was surveyed by A.W. Canning in 1906 and serviced by a chain of wells, each timber lined and provided with iron covers, a windlass, bucket and watering trough. The purpose was to provide an overland route for driving cattle from the Kimberley to market in the south, but it must have been a hazardous journey and the cattle must have lost a lot of condition on the way. In time the route went out of use, though the wells have remained usable. Contrary to popular impression, there was never a vehicle track along it, though in recent years one has been opened by private adventurers in 4WD vehicles.

The Commonwealth put in some graded tracks for survey purposes in 1958 and these were extended by the West Australian Petroleum Company (WAPET). The Company built a metalled road from the coast to Swindell Field in 1964 and extended this to Kidson in 1966. Other mineral surveys have provided a network of tracks and a major gold mine, Telfer, has been opened in the Paterson Range. The desert has thus become much more accessible.

The Great Sandy Desert occupies the greater part of the northern desert area north of the Tropic of Capricorn. It stretches from 18 to 20°S latitude, and from 120 to 128°E longitude. It is essentially a sandhill region and its boundaries have everywhere been drawn at the outer limits of sandhill country. In the northeast and east the sandhill desert is bounded by the desert sandplains of the Tanami Desert, in the north by the tree- and shrub-steppe formations forming the arid fringe of the Kimberley, and in the northwest by the pindan formation of Dampierland.

On the southwest the boundary skirts the mountainous country of the Pilbara block prolonged into the Throssel, Broadhurst, McKay and other ranges

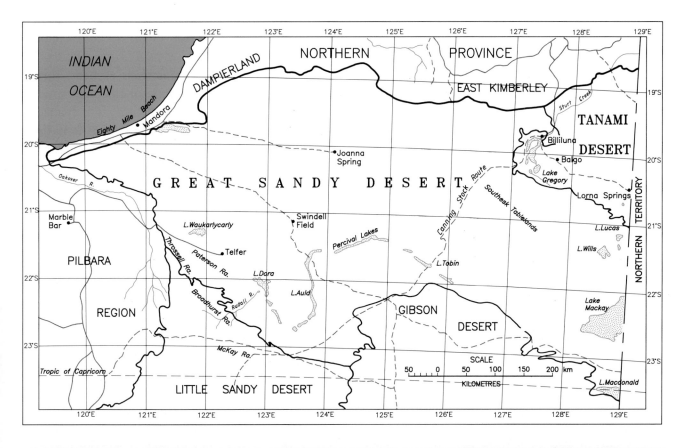

FIG 4.179

The Great Sandy Desert occupies the Canning Basin, an area of sedimentary rocks between the Pilbara and the Kimberley. The flat-lying rocks produce a mesaform topography, but not usually as hilly as seen here. The country consists mainly of sandy plains and sand ridges, with only occasional mesas. Callawa station.

which form a tongue partially cutting it off from the Little Sandy Desert. On the south the boundary lies at the edge of the laterite plains of the Gibson Desert.

There are no meteorological stations within the Great Sandy Desert and inferences as to the climate can be made only from stations outside its borders. Rainfall is probably very erratic and decreases from 300 mm per annum in the northwest to 250 mm or less in the southeast. Rain is received predominantly from thunder showers in the summer months and sometimes from cyclonic rains bringing heavy falls. The winter months are virtually rainless.

The physiography of the region consists basically of a gently undulating plain rising from the coast to some 450 m above sea level in the far interior. It is divided into two ancient and very extensive river basins, now inactive, divided by a low watershed which appears as the Anketell Ridge in the west and the Southesk Tablelands in the east. Both these upland areas are free of sand dunes and have a hard lateritic surface. The southern river basin is marked by a curving chain of salt lakes, Lake Tobin, the Percival Lakes and Lake Auld which represent the course of a former river. This would have continued to pick up drainage from the basin of Lake Disappointment, and turning to the north connected with the Rudall River at Lake Dora and flowed on towards the sea via Lake Waukarlycarly. The northern river was a continuation of the Sturt Creek. It has been established from satellite imagery that this once flowed on across the desert to reach the sea at Mandora on the 80-Mile Beach, though it cannot be detected on conventional aerial photography.

With the exception of the Paterson Range which is a solitary outcrop of the pre-Cambrian basement, all the surface rocks are more or less horizontally bedded sediments classified as fine to medium-grained micaceous sandstone, and coarse sandstone and conglomerate, of Permian to Cretaceous age.

The more recent rocks listed above frequently crop out to form small buttes and mesas throughout the area; however the linear dunes are the most conspicuous feature of all and sweep across the country in immense parallel lines. They are absent only from the uplands and salt lakes. The general trend of the dunes is at right angles to the coast, and conforms to the prevailing winds of the present day. Individual dunes can be very long, as much as 250 km without a break. Distance apart laterally is from 250 to 500 m (2 to 4 dunes per kilometre). It is therefore quite easy to traverse the country parallel to the dunes, as the WAPET road does, but arduous to travel across them as the Canning Stockroute does.

Since the dunes are a dominant element, their vegetation is the principal feature of the plant cover. It is simple in species and structure, is extremely consistent throughout the region, and differs from the vegetation of the interdune flats. The flanks of the dunes are normally thickly vegetated with hummock grass ('spinifex'), with scattered shrubs, but the crests are often somewhat bare and subject to wind action. Feather-top spinifex (*Plectrachne schinzii*) is the universal species of hummock grass. Among shrubs, *Grevillea stenobotrya* is common, very consistent, and confined to sandhills, *G. eriostachya* and *G. juncifolia* may be found, with many species of wattles, especially *Acacia victoriae, A. ligulata* and the soft shrub *Crotalaria cunninghamii*. In depressions *Melaleuca lasiandra* may become common. South of latitude 22° small shrubs of *Thryptomene maisonneuvii* may be locally frequent between the spinifex. In addition to the above there are often trees of a species confined to sandhills. Its identity was long in doubt but it has been recently named as *Eucalyptus chippendalei*, 'desert bloodwood'. Distribution is extremely irregular and it may be absent for miles, or else quite thickly distributed. It tends to grow on either side of the bare top of the dune, so much so that in many aerial photographs two rows of small dots can be distinguished along each crest. From the association of trees, shrubs and hummock grass this vegetation is termed 'tree-steppe'. The vegetation of the interdune may also be tree-steppe but is more commonly only shrub-steppe, i.e. with no trees present. The interdune is primarily influenced by the amount of sand present. On deep sand *Plectrachne schinzii* is dominant, but if this shallows or gives place to bare laterite gravel there is a change to *Triodia pungens* or, south of latitude 22°, to *T. basedowii*. The woody components are less completely dependent upon the soil constitution. In the northern part of the desert, that is northwest of the Percival Lakes, presumably under the influence of a higher rainfall, scattered trees of *Eucalyptus dichromophloia, E. setosa* and *E. aspera* are common in sandy ground, together with numerous shrubs, notably *Hakea suberea, Acacia pachycarpa, A. monticola, A. tumida, Grevillea wickhamii* and *G. eriostachya*. On hard lateritic ground the trees drop out. In the interior sector southeast of the Percival Lakes the trees are no longer seen and only shrub-steppe covers the interdune.

Fig 4.180

Seen from the air, the Great Sandy Desert is seen to be marked by *seif* dunes, long sub-parallel sand ridges, all aligned NNW–SSE and arranged 2 to 4 dunes per kilometre. Individual dunes may be 200 km long. They have been formed by the wind sweeping up the surface sandy topsoil of the country during a drier and more windy climatic phase than today's, perhaps during the last throes of the ice age about 15 000 years ago. At lower right, the lateritic subsoil can be seen exposed by deflation. The country can be seen to be generally covered by spinifex which becomes rather thin on the dune crests, with scattered trees and shrubs. Approaching Swindell Field.

Fig 4.181

Seen from one dune, the country stretches flat and hard surfaced to the next, which closes the view. It is quite easy to traverse this country parallel to the dunes, but very difficult to try to cross them. The distant dune is seen to bear trees along its crest, of which one appears at left foreground. This species is confined to the tops and sides of dunes, as far south as the fringe of the Great Victoria Desert, but does not occur between the dunes. It has been recently named as *Eucalyptus chippendalei*.

Fig 4.182

Typical crest of a *seif* dune. The flanks are thickly vegetated with spinifex, but the tops are often partly bare and the sand is still partly mobile. The *Eucalyptus* is seen here. Other typical dune plants are *Grevillea stenobotrya*, *Acacia ligulata* and other species, and *Crotalaria cunninghamii*.

Fig 4.183

Grevillea stenobotrya is no doubt the most typical of all sandhill plants, being found on every sandhill in Western Australia but in no other habitat. Average maximum size is shown here, and the white flowers. Dune near the Parry Range.

Fig 4.184

Flowers and foliage of *Grevillea stenobotrya*.

Fig 4.185

The flats between the dunes are vegetated with spinifex and eucalypts, but with different species. On deep sand the spinifex is *Plectrachne schinzii* as on the dune flanks, but on laterite it is *Triodia pungens* towards the coast, *T. basedowii* further inland. As seen here the *Plectrachne*, feathertop spinifex, is in full flower. The trees are *Eucalyptus aspera*.

Fig 4.186

The desert is interspersed with lateritic rises where the sandy covering has been removed by wind as in the Gibson Desert. There is a general cover of *Triodia pungens* in this case but no trees, only scattered shrubs. The commonest are acacias, but *Grevillea wickhamii* is certainly the most attractive. Both leaves and flowers are larger here than in the form of this species seen in Central Australia.

Fig 4.187

Within 100 km of the coast, perhaps in response to higher rainfall, the most conspicuous tree becomes the desert walnut, *Owenia reticulata*, with some *Gardenia keartlandii* and *Erythrophleum chlorostachys* (ironwood). It is unusual that the trees should not be eucalypts. The ground layer is composed of both *Triodia pungens* and *Plectrachne schinzii* with a minor occurrence of bunch-grasses such as *Eragrostis eriopoda*.

Fig 4.188

Fruit and foliage of desert walnut, *Owenia reticulata*. The fruit contains an edible kernel about the size of a pea, but it is well protected by a fibrous and indehiscent outer shell which needs an axe to open it.

Within 100 km of the coast the tree-steppe association changes, representing a transition to the pindan formation of the coastal strip which is included in the Northern Province (Chapter 5). The principal tree becomes the desert walnut, *Owenia reticulata* (so called because it produces edible nuts), with *Gardenia keartlandii* and the ironwood *Erythrophleum chlorostachys*. There are sparse shrubs mainly of *Acacia pachycarpa* with *A. monticola* and numerous others, *Hakea suberea*, *Grevillea refracta* and *G. wickhamii*. The ground cover is formed of spinifex, with two species, *Triodia pungens* and *Plectrachne schinzii*. The latter is dominant in sandy areas where *Eucalyptus setosa* may also be present.

The lateritic, sand-free uplands carry a type of shrub-steppe with *T. pungens* dominant (or *T. basedowii* south of latitude 22°) and numerous shrubs of the species listed above, frequently growing in clumps, or disposed along drainage. The vegetation of hills and mesas is normally very sparse, consisting of a thin cover of *T. pungens* or *T. basedowii* with some small shrubs of *Eremophila* and *Cassia* spp. Mulga *(Acacia aneura)* is to be seen on the tops of mesas in some cases.

The vegetation of salt lakes and their vicinity is distinctive. On approaching a lake there is frequently an appearance of travertine in the soil, a replacement of *Plectrachne schinzii* by *Triodia pungens*, and occurrence of the tea-trees *Melaleuca lasiandra* and *M. glomerata*, up to 2 m high and locally forming thickets, with *Acacia ligulata* and numerous large termitaria. Lake bed communities (studied at Lake Auld) show a zonation

Fig 4.189

The Percival Lakes, Lake Auld and others are salt lakes forming a chain along an ancient river channel. The bed of Lake Auld shown here carries the saltbush *Hemichroa diandra* and the grass *Eragrostis eriopoda*. Other plants identified were *Tephrosia arenicola* and *Dampiera candicans*. *Melaleuca glomerata* as larger bushes in the rear.

Fig 4.190

Depressions and drainage lines where calcrete has been deposited show a relatively dense growth of *Melaleuca lasiandra* and *M. glomerata*, filled in with *Triodia pungens*.

Fig 4.191

It does rain in the desert sometimes! On this occasion our party was trying to reach an oil-drilling rig in the northern part of the Great Sandy Desert during July, the dry season. Unfortunately freak rains in June had exceeded the total received in Perth that month, 250 mm! Some desert country can become very soft after rain but the more sandy stretches remain firm.

probably corresponding to the deposition of evaporites — lime, gypsum and salt. The lowest portions where salt accumulates carry samphire communities *(Halosarcia)*, surrounded by a saltbush zone *(Hemichroa, Sclerolaena, Frankenia)* and this in turn by a spinifex zone *(Triodia* sp. inedit.*)*.

4.11 *Tanami Desert*
MUELLER BOTANICAL DISTRICT

Shrub-steppe as general cover with *Triodia pungens*. Sparse tree-steppe on some ranges, locally plains with short-grass savanna.

Climate: Arid tropical with summer rain; precipitation 200-300 mm per annum.

Geology: Quaternary sandplain overlying Proterozoic and Permian rocks which are exposed locally.

Topography and soils: Gently undulating sandy plain, dominated by longitudinal dunes south of latitude 20°30'S and with occasional low rocky ranges of Proterozoics; laterite-crusted uplands on Permian.

Area: 35 192 km² (within Western Australia).

Boundaries: The western boundary with the Great Sandy Desert is the limit of desert sandplains and dunefields on Mesozoic rocks. The short southern boundary is the northern limit of mulga.

Like the Great Victoria Desert and the Central Ranges, the Tanami is a region shared with the Northern Territory, and in this case the Territory owns the major portion. The Tanami comes across the border southeast of Halls Creek as far as Bililuna, including the lower course of the Sturt Creek and its termination in Lake Gregory, and the Gardner, Lewis and Phillipson Ranges. The Aboriginal community at Balgo Hills and some pastoral stations fall within its borders. Near Balgo there is a hill named Mt Mueller after the famous botanist, who accompanied Augustus Gregory's North Australia expedition of 1856. This party followed the Sturt Creek in the hope that it would lead them to the great inland sea fabled to exist in the centre of Australia but were

FIG 4.192

Like the Great Victoria Desert and the Central Ranges, the Tanami Desert is shared with the Northern Territory and in this case the Territory owns most of it. The Tanami comes across the border southeast of Halls Creek as far as Bililuna. It consists mainly of flat sandy plains from which rise occasional abrupt low hills and ranges of Precambrian geology, making the structure of the country like that of the Little Sandy rather than the Great Sandy Desert. However the general appearance of the country is much the same.

FIG 4.193

A distinctive association is found on plains along the lower Sturt Creek, of *Hakea suberea* as often the sole woody species, with *Triodia pungens*. Billiluna.

FIG 4.194

Along the lower Sturt Creek groves of desert oak, *Casuarina decaisneana*, become quite common. As in the Central Ranges, it is supposed that these trees must have access to underground water, derived by seepage from the Sturt Creek.

disappointed and turned back leaving Lake Gregory named after their leader as a memento.

Physiographically this region consists mainly of flat sandy plains about 260-275 m above sea level, from which rise occasional abrupt low hills and ranges, much eroded and very stony, and not attaining much more than 90 m above the plains. The highest point, Mt Brophy in the Gardner Range, is about 550 m above sea level. The Sturt Creek and its tributary the Wolf Creek drain down into this region and are lost by evaporation. There are no active rivers other than minor flood channels draining the hills. Geologically the region is covered in the western part with sediments of Permian age, mainly sandstones. In the eastern part these have been eroded to expose harder underlying rocks of Proterozoic age which rise in hills and ranges of rounded outline forming 'whalebacks'. It was suggested in the geological survey that these show us a landscape of Permian age (270 million years old) which has been 'exhumed' — literally, 'dug up', meaning re-exposed by erosion after having been buried for millions of years.

Although most of the country consists of sandy plains, linear dunes are uncommon and the reason for this absence of sandhills is not clear. In many areas there is evidence that the surface sand is rather thin, overlying rock or calcareous hardpan. It appears that the red soil is firm and of a fairly loamy nature rather than loose and incoherent, so that it may possess binding qualities. The vegetation of the sand plains is shrub-steppe. As observed at Billiluna, this has a ground cover of soft

FIG 4.195

On its journey to its termination in Lake Gregory the Sturt Creek fills many billabongs and other flood-out zones. This one is known as the Stretch Lagoon, lined with river gums and paperbarks. Billiluna.

FIG 4.196

Flood-out zones of the Sturt Creek often appear rather desolate, with little but a few coolabah trees. Doubtless in time of flood they would appear more genial.

FIG 4.197

Lake Gregory where the Stuart Creek finally ends is a large dry lake with a bed of salt crystals. The North Australia Expedition led by Augustus Gregory in 1856 hopefully followed the Sturt Creek southward, spurred on by the current legend of a great inland sea in the centre of the continent, and were bitterly disappointed when it finally ran to nothing. There is evidence that in former times of higher rainfall the Sturt Creek continued to cross the Great Sandy Desert area and to reach the sea at the 80-Mile Beach.

FIG 4.198
The Tanami Desert is crossed by a track leading southeast from Billiluna eventually to reach Alice Springs. There are some heavier soils in this direction which support saltbush flats of various kinds. Texas Bore, Balgo.

FIG 4.199
Just before reaching the State boundary the Tanami Track passes through the Lewis Range (some low sandstone hills) and encounters this wonderful oasis, Lorna Springs.

spinifex, *T. pungens*, a variable species which here adopts a stoloniferous form. Numerous shrubs, up to 3 m tall, mainly include the following: *Hakea suberea, Acacia pachycarpa, A. monticola, Grevillea pyramidalis, G. wickhamii, Gardenia keartlandii, Cassia* spp. Very occasional trees of *Eucalyptus aspera* and *E. dichromophloia* are seen and the mallees *E. pachyphylla* and *E. odontocarpa*. *Melaleuca lasiandra* comes in, in depressions.

The vegetation of stony hills, where these occur, is very sparse, a thin cover of *Triodia pungens* with a few stunted bushes of *Grevillea wickhamii, Cassia* and *Eremophila* spp. or *Ptilotus*. Only on the northern border of the region do a few trees of *Eucalyptus brevifolia* appear, and some admixture of *Triodia intermedia*. These are typical elements of the tree-steppe of the Halls Creek area further north.

The course of the Sturt Creek, pouring down to terminate in this region, introduces an element of vegetation foreign to the desert. Along the Sturt Creek itself there is a flood plain mostly about 3 km in width, sometimes more, and there are also traces of a former course of the Sturt Creek cutting across to pick up the Lewis and Slatey Creeks and rejoining the present course below Billiluna station. These flood plains consist of a grey silt and are very sparsely vegetated with grasses (especially *Astrebla pectinata*) and soft spinifex *(Triodia pungens)* in between very numerous small bare claypans and occasional coolabah trees *(Eucalyptus microtheca)*. South of Billiluna the flood plain spreads out into the distributary zone where the river is lost in a maze of braided channels and pans, a very desolate region with the bare clay surfaces of the pans, low sandy mounds dividing them and covered with spinifex *(Triodia pungens)*, some flats with grass and herbs, small stunted trees of *Eucalyptus microtheca* mainly around pans, and shrubs of *Eremophila* and *Acacia* spp. Around some marginal claypans in sandy country there is a growth of tea-tree scrub, *Melaleuca lasiandra* and *M. glomerata*. In the adjacent sandplain there are also frequently extensive groves of desert oak *(Casuarina decaisneana)* over a ground cover of soft spinifex *(Triodia pungens)*. The casuarinas become quite large trees 9-12 m tall and 30 cm in diameter and appear to be dependent on supplies of underground water. The distributary zone is terminated at the south end by several extensive 'plains' which are just bare clay flats, and by Lake Gregory which is salt. There is otherwise little sign of salt along the Sturt Creek. The lake is ringed by a belt of a well-grown wattle (*Acacia* sp. unidentified) 6 m tall, in close, almost closed formation with a ground layer of short annual grass. This tree resembles mulga *(A. aneura)* with grey foliage but the leaves are much longer and pendant. The lake bed is vegetated with black samphire *(Halosarcia)* on grey silt.

BIBLIOGRAPHY

1 Beard, J.S. 1974–81. Vegetation Survey of Western Australia 1:1 000 000 Series. Univ. W.A. Press, Nedlands.
a. Sheet 1, Kimberley 1979
b. Sheet 2, Great Sandy Desert 1974
c. Sheet 3, Great Victoria Desert 1974
d. Sheet 4, Nullarbor 1975
e. Sheet 5. Pilbara 1975
f. Sheet 6. Murchison 1976
g. Sheet 7, Swan 1981
2 Lowry, D.C., 1970. Geology of the Western Australian part of the Eucla Basin. *Bull. Geol. Surv. West. Aust.* 122.

3 Mabbutt, J.A. *et al.*, 1963. General report on lands of the Wiluna-Meekatharra area, Western Australia, 1958. Land Research Series 7, CSIRO, Melbourne.
4 Semeniuk, V., Kenneally, K.F. & Wilson, P.G., 1978. Mangroves of Western Australia. W.A. Nat. Club Handbook 12. Perth.
5 Wilcox, D.G. & McKinnon, E.A., 1971. A report on the condition of the Gascoyne catchment. Lands Dept, Perth (unpub. report).
6 Craig, G.F., 1983. Pilbara Coastal Flora. Misc. Pub., Dept of Agriculture, South Perth.

5
THE NORTHERN PROVINCE

The Northern Province is roughly the same size as the Southwest Province, about 300 000 km², and comprises 12% of the State. It lies within the tropics, north of the 19th parallel, and enjoys a tropical climate with rain during the summer. However while the rainfall is higher and more reliable than in the desert to the south, it is by no means copious. This is the dry tropics with a very short rainy season of only four months so that crop production is only possible under irrigation and most of the area is utilised for grazing cattle on the indigenous grasslands. The climate favours grassland vegetation so that most of the area is covered by tropical savannas of one sort or another. Spinifex grasslands are found along the fringes of the desert, and elsewhere on shallow, sandy and rocky soils.

The province equates with the administrative Kimberley District of Western Australia, named in 1880 after the Earl of Kimberley who was then Secretary of State for the Colonies. He has left his name also on the South African diamond-mining centre of Kimberley. It is a coincidence that the Kimberley in Western Australia has become a significant diamond-mining centre in recent years.

The first European to land on Kimberley shores is thought to have been the English buccaneer William Dampier who careened his ship *Cygnet* in King Sound in 1688. He came again exploring for the British government in HMS *Roebuck* in 1699 and gave such an unfavourable report on the country that the British took no further interest for over a century, until the coast was surveyed for the British Admiralty by Lieut. King in 1818-21. Some disastrous attempts at settlement by sheep graziers were made at several points in 1863-65. Greater success followed the explorations of Alexander Forrest in 1875-78, when millions of acres were quickly taken up on pastoral lease. In general the west Kimberley (Dampierland) was settled by Western Australian sheep graziers from the south who occupied the short grass and spinifex country of the Fitzroy basin, while the east Kimberley, more of which is tall grass country, was settled by cattlemen driving herds overland from the Territory and Queensland. Ten thousand head had been introduced in this way by 1880, reaching as far west as Fossil Downs.

To this day, cattle raising remains the principal activity in the Kimberley. The area is divided into very large cattle stations which deliver their stock to the ports for slaughter and export. An active pearling industry based on Broome eventually died out, gold discovered at Halls Creek did not last, other mineral developments (until the relatively recent Argyle diamond venture) have been minimal. An attempt to develop irrigation farming at Camballin on the Fitzroy River has been unsuccessful and the same can be said, effectively, of irrigation on the Ord at Kununurra. The Argyle Dam on the Ord River has created an enormous lake with huge irrigation potential on fertile black soil plains below, but it has not so far been possible to develop economically successful crops.

Accessibility remains a problem in much of the Kimberley. The plains of the Dampierland and East Kimberley regions are easily traversed, and crossed today by the black-topped main highway. The rough country behind the King Leopold Ranges is quite another matter. It was as recently as 1965 that the first road was opened across the ranges, enabling the stations behind to move stock by road instead of droving. Still more recently this road has been extended through to Wyndham as the 'Gibb River Road' providing a scenic route for dry season use. Further north access is still very difficult. A track reaches the Aboriginal community at Kalumburu but the whole north and northwest coast is virtually uninhabited and accessible only by sea. Most of it is extremely rough country not even traversable by 4WD vehicles. Botanical exploration of this coast has been pushed forward in recent years based either on the Mitchell Plateau mining camp (now closed) or on visits by sea after a long interval since the first botanical collections were made by Alan Cunningham who accompanied the Admiralty survey of the coast in 1820.

One of the most exciting developments botanically has been the discovery of patches of vine thicket (rain forest in the broad sense) on and near this coast. Nothing of this sort had been previously known in Western Australia and the occurrences are now being actively examined and mapped by the Dept of Conservation and Land Management. The vine thickets usually occur only as small patches of a few hectares at a time and range

The Northern Province, equating with the Kimberley District of Western Australia, is the somewhat better watered tropical portion of the State lying north of the desert centre. Its tropical situation means that rain is received in summer, in the hot season. Winter is cooler and dry. This climatic regime favours grasslands, which are dominant in one form or another, and as tropical bunch grasslands rather than the desert spinifex which persists only on some dry substrates. The bunch grasslands are normally wooded with scattered trees and form true tropical savannas. Savanna with grey box trees, Mt House station.

from dense forest 10-15 m tall to mere scrub of 2 m, but have a rich flora of mixed tropical species quite distinct from the surrounding savannas.

Visitors to the Kimberley should understand that the climate is delightful in winter but unpleasant in the summer. Only the main trunk road is suitable for all-season use. Minor back roads become impassable during the 'wet' as the creeks fill with water and there are no bridges. Visitors interested in flora and vegetation may be assisted by the handbook by Petheram and Kok[1] which illustrates photographically 242 of the most common and conspicuous species — as usual there is the problem that the flora is too large to be illustrated in entirety. The vegetation has been mapped and described by Beard[2].

As we have noted above, the tropical summer-rain climate favours grassland development so that most of the Kimberley is covered by grasslands of one sort or another forming tropical savannas. A word about these may be helpful. In general such vegetation is two layered, with a tree layer and a grass layer. Locally the tree or the grass layer may be missing, or shrubs may replace the trees or form a third layer. On any site each layer tends to be characterised by one or a few prominent species which account for a high proportion of the biomass. Generally, the many other species together contribute little to the total biomass. Considering only the prominent species, each layer can be classified into reasonably homogeneous and distinct communities called *synusiae*. While it is common for a particular grass and a particular tree synusia to be associated over large areas, it is also common for several different grass synusiae to be associated with one tree synusia (and vice versa) on various sites. That is, the distribution of grass synusiae is independent of the tree synusiae (and vice-versa) and a large number of combinations occurs. The independent variation of the tree and grass layers is evidently due to their responding to different ecological factors.

The long winter dry period has had a dominant role in the selection of plant species in the area. According to their method of surviving the long dry period the species can be classified into three groups:

Perennial drought-resisting species — The leaves and stems of these plants remain in a growing condition through the dry period. The group includes most of the trees and shrubs and also the spinifexes (*Triodia* spp.) which are evergreen perennial grasses.

Perennial drought-evading species — This group includes those species in which at least the leaves die at the end of the wet season but new growth in the following wet season is initiated from vegetative organs. The group includes most of the perennial tussock grasses and deciduous trees and shrubs.

Annual drought-evading species — Plants in this group germinate from seed each growing season. It includes most of the forbs and short grasses but also a number of tall annual grasses.

In mapping the grassland vegetation of the Kimberley the following classification was adopted:

Grassland 1 — Savanna Grass layer closed and of bunch-grass form.
a High-grass savanna Grasses > 100 cm tall, drought evading.
b Tall bunch-grass savanna Grasses 50-100 cm tall, perennial, drought-evading.
c Short bunch-grass savanna Grasses < 50 cm tall, annuals, or short lived perennials.
d Curly spinifex communities Grasses 50-100 cm tall, perennial, drought-resisting.

Grassland 2 — Steppe Grass layer open and of hummock-grass form, grasses perennial, evergreen.

It was considered necessary to introduce a special classification for the curly spinifex communities in which *Plectrachne pungens* is dominant since this grass adopts a life form somewhat intermediate between the bunch-grasses and the hummock-grasses. It is a perennial, coarse, wiry grass in a genus in which many other, but not all, members are hummock-grasses, but strictly speaking it does not itself usually adopt a hummock-grass form. On the other hand as it is evergreen it is inappropriate to classify it among the tall bunch-grasses.

The primary classification above is based on the life-form and height of the grasses. A secondary classification arises from the fact that trees and shrubs are present also in most communities, and may be tall or short, dense or scattered. Two height classes for trees are recognised (medium height > 10 m; low < 10 m) and one class for shrubs. The density classes are as follows:

Savanna woodland: trees and shrubs forming a canopy which is generally light.
Tree savanna: trees scattered.

Sparse tree savanna: trees very scattered or sparse.
Grass savanna: trees and shrubs generally absent.

Table 5.1 shows the classification set out in tabular form. In addition woodland and shrubland types are recognised in which grasses are absent or play a minor role.

TABLE 5.1

Classification of tropical grassland communities

Tree and shrub layer	BUNCH GRASSLANDS			'CURLY SPINIFEX'	HUMMOCK GRASSLAND
	High >100 cm Drought-evading	Tall 50–100 cm Drought-evading	Short <50 cm Annual or short-lived perennial	50–100 cm Drought-resisting	Hummocks Evergreen
a Trees and shrubs forming a canopy, which is generally light	High-grass savanna woodland	Tall bunch-grass savanna woodland	Short bunch-grass savanna woodland *	Curly spinifex savanna woodland	Steppe woodland *
b Trees scattered	High-grass tree savanna *	Tall bunch-grass tree savanna	Short bunch-grass tree savanna	Curly spinifex tree savanna	Tree steppe
c Trees very scattered or sparse	High-grass sparse tree savanna	Tall bunch-grass sparse tree savanna	Short bunch-grass sparse tree savanna	Curly spinifex sparse tree savanna	Sparse tree steppe
d Shrubs only	High-grass shrub savanna *	Tall bunch-grass shrub savanna *	Short bunch-grass shrub savanna *	Curly spinifex shrub savanna *	Shrub steppe
e Trees and shrubs absent or virtually so	High-grass savanna	Tall bunch-grass savanna	Short bunch-grass savanna	Curly spinifex savanna *	Grass steppe

*Indicates no actual occurrences found in the Kimberley.
Note: If trees <10 m tall, insert the word 'low' appropriately.

5.1 Dampierland Region

DAMPIER BOTANICAL DISTRICT

Tree savanna of *Chrysopogon-Dichanthium* with scattered *E. microtheca* and *Lysiphyllum cunninghamii* on river plains; pindan on sandplains. The latter is a three-layered community, an open upper stratum of low trees, a closed middle layer of *Acacia* and an open ground layer of curly spinifex; hummock grassland with scattered trees on uplands.

Climate: Semi-arid to dry hot tropical; precipitation 250-800mm per annum, summer wet season 2-4 months.

Geology: Quaternary sandplain overlying Jurassic sandstones; Quaternary marine deposits on coastal plains. Devonian reef limestones and extensive alluvial river plains.

Topography and soils: Extensive riverine plains with grey and brown cracking clays; extensive sandplains on red earthy sands, low uplands of sandstone and limestone with shallow stony soils.

Area: 84 400 km².

Boundaries: The southern boundary is drawn according to vegetation at the edge of desert sandplains and dunefields; the northern and eastern boundaries at the vegetation change with the transition from Phanerozoic to Precambrian rocks.

The early visit by William Dampier is commemorated geographically in the Dampier Peninsula, the Buccaneer Archipelago and Roebuck Bay, and both province and botanical district are named after him. Dampier took

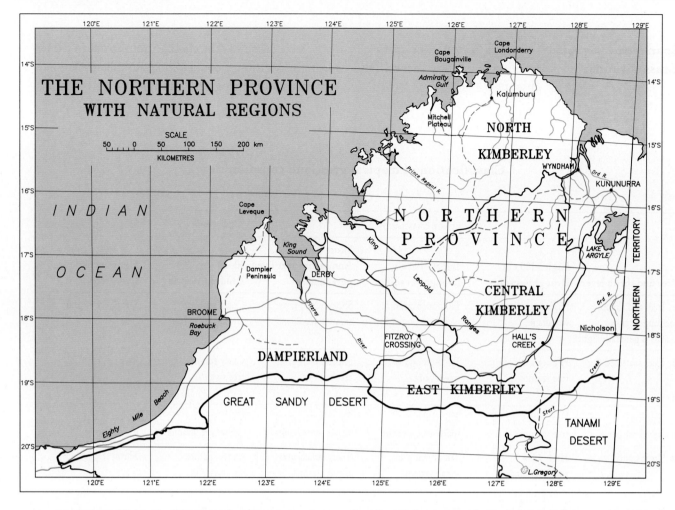

home the first collection of botanical specimens ever made in Australia but they were from further south on the coast, not from the Kimberley. The vegetation of Dampierland is characterised by a remarkable formation known as *pindan* which is unique to the province and occurs on all the sandy plains, covering 64% of the region. However it is entirely different from the sandplain vegetation of the Southwest Province: it has nothing in common with kwongan. Essentially the pindan is a grassland wooded by a sparse upper layer of trees from 5 to 15 m in height according to rainfall, and a dense, thicket-forming middle layer of *Acacia*. It has however many aspects, being subject to fire which periodically destroys the ground layer and the middle *Acacia* layer leaving the trees intact. The grasses regenerate from seed or rhizomes, the *Acacia* from seed. The grasses are quickly reestablished and for the first season or two after fire the pindan has the aspect of savanna, i.e. grassland with scattered trees. Gradually the *Acacia* shrubs grow up and eventually become dominant, suppressing the grasses, herbs and small woody plants. After a certain number of years the aspect is three-layered, with scattered trees, a shrub thicket and a sparse ground layer. Later still the acacias reach the height of the trees which disappear from view, giving the aspect of a tall thicket or low forest of *Acacia*. Eventually fire puts the succession back to the beginning. *Acacia eriopoda,* the principal pindan species, is known locally as 'seven years wattle' as it is expected to be burnt every seven years.

In the south, behind the 80-Mile Beach, the pindan is at its lowest and driest. The wattle is *Acacia pachycarpa,* the ground layer mainly *Triodia pungens,* trees are *Dolichandrone heterophylla, Erythrophleum chlorostachys, Gardenia keartlandii, Gyrocarpus americanus* and *Lysiphyllum cunninghamii.* Just for once, there are no eucalypts. From La Grange to Broome the *Acacia* layer consists almost entirely of *A. eriopoda* and reaches about 6 m in height at maturity. Other shrubs present include *Acacia holosericea, A. monticola, A. stipuligera, Calytrix exstipulata, Grevillea pyramidalis, G. refracta, G. wickhamii* and *Hakea macrocarpa.* Trees, which reach 8 m, include those listed above, now joined by *Eucalyptus dampieri* and *E. setosa.* The grass layer consists of relatively soft spinifex species not forming rigid hummocks, i.e. *Triodia pungens* and *Plectrachne schinzii.*

North of Broome along the length of the Dampier Peninsula the pindan gradually changes as the rainfall

FIG. 5.1

Dampierland or the West Kimberley is a low-lying region covered largely with sandy plains or alluvial river plains deposited by the Lennard and Fitzroy Rivers. The sandy plains which make up most of the western part of Dampierland are typically covered with an unusual formation known locally as pindan. Like the sandplains of the southwest it is frequently burnt, but there the resemblance ends. Pindan is 3-layered. When mature as shown here, it consists of a dense layer of wattles 5–6 m tall engulfing a number of scattered small trees of *Eucalyptus setosa*, *Gyrocarpus americanus* and others, and suppressing a ground layer of soft and curly spinifex *(Triodia pungens, Plectrachne pungens)*. Periodic fire destroys the wattle layer, after which the pindan has the aspect of a tropical savanna, but gradually the wattles regenerate and a return is made to the previous aspect.

increases. *Acacia tumida* replaces *A. eriopoda* while *Eucalyptus tectifica* and *E. grandifolia* replace the other trees. The grass layer is mainly of curly spinifex *Plectrachne pungens* with some ribbon grass *Chrysopogon fallax*. North of Pender Bay the trees are mainly woollybutt *Eucalyptus miniata* with *E. tectifica*, while *Acacia tumida* is joined by *A. platycarpa*; the ground layer remaining much the same. These pindan communities occur elsewhere in Dampierland on sandplains, e.g. on the Fitzroy Plains.

Between the pindan and the sea on the Dampier Peninsula and more especially further south behind both Roebuck Bay and the 80-Mile Beach there are extensive flat coastal plains of grey mud lying very little above sea level. They are partly saline and are thought to have been covered by the sea during a period of higher sea level about 6000 years ago. The plains are normally fringed by a zonation of mangroves, hypersaline bare mud flats and samphire communities of *Halosarcia halocnemoides*, *H. indica* and *Sesuvium portulacastrum* grading into short-grass plains of *Sporobolus virginicus* with few other plants other than an occasional samphire. The inland edge of the plain is usually marked by a steep declivity representing a former sea cliff, with a belt of *Melaleuca* at its foot. Inland of Mandora where the plain is at its widest there is a salt creek running out of the desert lined by the mangrove *Avicennia marina* although out of contact with the sea. As this area was tidal 6000 years ago the mangroves presumably became established at that time and survived after withdrawal of the sea perhaps 4000 years ago.

A belt of beach dunes has been built up along the 80-Mile Beach and on many other parts of the coast. *Spinifex longifolius* is the principal coloniser and normally

Fig. 5.2

For a few years after a fire, the pindan has the appearance of a savanna with scattered small trees of *Eucalyptus* and other species which are fire-resistant and not damaged by the burn, and a grassy ground layer of spinifex. This is a soft type of spinifex, not hummock grassland, formed of the soft spinifex *Triodia pungens* and the curly spinifex *Plectrachne pungens*. Wattle seed is stimulated to germinate by the fire, but the seedlings at first remain inconspicuous.

Fig. 5.3

After a few years the wattles have grown up sufficiently to begin to appear as a thicket once more, and will eventually equal the height of the longer-lived trees in the pindan.

Fig. 5.4

Pindan wattle *Acacia eriopoda*, known also as 'seven years wattle' as it is expected to be burnt about every seven years. There is a species zonation from south to north. Coming from the south, as the pindan is entered on Anna Plains the principal wattle is *Acacia pachycarpa* and the taller trees *Owenia reticulata, Dolichandrone heterophylla, Lysiphyllum cunninghamii, Gardenia keartlandii, Erythrophleum chlorostachys, Eucalyptus dampieri* and *E. setosa*. In the La Grange area *Acacia eriopoda* becomes established and continues to Broome and beyond. The two eucalypts become the commonest trees, joined by *Gyrocarpus americanus*. Twenty-five kilometres north of Broome *Acacia tumida* begins to replace *A. eriopoda*, the eucalypts become more varied and eventually settle to dominance of *E. tectifica* with a wattle understorey.

FIG. 5.5

North of Pender Bay the dominant eucalypt becomes *E. miniata* (woollybutt) recognisable by its black stocking. Taller grasses such as *Chrysopogon fallax* (ribbon grass) join *Plectrachne* in the ground layer, while *Acacia platycarpa* and *Grevillea pyramidalis* are common shrubs. This assemblage is found also on the sandy plains to the northeast of Derby.

FIG. 5.6

Along the west coast, principally behind the 80-Mile Beach and Roebuck Bay, there are swampy clay plains lying little above sea level. Fronting the sea these are bordered by dunes or mangroves and bare mud flats, and inland of these a samphire zone of *Halosarcia* spp. Most of these flats consist however of saline short-grass plains, with *Sporobolus virginicus* forming a dense grassland 15–30 cm tall. Other plants are occasional samphire, *Sclerolaena*, and *Eragrostis falcata*.

FIG. 5.7

Seasonally swampy bottomlands representing former river channels are seen here and there in the Dampier Peninsula and have a scattered growth of paperbark trees *(Melaleuca nervosa)* over a ribbon grass-bluegrass-white grass ground layer *(Chrysopogon-Dichanthium-Sehima)*.

FIG. 5.8

The Lacepede Islands are a low sandy group off the west coast of Dampierland which are a prime breeding ground for seabirds. Here brown boobies are nesting among the thin vegetation of *Spinifex longifolius* (large clumps) and the trailing vine *Ipomoea pes-caprae*. Note booby chick centre, whose mother has been disturbed by the photographer. Such unguarded chicks quickly fall prey to gulls. The genus *Spinifex* which occurs on coastal dunes is not to be confused with the hummock grasses popularly known as spinifex which belong to the genera *Triodia* and *Plectrachne*.

FIG. 5.9

In the eastern part of Dampierland, river flood plains become important. Termite mounds are often a conspicuous feature. In this example near Fitzroy Crossing, scattered trees of *Eucalyptus papuana* (ghost gum) are associated with grasses of the 'frontage grass' type, i.e. *Dichanthium* spp. (blue grass), *Chrysopogon* spp. (ribbon grass) with *Sorghum plumosum*, *Themeda australis*, *Bothriochloa bladhii* and *Heteropogon contortus*.

covers the foredunes. In the south *Triodia pungens* covers the inner dunes but in the north higher rainfall brings more mixed communities of *Acacia, Terminalia, Santalum* and *Ficus* with herbaceous species such as *Ptilotus exaltatus, Salsola kali, Crotalaria cunninghamii* and *Euphorbia* spp. On the Dampier Peninsula the leeward side of the dunes is characterised by thickets of *Acacia ampliceps*, and an exciting recent discovery[3] has been that these include distinct pockets of mixed vine thickets. They are best developed at the north end of the peninsula but extend to the outskirts of Broome where typical components are *Terminalia petiolaris, Grewia breviflora, Pouteria sericea* and *Celtis philippinensis* mixed with pindan species and dune acacias. The flora is richer further north and is fully listed in (3).

Further inland, Dampierland includes the spreading plains along the Fitzroy and Lennard Rivers. To a large extent these plains are sandy and covered with pindan which is generally similar to the pindan of the Dampier Peninsula. A novel and conspicuous element however, usually in lower-lying parts, is the Australian baobab or boab as it is called locally, *Adansonia gregorii*. Other species of *Adansonia* occur in Africa and Madagascar. The Australian one is found in the Kimberley and in the Northern Territory on the Victoria River. It does not extend to Queensland, where the somewhat similar Queensland bottle tree is a kurrajong, *Brachychiton rupestris*. Baobabs are huge trees with a vastly swollen trunk consisting of spongy water storage tissue. Their great size does not imply great age but it does

FIG. 5.10

The baobab or boab tree *(Adansonia gregorii)* is often a conspicuous feature of the river plains owing to its striking shape and enormous size. The huge swollen trunk consists of spongy water-storage tissue, and when a tree dies it collapses into a rotting mass of pulp. There are numerous species of *Adansonia* in Africa and Madagascar but only this one in Australia which is localised in the Kimberley and on the Victoria River in the Northern Territory, not extending to Queensland. It frequently associates with beefwood *(Grevillea striata)* seen here on right and with bauhinia *(Lysiphyllum cunninghamii)*, and conspicuous termite mounds. The ground layer is ribbon grass.

FIG. 5.11

It is frequently supposed from the huge size of baobab trees that they must be very old, but actually this is not the case as they quickly develop their swollen trunk and young ones are often seen. In this photograph a range of sizes is seen, some obviously quite young. Studies of African baobabs have shown that having attained a mature size they fluctuate with available moisture, and may actually shrink in diameter during a drought period!

make them very conspicuous. A typical community often seen on alluvial flats is an association of baobab with bauhinia *(Lysiphyllum cunninghamii)* and beefwood *(Grevillea striata)* with a ribbon-grass *(Chrysopogon)* ground layer.

Near the rivers there are floodplains of heavier alluvial soil — grey and brown cracking clays — which may be 22 km wide, and these are generally grassland with sparse trees except for fringing woodland along the riverbanks. The latter consists typically of *Eucalyptus camaldulensis* with *Terminalia platyphylla* but may be associated with *Ficus coronulata*, *F. racemosa*, *Lophostemon grandiflorus*, *Adansonia gregorii* and the 'Leichhardt pine' *Nauclea orientalis*. Many smaller trees such as *Melaleuca* spp., *Lysiphyllum cunninghamii*, *Acacia* spp., *Brachychiton* spp., *Planchonia careya* and *Pandanus* spp. assist in forming a more or less dense fringe to the channel. The ground layer is characterised by patches of perennial 'frontage grasses', *Dichanthium* spp. and *Chrysopogon* spp. On levee crests and backslopes, the trees change to *Eucalyptus papuana* or *E. microtheca* over these same grasses.

On the back plains with their 'black soil' there are essentially two communities, Mitchell grass and ribbon grass-blue grass. In Mitchell grass plains tussocky perennial grasses 60-120 cm tall form a moderately dense grass layer distinguished by *Astrebla squarrosa*, *A. pectinata* and *A. elymoides*, 'bull', 'barley' and 'weeping' Mitchell grasses. Ribbon grasses *Chrysopogon* spp., whitegrass *Sehima nervosum* and feathertop wiregrass *Aristida latifolia* are commonly also present. The community is further

FIG. 5.12

Back plains away from the main river channels may be treeless or almost so, perhaps with a few scattered *Eucalyptus microtheca* (coolabah) trees. The grasses of such plains may be of the blue grass-ribbon grass type, or of Mitchell grass. In the latter case the three Mitchell grass species *Astrebla squarrosa*, *A. pectinata* and *A. elymoides* form a moderately dense grass layer 60–120 cm tall with some ribbon grass, white grass and wiregrass (*Aristida latifolia*).

FIG. 5.13

The Napier Hills extend in a narrow band across the northeast of the plains of Dampierland and contain the fossilised remains of a barrier reef similar to today's Great Barrier Reef of Queensland but of Devonian age, 370 million years old. The reef limestones contain excellent fossils of Devonian armoured fishes. Where exposed, the reef limestones rise like a wall from the plain, winding across it over a length of 300 km. The reef may be only 1–2 km wide and traversed by some spectacular gorges.

FIG. 5.14

Much bare rock is exposed in the reef limestones of the Napier Hills, carved into fantastic shapes by solution-weathering where the limestone is slowly dissolved by the acid in rain water. Vegetation is sparse, consisting of scattered trees including baobabs and *Eucalyptus dichromophloia* and the spinifex *Triodia wiseana*.

FIG. 5.15

One of the many spectacular gorges cut through the reef limestones, Brooking Gorge near Fitzroy Crossing, is well known. Permanent streams in the gorges are well lined with river gums, paperbarks, wild plum *(Terminalia platyphylla)* and Leichhardt pine *(Nauclea orientalis)*.

FIG. 5.16

Any plateau on top of the reef limestones is apt to present a weird landscape of eerie shapes of weathered rock. This site above Geikies Gorge suggests an ancient Buddhist temple, or the Ming tombs of China! Vegetation is a few scattered *Ficus* trees and the spinifex *Triodia wiseana*.

FIG. 5.17

An interesting small tree on the limestone ranges is this red-flowered *Sterculia viscidula*.

FIG. 5.18

Flowers of *Sterculia viscidula*.

FIG. 5.19
One of the most beautiful of the gorges, Windjana Gorge, at sunset.

enriched by annuals and herbs including a number of leguminous species of *Flemingia, Rhynchosia* and *Neptunia*. Sparse low trees of *Lysiphyllum cunninghamii, Eucalyptus microtheca* and *Acacia suberosa* and some shrubs dot the grassland. The ribbon grass-blue grass community is structurally and floristically similar, with the same scattered trees and shrubs, but the Mitchell grasses are missing and replaced by a rich assemblage in which *Chrysopogon* spp. and *Dichanthium fecundum* (blue grass) are dominant.

The Fitzroy-Lennard plains are bounded on the north by the Napier Hills, an extraordinary feature which represents a former barrier reef similar to the Great Barrier Reef of Queensland today, except that this one is of Devonian age, about 370 million years old! It seems that during the Devonian the Canning Basin to the south was submerged but the highland area of the present Kimberley to the north was dry land, and a typical barrier reef 300 km long was formed. It was subsequently buried by other sediments which preserved it, and in geologically more recent time has been 'exhumed', exposed by erosion of these sediments on uplift of the area above sea level. The reef limestones are noteworthy for a wonderful assortment of fossils of armoured fishes of the Devonian period. The topography can be very striking where these reef limestones are exposed. On both sides of the formation limestone walls rise abruptly from the plains, cut by narrow gorges. Wandering off across the plains on both sides into the far distance the cliffs appear like the Great Wall of China. The well known Geikie Gorge has been formed by the Fitzroy River cutting through the formation and there are others, e.g. the Brooking and Windjana Gorges. One creek east of Windjana Gorge passes through the limestone in a tunnel.

On the summits the limestone is normally devoid of soil and is sparsely vegetated, but where the formation is broad, cracking clay plains have developed. In the former case scattered hummocks of *Triodia wiseana* perch on rocks and ledges, with some other sparse short grasses and herbs such as *Achyranthes aspera, Cleome viscosa, Indigofera linifolia* and *Portulaca* sp. Irregularly scattered baobabs as trees 5-10 m in height occur rooted in crevices with a number of shrubs or small trees such as *Sterculia viscidula*, *Celtis philippinensis*, *Cochlospermum fraseri*, *Ficus opposita*, *Terminalia* sp. and *Wrightia* sp. The baobabs drop out towards the southeast and are no longer present at Geikie Gorge, where *Ficus* is the principal woody plant. 'Black soil' plains, either as foot plains to the ranges or high plains upon them are covered with Mitchell grass and ribbon grass-blue grass communities as previously described.

5.2 *East Kimberley Region*
HALL BOTANICAL DISTRICT

Hummock grassland of *Triodia* spp. with sparse trees on ranges; short grass of *Enneapogon* spp. on dry calcareous plains, medium-height grasses *Astrebla* and *Dichanthium* on cracking clays.

Climate: Semi-arid to dry hot tropical, precipitation 350-500 mm, summer wet season 2-4 months.

Geology: Very varied Archaean, Proterozoic and Phanerozoic rocks both acid and basic.

Topography and soils: Harder siliceous rocks form abrupt parallel ranges with shallow stony sand and loam soils; volcanics and limestones form extensive plains with neutral red earths and red loams or grey and brown cracking clays.

Area: 50 510 km².

Boundaries: Drawn according to vegetation changes as noted above in bounding with the other districts. The southern boundary is the edge of desert sandplains and dunefields.

The botanical district is named after Charles Hall whose name survives in Halls Creek where he discovered gold in 1884. The region extends round the southern and eastern flanks of the Kimberley highlands and has more of the character of a desert fringe than Dampierland. Seventy per cent of the region is covered by spinifex grasslands which replace the pindan on sandplains and

FIG. 5.20

In the south, the East Kimberley Region forms a desert fringe. From this vantage point in the Shore Range southeast of Fitzroy Crossing we look south into the empty wastes of the Great Sandy Desert. Bare rocky hills are covered with little but spinifex, while the flats have some scattered trees, principally *Lysiphyllum cunninghamii* and *Ventilago viminalis* in short grass plains.

also cover the numerous rocky ranges. However bunch-grass savannas make up most of the other 30% on plains with heavier soils, both short grass and medium height grass types being present.

Along the Great Northern Highway the region is entered just east of Fitzroy Crossing and traversed as far as Louisa Downs, after which the highway lies just within the Central Kimberley Region with the regional boundary running parallel and a few kilometres to the east. The township of Halls Creek is just on the boundary. The road south to Billiluna and the Duncan Highway through Nicholson both traverse the region. Shortly beyond the Bow River on the Great Northern Highway and Spring Creek on the Duncan, the North Kimberley Region is entered as the plant cover responds to increasing rainfall nearer the coast.

The East Kimberley Region is considerably varied in geology, topography and soils so that the vegetation is correspondingly varied within the general limits of spinifex and other dry grasslands. East of Fitzroy Crossing the continuation of the Napier Hills carries similar vegetation to that further west where the outcrop is of limestone, but much of it here is of conglomerate forming bare rounded hills with a general cover of *Triodia intermedia* but few trees or shrubs. Further south and east a complex of sandstone ranges and plateaux alternates with river plains. In the former case in rocky areas there are sparse trees of *Eucalyptus dichromophloia* and *E. setosa* standing over *Triodia intermedia* and *T. pungens*, in more rolling country *Eucalyptus brevifolia* with *T. intermedia*. Sandy valleys contain short-grass plains of *Aristida browniana* with *A. hygrometrica*, *Eriachne obtusa*, *Perotis rara* and *Enneapogon polyphyllus*. Scattered low trees are principally *Lysiphyllum cunninghamii* and *Ventilago viminalis*. Plains of red earth are covered in general with *Triodia pungens* and a variety of tree and shrub communities. One of these which is widespread is the *Acacia pyrifolia-Grevillea pyramidalis-Hakea suberea* association as found in the Pilbara Region. On the river plains of Christmas Creek the ribbon grass-blue grass community with *Eucalyptus microtheca* reappears. On Bohemia Downs and to the east for 100 km there is much dissected laterite country with scattered low trees of *Eucalyptus brevifolia*. *Triodia pungens* provides the ground cover on sandy areas, *T. intermedia* on stripped laterite or stony surfaces. There is a range of shrubs, mainly of *Acacia* and *Grevillea*.

Running through the township of Halls Creek with a strike of SW to NE are the Halls Creek Ridges, a belt of tightly folded rocks with a rough, hilly relief, much outcropping rock and little soil.

The general vegetation consists of sparse tree-steppe. There are scattered stunted trees, mainly *Eucalyptus brevifolia*, though *E. dichromophloia* occurs in variable proportions on sedimentary rocks and *E. terminalis* on basalt. The ground cover is of hummock grasses varying again according to rock type: *Triodia intermedia* and *T. inutilis* are associated with sedimentary rocks, *T. wiseana* and some *T. intermedia* with basalt. The tall grass *Aristida inaequiglumis* is very common and there are shorter grasses including *Enneapogon* sp. and *Tragus australianus* growing between the spinifex hummocks. *Plectrachne schinzii* appears in sandy patches. Scattered shrubs belong to numerous species including *Acacia bivenosa*, *A. holosericea*, *A. monticola*, *A. lysiphloia*, *Cassia chatelainiana*, *Grevillea dimidiata*, *G. pyramidalis*, *G. wickhamii*, *Hakea arborescens*. The subshrubs *Anisomeles malabaricum* and *Ptilotus calostachyus* were noted, and the forbs *Trachymene villosa* and *Trichodesma zeylanicum*. *Eucalyptus setosa* appears in sandy patches and *E. grandifolia* on drainage.

The much narrower parallel Albert Edward Range on the east is similar but with *Triodia pungens* replacing *T. intermedia* and *T. inutilis*. The Osmond Range again, further to the northeast is similar to this but influenced by higher rainfall so that *Plectrachne pungens* joins the *Triodia*. On the lower gentle slopes and drainage lines there is low-tree savanna of *Eucalyptus terminalis*, *E. argillacea* and *E. confertiflora* with a ground layer of *Aristida pruinosa*, *A. browniana* and *Chrysopogon fallax*.

East and south of Halls Creek beyond the 30 km wide Halls Creek Ridges lies the Sturt Plateau, a flat elevated plain drained to the south by the Sturt and Wolf Creek systems. The principal underlying rock is basalt which produces 'black' cracking clay soils and treeless grass plains, but elsewhere there is a laterite crust which has been partly dissected to form mesas but mainly underlies extensive desert sandplains.

The open grasslands which cover the Denison Plains and are such a conspicuous feature, are dominated by barley Mitchell grass, *Astrebla pectinata*. This is a perennial bunch-grass 30-75 cm high and 20-30 cm in diameter. The tussocks are 15-90 cm apart. The few other perennial bunch-grasses such as *Astrebla squarrosa*, *A. elymoides*, *Dichanthium fecundum*, *Aristida latifolia* and *Chrysopogon fallax* are scattered and unimportant. *Themeda avenacea* grows in small depressions. In low rainfall years and towards the end of the dry season the

FIG. 5.21

Slightly further north as rainfall increases, hard lateritic plains are covered with *Eucalyptus brevifolia* (snappy gum) forming a low savanna woodland over a ground layer of *Triodia intermedia*.

FIG. 5.22

Running through the township of Halls Creek with a strike of SW to NE are the Halls Creek ridges, a 30 km wide belt of tightly folded rocks with a rough, hilly relief, much outcropping rock and little soil. The general cover is of stunted eucalypts, *E. dichromophloia*, *E. brevifolia*, *E. terminalis*, and spinifex of various species according to rock type, interspersed with short bunch grasses.

FIG. 5.23

Palm Springs is a beauty spot near Halls Creek where there are permanent water holes lined with cadjuput trees and ancient cycad palms. The taller fan-palm *Livistona loriphylla* also occurs, although as at so many other 'palm springs' in the area these have been cut out in the past for house construction.

FIG. 5.24

Large termite mounds are a feature of the spinifex plains, sometimes reaching enormous sizes. Termites play an important role in these ecosystems by contributing to the recycling of nutrients otherwise locked up in dead wood and other plant material.

FIG. 5.25

East of the Halls Creek ridges lies the Sturt Plateau, a flat elevated plain drained to the south by the Sturt and Wolf Creek systems. The principal underlying rock is basalt which produces 'black' cracking clay soils and treeless grass plains known as the Denison Plains. These stretch over the border into the Northern Territory, and are normally treeless except on drainage where the creeks are lined with *Eucalyptus microtheca*, *E. tectifica*, *Lysiphyllum cunninghamii* and *Terminalia arostrata*.

FIG. 5.26

The Denison Plains are composed typically of the Mitchell grasses *Astrebla pectinata*, *A. squarrosa* and *A. elymoides* interspersed with blue grass, wire grass and ribbon grass. Native legumes, especially the *Neptunia* sp. seen here as the darker green, are important components for grazing.

spaces between the perennial bunch-grasses are almost bare but with high rainfall *Sorghum* spp. are prominent in patches, especially where water lies for several weeks. Under average to good rainfall the spaces are almost completely covered by short grasses and forbs 8-30 cm high, including *Iseilema* spp., *Echinochloa colona*, *Eragrostis japonica*, *Brachyachne convergens*, *Malvastrum americanum*, *Neptunia* spp., *Crotalaria medicaginea*, *Sida fibulifera*, *S. spinosa*, *Alysicarpus rugosus*, *Rhynchosia minima* and many others. It will be noted that there is a number of native legumes, which are an important element.

The plains are virtually treeless except on drainage, where the creeks are lined with low, open savanna woodland of *Eucalyptus microtheca*, *E. tectifica*, *Lysiphyllum cunninghamii*, *Terminalia arostrata*, with a grass layer of *Themeda avenacea* and *Eulalia fulva*. Shrubs such as *Acacia stenophylla* and *Eremophila bignoniiflora* are also present. Along the main course of the Sturt Creek this community occupies linear tracts up to 1 km wide with an intense braided pattern of small stream channels. Downstream, where the Sturt Creek is flowing through the desert sandplains, the savanna woodland thins mainly to *Eucalyptus microtheca*, and there are numerous low-lying distributary zones which have a vegetation of chenopodiaceous succulent steppe. *Chenopodium auricomum* and *Muehlenbeckia cunninghamii* are the principal species.

The red soil plains are contrastingly vegetated with tree-steppe of low trees, spinifex and bunch-grasses. *Eucalyptus brevifolia* is the principal species, reaching

FIG. 5.27

Red soil plains on the Sturt Plateau are much more poorly vegetated: a scattered ground cover of spinifex and short grasses (Flinders grass, wire grass), and occasional trees of *Terminalia* or *Eucalyptus*.

FIG. 5.28

Heading north from Nicholson on the Duncan Highway, a belt of limestone country is crossed. This is very stony and inhospitable, covered with the spinifex *Triodia wiseana* and very sparse trees and shrubs.

FIG. 5.29

The limestone country is succeeded by plains developed on softer limestone and shale which are largely treeless, covered with short bunch-grass savanna of *Enneapogon* spp. (nineawn grass) and other grasses and herbs. Trees are *Terminalia canescens* and *Eucalyptus pruinosa*.

FIG. 5.30

A shrub attractive when in flower is the caustic bush, *Grevillea dimidiata*. The bush itself is usually straggly, affected by the frequent grass fires.

FIG. 5.31

Kapok bush, *Aerva javanica*, is a soft shrub or weed introduced in the Kimberley in the 1880s. It has been used widely with success, as shown here, for the revegetation of denuded pastoral country in the Ord River catchment.

6-8 m in height with *E. dichromophloia* more rarely. Numerous shrubs include *Acacia monticola*, *A. tenuissima*, *A. pachycarpa*, *A. tumida*, *Cassia* spp., *Dolichandrone heterophylla*, *Grevillea pyramidalis*, *G. wickhamii*, *Gossypium australe*. *Triodia pungens* forms an open hummock grassland, with numerous short grasses (*Iseilema* spp.), a taller grass (? *Aristida inaequiglumis*) and forbs filling the spaces between the hummocks. The latter include *Gomphrena brachystylis*, *G. canescens*, *Ptilotus calostachyus*, *P. exaltatus*, *P. fusiformis* and *Trichodesma zeylanicum*.

After passing through Nicholson on the Duncan Highway one descends from the Sturt Plateau to traverse the Ord Plains where the Ord River has formed widespread low lying plains and low hilly tracts on the relatively soft rocks. The plains are about 80 m above sea level in the north and rise gradually south and east to about 300 m to merge with the Sturt Plateau. The plant cover is primarily controlled by the underlying rocks which may be basalt, limestone, shale or sandstone from Cambrian to Devonian in age. After leaving Nicholson one enters basalt country, a jumble of small rocky, stony hills and valleys with better soil. On the former, scattered *Eucalyptus brevifolia* make up the tree layer with some *E. terminalis* and *Terminalia* spp., standing over a cover of spinifex of *Triodia intermedia* and *T. pungens* mixed with tall grasses, e.g. *Aristida* spp., *Chrysopogon fallax*, and herbaceous species between the hummocks. These include *Jacquemontia browniana*, *Pterocaulon glandulosum*, *Ptilotus spicatus*, *Trichodesma zeylanicum* and *Wedelia asperrima*. Scattered shrubs include

FIG. 5.32

On the route north by the main highway a point of attraction is Pompey's Pillar, a boulder-capped tor east of the road. The surrounding vegetation is low-tree savanna of *Eucalyptus brevifolia* and *E. pruinosa*, the latter conspicuous from its silvery leaves as the name implies.

FIG. 5.33

In the East Kimberley 'pindan' means something different from Dampierland. It is still the vegetation of sandy ground, but in this case it is a much more open community of typically sand-loving plants. The characteristic dominant is the small tree *Grevillea pteridifolia* with its light foliage and conspicuous orange-yellow flowers.

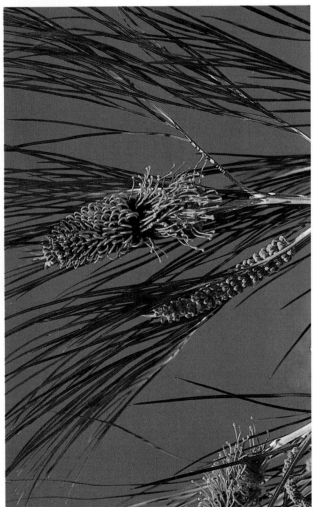

FIG. 5.34

Flowers and foliage of *Grevillea pteridifolia*. The specific name implies that the leaves are fern-like which is hardly the case except for a superficial impression at a distance. They are deeply divided into very narrow lobes. The inflorescence makes this one of the most striking of grevilleas.

FIG. 5.35
Another pindan plant found in the northeast Kimberley is the northern morrison or feather flower, *Verticordia cunninghamii*. *Verticordia* is a genus endemic to Western Australia and generally considered typically southwestern where the numerous species are conspicuous in kwongan. However this species and another, *V. verticillata*, are found in the Kimberley.

Tom Muir

FIG. 5.36
Three species of star flowers, *Calytrix*, are found in the Kimberley in pindan or on sandstone. *Calytrix exstipulata*, known as Kimberley heather, has pink flowers and is a common pindan element.

Cassia venusta, *Grevillea dimidiata*, *G. pyramidalis*, *G. wickhamii* and *Hakea lorea*. On the other hand, on the lower slopes and flats in this country the vegetation is quite different, the trees being *Terminalia* spp. and *E. argillacea*, while the ground layer consists of short grasses less than 30 cm tall. The commonest grasses are *Enneapogon* spp., *Aristida contorta*, *Sporobolus australasicus*, *Tragus australianus* and *Oxychloris scariosa*, with some herbaceous species such as *Cleome viscosa*, *Portulaca oleracea* and *Sida fibulifera*.

The succeeding formation is a belt of Headley's Limestone, extremely barren country consisting of steep rocky hills with little soil, covered with the spinifex *Triodia wiseana* and very sparse trees and shrubs. This is followed by plains developed on softer limestone and shale which are almost completely treeless and under short bunch-grass savanna of the *Enneapogon* community detailed above. On crests and low scarps *E. terminalis* and *E. argillacea* appear with some *Triodia wiseana*. Down slope the community grades into Mitchell grass plains on cracking clay. Further north still towards Lake Argyle there is more basalt country, this time — with higher rainfall — under *Eucalyptus terminalis* and *E. argillacea* with a different grass association known as the tippera tall grass type. *Themeda australis* (kangaroo grass) and *Sehima nervosum* (white grass) are generally dominant with *Sorghum plumosum*, *Chrysopogon fallax*, *Aristida pruinosa*, *Heteropogon contortus* and other species.

'Black soil' plains on brown cracking clays tend to occur on all the flattest low lying ground where there is a calcareous influence, and these again are largely treeless with a cover of Mitchell and other grasses. The few sparse low trees consist of *Terminalia arostrata* and *T. volucris*. In the ground layer the grasses are spaced 30-90 cm apart and reach 30-75 cm in height. The dominants are the three Mitchell grasses with *Dichanthium fecundum* and *D. annulatum*. Other tall grasses include *Aristida latifolia*, *Chrysopogon fallax*, *Panicum decompositum* and *P. whitei*. Short grasses comprise *Iseilema* spp. The numerous herbs include a number of native legumes; important species include *Boerhavia ? coccinea*, *Crotalaria medicaginea*, *Abelmoschus ficulneus*, *Ipomoea* spp., *Flemingia pauciflora*, *Neptunia* spp., *Ptilotus spicatus*, *Rhynchosia minima*, *Sida fibulifera*, *S. spinosa* and *Trichodesma zeylanicum*. This community, described above in its pristine state, had been substantially degraded by faulty grazing practices prior to the construction of the Ord River dam when the State government resumed the cattle stations. Rehabilitation work has been in progress.

5.3 *Central Kimberley Region*

FITZGERALD BOTANICAL DISTRICT

Curly spinifex *(Plectrachne pungens)* with low trees of *E. phoenicea-E. ferruginea* or *E. brevifolia-E. dichromophloia* on sandstones, ribbon grass *(Chrysopogon)* with *E. tectifica* on basalt.

Climate: Dry hot tropical; precipitation 400-800 mm per annum, summer wet season of four months.

Geology: Siliceous sedimentary and basic volcanic rocks of Proterozoic age, overlying locally exposed basement of Archaean granite, acid volcanics and siltstones.

Topography and soils: Hilly to mountainous country with the harder siliceous rocks forming parallel ranges; volcanics in the valleys.

Area: 83 330 km².

Boundaries: Boundaries were drawn to encompass the curly spinifex communities. The northern boundary is the transition to high-grass savanna, the southern follows the boundary of Precambrian rocks from King Sound to the Margaret River, thereafter the vegetation boundary between curly spinifex and semi-desert spinifex steppe.

The name of the botanical district commemorates W.V. Fitzgerald for his pioneer botanical exploration of the area in 1905-06. Together with the North Kimberley Region, the Central forms part of the highland area of the Kimberley whereas Dampierland and East Kimberley are situated on the plains and lowlands peripheral to this massif. The highland area comprises the Kimberley Plateau and the Kimberley Foreland. The former is a broad dissected plateau formed upon gently dipping Proterozoic sedimentary and volcanic rocks whereas the latter is a marginal belt on the south and east where the rocks become intensely folded and faulted and dip steeply, eroded into high, almost vertical scarps facing outward from the plateau. The King Leopold and Durack Ranges are the innermost and highest of the ridges so formed, reaching heights of 950 m at Mt Broome and Mt Ord. To the south and east the country is very broken, progressively lower, and exposes more ancient rocks lying beneath those of the Kimberley Plateau.

Since the boundaries of the Central Kimberley Region are determined for biogeographical purposes by vegetation, which is influenced but not entirely controlled by the geology, the region does not altogether conform with the above geomorphic units. It in fact includes the whole of the Kimberley Foreland but only the southeastern part of the Kimberley Plateau. Changes in vegetation in response to higher rainfall place the northern and western parts of the plateau within the North Kimberley Region.

The characteristic feature of the Central Region is the occurrence of 'curly spinifex' *(Plectrachne pungens)* communities which cover 70% of the area. Spinifex grasses belong to two genera, *Triodia* and *Plectrachne*, and while all of the species are hard, wiry, perennial and drought-resisting, not all of them adopt the hummock-grass form. Some are bunch-grasses and not therefore strictly spinifex. *Plectrachne bromoides* and *P. danthonioides* show by their names (from *Bromus* and *Danthonia*, two grasses) that they belong to this category whereas *P. rigidissima* is obviously a hummock-grass. *P. pungens* is in a sense intermediate. Under adverse conditions it may become more hummocky and prickly, under better conditions more lax and luxuriant. This has inspired the local name curly spinifex given by the pastoralists, and it is perhaps best regarded as forming a special category somewhere between bunch and hummock grassland.

The Central Kimberley is not well provided with access roads. The main highway skirts it, passing just inside the regional boundary from Louisa Downs to Bow River. The Gibb River road crosses the King Leopold Range at Inglis Gap, and beyond Mt Barnett also travels close to the regional boundary. For the rest only station

FIG. 5.37
The Central Kimberley Region covers a belt of broken country known as the Kimberley Foreland, and part of the Kimberley Plateau lying behind, the latter consisting of dissected plateaus and scarps. The typical cover, shown here, is of curly spinifex, *Plectrachne pungens*, a hard wiry grass which is perennial and drought-resisting but does not form hummocks like more typical species of spinifex.

tracks are available and in the most rugged country not even these. Much of the region is devoid of access by wheeled vehicles.

Where the Gibb River road enters the foothill country it consists of bouldery tors and whaleback hills of granite separated by narrow belts of alluvium and sandy pediments, and rocky hills of stratified siltstone dipping at high angles. This country is mapped as low-tree savanna of the *Eucalyptus brevifolia-E. perfoliata* community with a curly-spinifex ground layer, as this is the most important component, occurring characteristically on rocky hill slopes of granite, gneiss and schist. A great deal of the country with outcropping granite consists however of bare rock, vegetated only in cracks, crevices and pockets of soil, with scattered trees and shrubs and the hummock grass, *Plectrachne bynoei*, creating a unique and interesting community. Scattered trees of *Eucalyptus perfoliata*, *E. dichromophloia* and *Adansonia gregorii*, less prominently *Gyrocarpus americanus*,

Lysiphyllum cunninghamii and *Petalostigma pubescens*, grow where they can get a place on their rocky habitat. The sporadic shrub layer contains also a rich and distinctive assortment of species, including *Sterculia viscidula*, *Buchanania obovata*, *Cochlospermum fraseri*, *Gardenia resinosa*, *Ficus* spp., *Terminalia ferdinandiana*. The ground-layer species *Plectrachne bynoei*, which is restricted to this habitat, is a coarse tussocky perennial with long, resinous, shiny green blades. It forms hummocks up to 1.5 high when in flower and up to 75 cm in diameter. Associated tall grasses such as *Cymbopogon procerus*, annual *Sorghum* and *Heteropogon contortus* grow in the cracks and crevices wherever there are small pockets of soil. Short grasses and forbs such as *Enneapogon polyphyllus*, *Eriachne ciliata*, *E. mucronata*, *Fimbristylis dichotoma* and *Portulaca filifolia*, are sparse and patchy.

This granite country only extends as far east as the Fitzroy River, beyond which the foothills country of the Kimberley Foreland is essentially similar to the

Fig. 5.38

The frontal scarp of the King Leopold Range seen from Inglis Gap. This consists of quartzite overlying granite and is part of the Kimberley Foreland. The cover here is low-tree savanna with *Terminalia canescens* (grey foliage) and *Eucalyptus brevifolia* over a ground layer of curly spinifex and taller annual sorghum.

Fig. 5.39

Flowers of kapok tree, *Cochlospermum fraseri*. The bush is deciduous, losing its leaves in winter.

Fig. 5.40

An attractive small tree common on rocky slopes and hillsides is the kapok tree, *Cochlospermum fraseri*. This is not the kapok of commerce, but is related to it and has large kapok-like fruits which split to release numerous seeds on silky parachutes. The green bark is very fibrous and can be used to make rope.

Fig. 5.41

Hibiscus spinulosus, one of about 10 species of hibiscus in the Kimberley. Most of them are soft-leaved shrubs which appear occasionally, often on disturbed ground.

King Leopold and Durack Ranges. The basic pattern is an alternation of low-tree curly spinifex savanna dominated by *Eucalyptus brevifolia* (snappy gum) on the sandstone ridges with savanna woodland of *E. tectifica* (grey box) in the valleys, which tend to be excavated in basalt country and have quite different soils. *E. brevifolia* as a stunted tree 3 m to 6 m high characterises a range of low, open, grassy woodland communities.

Except for small areas, where *E. pruinosa* is co-dominant, *E. brevifolia* tends to form pure stands in the drier parts of its range, but in the wetter parts several other eucalypts occur in the tree layer. Similarly a shrub layer of *Grevillea striata*, *Acacia* spp., *Cochlospermum fraseri* and *Carissa lanceolata* is important in the wetter areas but thins out in the drier. *Eucalyptus brevifolia* may occur over a variety of ground layers, curly spinifex being normal in the wetter areas.

In these ranges *E. brevifolia* may occur as sole species or associated with *E. perfoliata*, *E. dichromophloia* and *E. collina*. *E. perfoliata* occurs mostly as a shrub or small tree, but the other two species are trees of equivalent size to *E. brevifolia*. *Callitris intratropica* occurs sporadically, usually among rocks, where it is fire-protected. At Mount Bell in the King Leopold Range the writer noted a rich shrub-small tree flora including *Sterculia viscidula*, *Calytrix* sp., *Gardenia resinosa*, *Grevillea refracta*, *G. heliosperma*, *G. pyramidalis*, *G. wickhamii*, *Eucalyptus mooreana* and *Terminalia* spp. Small plants included *Acacia lycopodifolia* and *Gomphrena canescens*.

The ground layer consists predominantly of *Plectrachne pungens* in its curly-spinifex form. Patches of annual *Sorghum* are normally associated with the presence of *Eucalyptus dichromophloia* and *E. collina*.

On many of the gentler slopes and valley floors *Eucalyptus brevifolia* occurs with short-grass savanna, < 50 cm tall, of *Enneapogon* spp. with some *Aristida*, *Chloris*, *Sporobolus* and *Fimbristylis* spp. Tall perennials are scattered or absent, but there may be some admixture of curly spinifex. This community becomes more prevalent towards the eastern end of the unit, where it is mapped.

Most of the broader and lower valley floors, however, are under *Eucalyptus tectifica-E. grandifolia* savanna woodland. In response to the lower rainfall the ground layer is of the ribbon grass type rather than white grass, i.e. of *Chysopogon latifolius* and *C. fallax*, with *Aristida latifolia*, *A. inaequiglumis*, *Sehima nervosum*, annual *Sorghum* and *Plectrachne pungens*. The tree layer includes *Eucalyptus dichromophloia*, *E. polycarpa*, *Adansonia gregorii* and *Lysiphyllum cunninghamii* sparsely present in places. The shrub layer is moderately dense, but patchy.

On the Kimberley Plateau there are basically two major types of country, recognised by Charles Gardner in the early 1920s when he wrote of the 'sandstone savanna' and the 'basalt savanna'. Sandstone beds being harder and more resistant form relatively high plateau surfaces, prominent mesas and escarpments with thin siliceous soils. The more erodible basalt forms extensive plains and low hills with richer and deeper soils. These two types are distinguished in this region as the Karunjie Plateau and Gibb Hills surfaces respectively, with the addition of a minor unit, the Glenroy Plains, developed on shale. Typical 'sandstone savanna' on the Karunjie Plateau is dominated by *Eucalyptus phoenicea* (a very ornamental species with yellow flowers) and *E. ferruginea* and classed as curly spinifex low tree savanna woodland. The upper tree layer is typically 8-10 m tall, the trees of poor form, gnarled and twisted with trunks seldom more than 30-50 cm in diameter. *Callitris intratropica* occurs sparsely. There is a shrub layer, generally sparse, but in some areas *Acacia* spp. develop into thickets reminiscent of pindan. Principal components are *Acacia sericata* and other spp., *Gardenia* spp., *Grevillea agrifolia*, *G. cunninghamii*, *G. pteridifolia*, *Jacksonia thesiodes*, *Petalostigma pubescens*, *Planchonia careya* and *Ventilago viminalis*.

The curly-spinifex ground layer is characteristic of so many communities in the Fitzgerald District, and this is the dominant species and grows in straggly tussocks up to 100 cm wide, with vegetative tillers usually less than 60 cm high, and flowering culms attaining 100 cm. Commonly in rugged habitats the community is seasonally characterised by patchy, dense stands of annual *Sorghum* up to 2 m tall. Associated perennials, though rare, include *Aristida hygrometrica* in the north, merging to *A. pruinosa* and *A. inaequiglumis* in the south. The tussock interspaces, usually bare in the dry season, carry in favourable periods a sparse cover of annuals and ephemerals such as *Brachyachne convergens*, *Enneapogon* spp., *Fimbristylis dichotoma*, *Polycarpaea* spp. and *Gomphrena* spp. The associated shrub layer is sparse to moderate in density and varies with the habitat.

In the southern marginal portion of the Karunjie Plateau due to lower rainfall *Eucalyptus brevifolia*, which becomes extremely widespread further south still, makes its appearance. It associates with *E. ferruginea* on broad rocky surfaces in the tableland area, occurs widely on plains and interfluve crests throughout the south, and

FIG. 5.42

Cypress pines, *Callitris intratropica*, are a feature of sandstone country but as they are tender to fire usually require some degree of natural protection. In this case, on the top of the King Leopold Range escarpment, the trees are growing between rocks where grass growth is reduced, diminishing fire intensity.

FIG. 5.43

There are basically two types of vegetation on the Kimberley Plateau—the sandstone savanna and the basalt savanna. Both of these types are savanna woodlands with tall bunch-grass understoreys, the component species varying with rock, soil and locality. In this example of the sandstone savanna from Beverley Springs station the principal tree is *Eucalyptus miniata* (woollybutt) distinguished by the black stocking on the trunk. The grass layer is formed of curly spinifex and annual sorghum.

FIG. 5.44

In other examples *Eucalyptus collina* with its silvery leaves may be dominant usually on rocky mesas as in this case on the Phillips Range. A dead stem of cypress pine, killed by fire, is seen on the right.

Tom Muir

FIG. 5.45

In the central Kimberley the most typical dominant of the sandstone savanna is *Eucalyptus phoenicea*. It is a smaller tree than the woollybutt and has flaky yellowish red bark on the trunk without the black stocking, but like it has showy orange flowers.

FIG. 5.46

A conspicuous understorey tree is the quinine tree, *Petalostigma pubescens*. The flowers are not showy, but the fruits are about 2 cm in diameter, turning orange when ripe. They have a very bitter taste and hence give their name to the tree.

associates with *E. collina* and *E. dichromophloia* on rocky mesas such as Mount Clifton. The ground layer is normally curly spinifex.

The basalt savanna is of the grey box type, high-grass savanna woodland with a white grass ground layer. Curly spinifex is not found on these soils. On hills and rocky slopes *Eucalyptus tectifica* is the principal and frequently the only taller tree species. Smaller trees found sparsely include *Cochlospermum fraseri*, *Gyrocarpus americanus*, *Adansonia gregorii*, *Terminalia canescens*, *Dolichandrone heterophylla* and *Atalaya hemiglauca*. The dominant grasses are *Sehima nervosum* (white grass) and the annual sorghums (*S. stipoideum* and *S. timorense*) and *Heteropogon contortus*, *Cymbopogon bombycinus*, *Aristida pruinosa*. There is a well-developed layer of shorter grasses including *Eriachne glauca*, *Panicum mindanaense*, *Enneapogon polyphyllus* and *Brachyachne convergens*. Forbs are abundant and include species of *Tephrosia*, *Corchorus*, *Dicliptera*, *Cleome*, *Heliotropium*, *Polycarpaea* and *Borreria*.

On lower slopes *Eucalyptus dichromophloia* joins the tree layer, and the grass layer is floristically richer. The annual sorghums are replaced by *Dichanthium fecundum* (blue grass), and associated grasses now include *Chrysopogon* spp., *Aristida inaequiglumis* and *Themeda australis*.

The cracking-clay plains carry a sparse low-tree savanna with a ribbon grass-blue grass ground layer. The scattered low trees are principally *Acacia suberosa* and *Eucalyptus microtheca*, but *Lysiphyllum cunninghamii*, *Terminalia volucris* and *Atalaya hemiglauca* can be seen, with the shrubs *Acacia farnesiana* and *Carissa lanceolata*. The dominant grasses are *Chrysopogon* spp. (ribbon grasses) and *Dichanthium fecundum*, with *Eulalia fulva*, *Sorghum plumosum*, *Dichanthium tenuiculum*, *Bothriochloa ewartiana* and *Heteropogon contortus*. There is a lower layer 30-50 cm tall of shorter perennial or annual grasses and forbs including *Iseilema* spp., *Brachyachne convergens*, *Eriachne glauca*, *Eragrostis setifolia*, *Echinochloa colona*, *Paspalidium rarum*, *Xerochloa laniflora*, *Dactyloctenium radulans*, *Polycarpaea* spp., *Ipomoea* spp., *Desmodium* spp., *Boerhavia ?coccinea*, and many others, including the succulents *Calandrinia* and *Portulaca* spp.

The Glenroy Plains extend between Mt House and Mt Clifton, developed upon shale and siltstone with a very thin soil. This lack of soil is responsible for a poor, open vegetation of a more xeromorphic character than is normal in this region, over much of the plains. About half of the area is covered by tree-steppe, scattered low trees of *Eucalyptus argillacea* (Mt House box) and *E. brevifolia* over a hummock-grass floor. According to the writer's collections, the latter comprise *Triodia pungens*, *Plectrachne pungens* and *Plectrachne* sp. inedit. (JSB 4201). A number of small bunch-grasses grow between the hummocks, of which *Eriachne sulcata* and *Eragrostis japonica* were collected. *Melaleuca minutifolia* is a common shrub reaching 2 m, rarely 3 m. Other shrubs include *Acacia monticola* locally, *A. translucens* locally abundant, *Cassia desolata*, *Grevillea pyramidalis*, *Gossypium sturtianum*, *Hibiscus panduriformis*. Numerous herbs include *Goodenia scaevolina*, *Ptilotus calostachyus* and spp. The tree-steppe covers the bulk of the interfluves between drainage, and merges in places into hard plains which are treeless and carry the hummock grassland with scattered *Lysiphyllum cunninghamii, Acacia pyrifolia, Grevillea pyramidalis* and *Hakea lorea*.

At the other extreme the steppe merges into tree savanna of the *Eucalyptus polycarpa-E. grandifolia* community. The upper tree layer 6-9 m high consists of *E. grandifolia* and *E. polycarpa*. A smaller tree layer 3-6 m containing *Terminalia canescens, Melaleuca minutifolia* and *Dolichandrone heterophylla* is commonly present. The shrub layer of *Carissa lanceolata* and *Atalaya hemiglauca* is sparse. The ground layer consists of an assortment of 'frontage grasses' including *Dichanthium* spp., *Chrysopogon* spp., *Sorghum plumosum, Themeda australis, Bothriochloa bladhii* and *Heteropogon contortus*.

FIG. 5.47

The basalt savanna is seen typically on the plains country in the Kimberley Plateau, savanna woodland where the typical tree is the grey box, *Eucalyptus tectifica*. Curly spinifex is not found on the red loam soils of these plains, where the typical constituent is white grass, *Sehima nervosum*, with the annual sorghums *S. stipoideum* and *S. timorense*.

FIG. 5.48

The basalt savanna extends widely over red soil plains, dominated by the occasional sandstone mesa such as Mount House seen here in the distance. More open patches at the right rear are less well-drained, black soil areas.

Fig. 5.49

Where water accumulates in the rainy season, the soil changes to dark, cracking clay of so-called black soil plains. This clay dries out and cracks deeply during the dry season but can absorb more than its own weight of water in the rainy season, becoming very soft and impassable. Such soil can only carry a few specialised species of trees such as the coolabah (foreground), *Eucalyptus microtheca*, and the extraordinary spindly *Acacia suberosa*. Ribbon grass *(Chrysopogon)* and blue grass *(Dichanthium)* typically form the ground layer.

Fig. 5.50

Between Mount House and Glenroy there are extensive shale plains where soil is very shallow and is responsible for a return to a spinifex formation. The scattered trees are of *Eucalyptus brevifolia* (snappy gum) and *E. argillacea* (Mt House box) while the ground layer is principally of *Triodia pungens, Plectrachne pungens* and another *Plectrachne* sp.

5.4 *North Kimberley Region*

GARDNER BOTANICAL DISTRICT

High-grass savanna woodlands. *Eucalyptus tetrodonta-E. miniata* alliance with *Plectrachne pungens* and annual *Sorghum* spp. on sandstone; *E. tectifica-E. grandifolia* alliance with *Sehima nervosum* and *Sorghum* on basalt.

Climate: Dry hot tropical; precipitation > 700 mm per annum, summer wet season 4-6 months.

Geology: Siliceous sedimentary and basic volcanic rocks of Proterozoic age.

Topography and soils: A dissected plateau with mesa shape hills and ranges. Chiefly shallow stony and sandy soils on sandstones; neutral red and yellow earths on basalt.

Area: 99 100 km².

Boundaries: The southern boundary is the mapped boundary of the high-grass savanna and corresponds approximately to the 700 mm isohyet.

The name of the botanical district commemorates C.A. Gardner for his pioneer botanical exploration of the area in 1921. The region is defined as usual by vegetation, in this case by the predominance of high-grass savanna, which covers 90% of the area. This more luxuriant vegetation type appears in response to rainfall which exceeds 700 mm per annum in the east, 800 mm in the centre and 900 mm in the west. The region comprises the whole of the northwestern and northern coast of the Kimberley above Walcott Inlet round to the Cambridge Gulf and a further substantial sector east of Wyndham. The Cambridge Gulf divides the region into two distinct parts, one on the Kimberley Plateau and the other in a more topographically varied sector including Wyndham, Kununurra and Lake Argyle. The former is remote, inaccessible, almost uninhabited, while the latter contains two towns, the Ord River irrigation scheme and numerous tourist attractions including the Bungle Bungle National Park.

The former, the northwest Kimberley, contains a handful of cattle stations, a handful of Aboriginal settlements and the Japanese cultured pearl centre at Kuri Bay. There are no permanent roads and the coastline is only accessible by sea. The section east of the Carson escarpment through to the Cambridge Gulf is particularly unvisited. As far as is known no professional botanist has ever entered this area, even today. The reason for this state of affairs is principally that sandstone country predominates and not only produces poor herbage for stock but is excessively rocky and hard to traverse. Most of such country cannot even be penetrated by 4WD vehicles. The northeast Kimberley, that is the Wyndham-Kununurra area, is in complete contrast, varied country comprising plains, hills and rough ranges. Being on the through route from the Kimberley to Darwin it is served by sealed roads. Even here however the bizarre scenic attractions of the Bungle Bungle were only quite recently realised as it lies in such rough inaccessible country that hardly anyone had ever been there.

In the northwest Kimberley, the portion of the region lying on the Kimberley Plateau, there are three

FIG. 5.51

The North Kimberley Region receives a higher rainfall so that all vegetation tends to be taller and denser and new tropical forms appear. We still have the sandstone and basalt savanna types but they are differently developed. In this example of sandstone savanna from Mitchell River station, the two dominants are seen, at left messmate *Eucalyptus tetrodonta* with fibrous grey bark, and at right woollybutt *E. miniata* with its black stocking trunk. Annual sorghum is dominant in the ground layer and has dried off and collapsed during the dry season. Several plants of screw-pine, *Pandanus*, are visible at left.

FIG. 5.52

Eucalyptus miniata is actually a very decorative species and is worthy of planting as an ornamental in northern Australia and other tropical countries. Although much use has been made of eucalypts as both ornamental and economic trees throughout the world, the focus has so far been almost entirely on species of southern origin. Tropical eucalypts are a neglected field.

topographic units, the Prince Regent Plateau in the west and Karunjie Plateau in the east formed upon sandstone country, and the Gibb Hills on basalt. The first two are substantially similar as far as vegetation is concerned so that for practical purposes we may speak of sandstone country and basalt country.

The vegetation of sandstone country varies considerably with the amount of soil present upon the rock. The two trees *Eucalyptus tetrodonta* (messmate) and *E. miniata* (woollybutt) are characteristic dominants but vary in height, density and abundance. On gentler and lower-lying country where the soil is deep, the trees are tall and straight reaching 20-25 m in height, though commonly about 12-15 m, with an incomplete canopy, forming woodland, or open forest in the best stands. Height and density are very variable.

Nine upper storey species have been recorded. *E. tetrodonta* (messmate or stringybark) is the commonest species and occurs most consistently. *E. miniata* (woollybutt) is second in importance and less consistent. In the southwest it extends beyond the range of *E. tetrodonta* and dominates the sandstone areas extending also onto the Fitzroy Plains. *E. polycarpa* associates on the deep sandy soils of the slight depressions, *E. dichromophloia* conversely comes in as the soil shallows, *E. bleeseri* and *E. latifolia* may appear on laterite, *E. confertiflora* and *E. foelscheana* are minor casual species. *Callitris intratropica* occurs scattered throughout but may form pure stands on deep red sands.

A number of species of smaller trees occurs but these are normally somewhat scattered. They include *Brachychiton diversifolius, Buchanania obovata, Gardenia pyriformis, Grevillea cunninghamii, G. heliosperma, G. pteridifolia, Melaleuca* sp., *Owenia vernicosa, Persoonia falcata, Petalostigma pubescens, Planchonia careya, Syzygium suborbiculare, Terminalia* spp. and *Ventilago viminalis*. Scattered shrubs are also common and include several *Acacia* spp., notably *A. monticola, A. sericata* var. *dunnii, A. tumida, Bossiaea bossiaeoides, Calytrix leschenaultii, Grevillea agrifolia, Jacksonia argentea, J. thesiodes* and *Verticordia cunninghamii*.

The grass layer is relatively poorly developed and consists mainly of the annual species of *Sorghum, S. stipoideum* and *S. timorense*, with sparse *Plectrachne pungens*. The sorghums grow usually to a height of over 2 m during the wet season. There are a few associated species of forbs and other annual grasses of which little is known.

On the more rocky country the tree cover becomes sparser and more open, with heights not above 10-12 m. *Eucalyptus dichromophloia* may tend to become dominant, with the recession first of *E. tetrodonta* and in the most extreme cases of *E. miniata* also, while *E. herbertiana* comes in. The botanical exploration of recent years has revealed that rainfall declines towards the coast while at the same time the sandstone country becomes even more rocky and devoid of topsoil. All vegetation then becomes less luxuriant, plants shorter and more scattered. In extreme cases *E. herbertiana* may be the commonest large tree, with some *E. dichromophloia*, very scattered and irregular, height about 10-12 m. Many species of smaller trees are characteristic of this habitat,

Fig. 5.53

Sandstone country in the north Kimberley tends to be very rough and rocky, the more so the nearer to the coast. This aerial view of the mainland east of the Admiralty Gulf shows mostly bare outcropping rock with little soil or vegetation cover in the top half of the picture. The darker area at the bottom is a plain with deep sandy soil and a cover of good quality *E. tetrodonta–E. miniata* savanna woodland.

Fig. 5.54

Under not too extreme conditions, it is possible for the sandstone to carry a reasonably dense tree growth rooting in crevices. *E. miniata* is normally more tolerant of these conditions than *E. tetrodonta* which drops out, and is replaced by other species such as *Owenia vernicosa* (front left). *Plectrachne pungens* makes up the bulk of the grass cover. It can readily be seen that country of this kind is not traversable by vehicles, even with four-wheel drive. On the Prince Regent River.

Fig. 5.55

In many places on the rough sandstone, thickets of the wattle *Acacia delibrata* are seen. It is possible that these may be due to mass regeneration after fire.

FIG. 5.56
This small palm which probably belongs to an undescribed species adds character to the landscape in rough sandstone country on the lower Prince Regent. It is a *Livistona* but does not closely resemble *L. eastonii* at the Mitchell Plateau.

though very scattered, and include *Acacia delibrata, Grevillea pteridifolia, Owenia vernicosa*, several *Terminalia* species, *Xanthostemon eucalyptoides* and *X. paradoxus*. Small trees of vine thicket species such as *Canarium australianum, Pouteria sericea* and *Vitex glabrata* are not uncommon but appear normally to be destroyed by fire before maturity. Understorey shrubs are of much the same species as before but are scattered in patches where there is deeper soil, as are the herbaceous and ground plants. The spinifex *Plectrachne melvillei* and an undescribed *Triodia* form mats on exposed bare rock.

Thickets of the wattle *Acacia delibrata* are often seen, especially on offshore islands, and appear to be a seral stage after fire which causes copious germination.

A mixed fringing forest occurs along the rivers, composed of *Terminalia platyphylla, T. grandiflora Ficus coronulata, F. racemosa, Melaleuca leucadendra*, and *Acacia holosericea*. The banks of the larger rivers are usually lined with *Eucalyptus camaldulensis* together with *M. leucadendra*, often magnificent trees 20-25 m tall. *Pandanus aquaticus* is very common.

Mangrove communities occur on tidal mud in the more sheltered bays and inlets of the coast. Bare expanses of hypersaline mud occur here where the mudflats are of large area but not to the extent that they do further south in the Kimberley and Pilbara, while the fringing mangroves are correspondingly more extensive. Species richness in mangroves is much enhanced here and reaches a maximum for Western Australia with 15 species, which may occur in pure stands or form mixed associations. It is difficult to indicate any typical patterns or zonations. The 15 species recorded are *Aegialitis annulata, Aegiceras corniculatum, Avicennia marina, Bruguiera exaristata, B. parviflora, Camptostemon schultzii, Ceriops tagal, Excoecaria agallocha, Lumnitzera racemosa, Osbornia octodonta, Pemphis acidula, Rhizophora stylosa, Scyphiphora hydrophylacea, Sonneratia alba, Xylocarpus moluccensis*.[4] As may be expected of these plants whose propagules are distributed by ocean currents, the mangroves found here occur widely on the coasts of Australia and New Guinea and in most cases still further afield in Southeast Asia. A local species *Avicennia*

Fig. 5.57

The northwest coast of the Kimberley is immensely beautiful with a mass of islands, estuaries and inlets, and bold bluffs with sandstone cliffs overlying basalt. It can only be approached from the sea as there are no roads, and the coast is uninhabited except for the Japanese cultured pearl settlement at Kuri Bay. This view is from St Patricks Island in the St George Basin on the Prince Regent, looking north towards Mt Waterloo. *Calytrix exstipulata* in flower in the foreground.

Fig. 5.58

One of the principal attractions of the Prince Regent is the King Cascades, named after Lt King who charted this coast for the British Admiralty in 1819–20. He was accompanied by the botanist Alan Cunningham who made pioneer collections here and has left his name attached, as *cunninghamii*, to many local species which he discovered.

Fig. 5.59

Sunset at Bigge Island, on the northwest coast. Will the North Kimberley ever be better known, better visited?

eucalyptifolia listed in the mangrove book[4] has since been concluded to be only a variety of *A. marina*.

Mangroves of course occur indiscriminately on the coasts of both sandstone and basalt country, but the land vegetation is another matter. On basalt it is entirely different from the sandstone. Basaltic rock known as the Carson Volcanics was originally formed by lava flows and occurs overlying the sandstone which forms the western Prince Regent Plateau, and underlying the sandstones of the Karunjie Plateau on the east. It is exposed mainly in a north-south belt from Gibb River to Kalumburu but also in local pockets along the northwest coast. Basalt country is relatively easily traversed; it may be hilly and bouldery but there are no inaccessible masses of rock. It can also be utilised for running cattle, and is therefore the best known part of the Kimberley Plateau. The general vegetation is high-grass savanna woodland of the *Eucalyptus tectifica-E. grandifolia* alliance.

The upper tree storey is open, usually one to two crown widths apart and approximately 9-12 m tall, varying from 6-15 m. *E. tectifica* (grey box) is the principal tree species occurring almost throughout either singly or with *E. grandifolia* (cabbage gum). On the areas of more rugged topography, *E. grandifolia* is usually replaced by *E. confertiflora*. In this habitat the species in the low tree layer are more numerous and the ground layer is dominated by annual *Sorghum*. On levees and gentler slopes *E. foelscheana* joins the assemblage. *E. jensenii* comes in as an associate, or forms pure stands on sandy red earths. The shrub layer is then apt to be more strongly developed and to contain many species from adjoining sandstone country. A minor association of *E. tectifica* with *Terminalia canescens* and spp. occurs on stony ridges with cherty rock outcrops near the junction of the volcanics and sandstones. It forms much denser stands of trees.

The low-tree layer is very variable and consists usually of a small number of species, many deciduous. The most important are *Hakea arborescens, Cochlospermum fraseri, Terminalia canescens, Erythrophleum chlorostachys, Atalaya hemiglauca*. The shrub layer is normally very poorly developed or non-existent, except in association with *E. jensenii*. Patches of *Melaleuca viridiflora* occur locally in swampy places.

The ground layer consists of dense tall perennial grasses > 1.5 m. The principal species are *Sehima nervosum* (white grass) and *Chrysopogon fallax* (beard grass), together with *Sorghum plumosum, Heteropogon contortus, Themeda australis, Dichanthium* spp. and *Bothriochloa* spp. There is a tendency for *Plectrachne pungens* and annual *Sorghum* to come in and to become dominant in habitats where *Eucalyptus foelscheana* and *E. jensenii* are present, and in the *E. tectifica-Terminalia* community. *Aristida hygrometrica* may be present also in these and other cases, and becomes increasingly important towards the drier south.

An *Eucalyptus latifolia* community may replace the above in flats or depressions. *E. latifolia* is the principal species either singly or associating with *E. tectifica, E. tetrodonta, E. grandifolia* and *E. tetrodonta*, or *E. polycarpa* and *E. oligantha*. Trees are normally scattered so that the community tends to present an open, park-like appearance, with little or no low-tree or shrub layers. A correspondingly well-developed ground layer occurs of *Chrysopogon fallax, Sehima nervosum, Themeda australis*, and *Plectrachne pungens*. Other grasses include *Alloteropsis semialata, Eriachne obtusa, Heteropogon contortus* and *Sorghum plumosum*.

Eucalyptus papuana dominates a community on river flats. The tree layer is open, with widely-spaced shapely trees 12-24 m high. Low-tree and shrub layers are sparse or absent and the communities generally have an open, park-like appearance. *E. papuana* is the principal tree species and may occur singly or associated with *E. apodophylla, E. alba* and *E. tectifica*. The ground layer consists mainly of *Sorghum plumosum* with lesser patches of other grasses.

An interesting feature of the basalt country is the occasional appearance of flat-topped, laterite-capped small plateaux and mesas which preserve remnants of a very old landsurface formed probably in the early Tertiary 50 million years ago or more. This surface is evidently the same as we see commonly in the south, in the Darling Range for example, and in the same way is capped by bauxitic laterite overlying a thick 'pallid zone' of kaolinised and decayed rock. The heavier tropical rains in the north have almost entirely destroyed the old surface which persists only in remnants protected by the hard laterite capping. As this is gradually removed by erosion at the edges the underlying soft pallid zone goes too, leaving abrupt upstanding mesas bounded by breakaways. This formation is best known at the Mitchell Plateau which was for a long time accessible through mining exploration, now discontinued.

The surface of such plateaux is almost flat and consists of hard laterite overlain by a variable depth of a red clay-loam. Another interesting feature is that the

FIG. 5.60

Further inland, the abundant rainfall of the North Kimberley feeds numerous permanent rivers, lined with paperbark trees *(Melaleuca leucadendra)* and *Pandanus*. Here we have the King Edward River on the way in to the Mitchell Plateau.

FIG. 5.61

The lower lying parts of the country become flooded or swampy during the wet season, making travel difficult or impossible. Black soil flats in this region are usually heavily wooded, with *Eucalyptus papuana* often as sole species as seen here with a ground layer of *Sorghum plumosum* and some *Pandanus*.

FIG. 5.62

Tidal creeks are all thickly lined with mangroves in spite of the enormous rise and fall of tide which amounts to 10 metres daily. Some mangroves disappear under water at high tide! The high rainfall, by diluting sea water and diminishing evaporation, favours a high number of mangrove species represented here. Fiften have been recorded.

FIG. 5.63

Basalt savanna country is characterised as further south by *Eucalyptus tectifica* (grey box) and a sward of mixed tall grasses. Successive lava flows which built up the basalt formation are now to be seen forming prominent benches and small scarps which give protection to patches of vine thicket, seen here in a colour contrasting with the savanna. A laterite-capped plateau is present at the top. In the bay below there is a large mangrove swamp fronting the sea on the left, and backed by a bare mudflat. Port Warrender, on the Admiralty Gulf.

FIG. 5.64

The nine Kimberley species of *Polycarpaea* (family Caryophyllaceae) are attractive annuals appearing in the grassland. The flowers are stiff and papery.

FIG. 5.65

Other attractive herbs are the mulla-mullas (*Gomphrena*). This one is a perennial, *Gomphrena canescens*, growing to about 40 cm. They are related to the genus *Ptilotus* which is more common in the south.

plant cover is typical not of the basalt but of the sandstone country, since the surface materials are highly leached and siliceous. *Eucalyptus tetrodonta* and *E. miniata* are the commonest trees, the former replaced by *E. nesophila* at the drier north end of the Mitchell Plateau. These trees are of impressive straight and tall growth on deep soil, reaching 18 m in height, and form dense stands, but become more open and stunted where the soil is shallow. On sheets of bare laterite they are replaced by scattered *E. latifolia*. An associate confined to these laterites is a spindly palm, *Livistona eastonii*, with the 'fan-palm' type of leaf. It reaches 3 to 5 m in height, occasionally 10 m and has a thin trunk only 10 cm in diameter. It is extremely abundant and gives a distinct character to this community. There are a few small trees in the assemblage, rare shrubs and a grassy but not very dense understorey 1-2 m tall notably of *Heteropogon contortus*, *Themeda australis* and *Cymbopogon bombycinus*. A cycad, *Cycas armstrongii*, occurs occasionally at the southern end of the plateau.

Another very interesting feature of the basalt country is provided by the small patches of vine thicket which, as noted at the beginning of this chapter, were first discovered in the Kimberley in the Mitchell Plateau area in 1961. This vegetation is very common in Queensland and occasional in the Northern Territory but was thought not to occur in Western Australia. When found at the plateau the occurrence was at first attributed to the local high rainfall (the mining camp has the highest average rainfall in the State), but vine

FIG. 5.66

An interesting feature of the basalt country is the occasional appearance of flat-topped, laterite-capped small plateaux and mesas which preserve remnants of a very old land surface formed perhaps 50 million years ago. When eroded, the edges form scarps which are host to patches of vine thicket. These are rainforest in the broad sense and consist of a mixed tropical flora of trees, shrubs and above all vines quite distinct from that of the surrounding savanna. Mitchell Plateau.

FIG. 5.67

Plateau surfaces are capped with laterite, some highly bauxitic and a potential mining resource. The vegetation of such surfaces is similar to that of the sandstone savanna, with *E. tetrodonta* and *E. miniata* as principal dominants. At the Mitchell Plateau an important component, which is very abundant, is a small palm, *Livistona eastonii*, described by C.A. Gardner in 1923 and named after his expedition leader.

thicket has since been discovered to be common all along the northwest coast even where rainfall is much less. It takes the form of small patches up to about 20 ha each of mixed tropical evergreen and deciduous species forming forest and shrubland. The canopy is closed 5-10 m above ground but tall emergent crowns of some tree species reach 10-15 m or more. These are typically deciduous in winter whereas the lower canopy is evergreen. The understorey consists of low trees, shrubs and vines. There is virtually no ground flora apart from a few young seedlings. The contrast with the surrounding eucalypt-dominated communities is therefore extreme. Here we have no eucalypts and no grass; *Acacia* is rare, Proteaceae absent; many of the creepers are very prickly but sclerophylly is not in evidence. This is a tropical woodland community, 'rainforest' in the broad sense. Trees are orthophyllous ('ordinary' broad soft leaves), many have succulent fruits attractive to birds which propagate them. The presence of creepers is an important feature which has led to the distinguishing term 'vine thicket'. An extensive flora list is given in (5) while some idea of the size and distribution of the thickets can be obtained from (6).

The biggest and most important difference of all is that the vine thickets are not fire-prone. They appear to owe their survival to some degree to topographic protection from fire, on islands and peninsulas, in pockets against the breakaways of the Mitchell Plateau, below basalt ledges and so on. Rainfall is not the significant factor at first thought, except in so far as higher rainfall

FIG. 5.68
Owing to the difficulty of moving about in the Kimberley during the wet season, most photographs are taken during the dry, when vegetation appears in its winter colours, the grass dry and brown. The wet season colouring is entirely different, with both grass and tree layers in new leaf and growing actively. This wet-season rendering shows the plateau woodland on the Mitchell Plateau, at the drier northern end where the *Livistona* palm becomes rare, and *Eucalyptus nesophila* replaces *E. tetrodonta* (dark full-barked trees in contrast to the stockinged *E. miniata*).

may be associated with lower inflammability. The vine thickets are of special interest as they represent a biome or life zone completely distinct from the eucalypt savannas, the one fire free and the other fire-controlled.

Vine thickets have not been found in the smaller northeastern sector of the North Kimberley Region between Lake Argyle and the Cambridge Gulf. This sector is extremely varied, including rugged ranges, hilly and undulating country, sandy plains, alluvial plains and coastal mudflats. Broadly, the distinction between sandstone savanna and basalt savanna continues. Sandstone ranges are covered as on the Kimberley Plateau with high-grass savanna woodland of the *Eucalyptus tetrodonta-E. miniata* alliance, with *E. dichromophloia* as the principal tree species on the rugged country and *E. tetrodonta* on the deeper soils. The ground layer consists of annual *Sorghum* and *Plectrachne pungens*. On the extremely rocky and precipitous terrain of the Bungle Bungles plant cover of any kind may be somewhat sparse. On the other hand this same community, more luxuriant and well-grown, occurs on sandy plains (unless they are underlain by limestone, see below) with all three tree dominants occurring together.

Hilly and undulating country on basalt and dolerite is covered as elsewhere in this region by high-grass savanna woodland of the *E. tectifica-E. grandifolia* alliance. The principal tree species here are *E. tectifica* (grey box) and *E. foelscheana* (smooth-barked bloodwood) with *E. confertiflora, E. latifolia, E. parellaria* and *E. grandifolia* less commonly. There are only scattered shrubs. The ground layer is predominantly of the same 'upland tall grass' type with *Sorghum* and *Plectrachne* as on the sandstone. The flatter country on basalt and dolerite carries instead high-grass low-tree savanna, the tree layer shorter (less

FIG. 5.69
At the southern end of the Mitchell Plateau, the assemblage is joined by this cycad palm named *Cycas lane-poolei* by C.A. Gardner in 1923 after his chief, Charles Lane-Poole, the conservator of forests. It has since been referred to *C. armstrongii*. An unusual feature of this species is that it is deciduous, and it is shown here with its leaves beginning to dry off for their annual winter shed.

FIG. 5.70
The Mitchell Plateau mining camp had (when operating) the highest rainfall of any meteorological station in the State with 1500 mm annually. As a result water is plentiful and this permanent stream, 'Camp Creek', flows to join the Mitchell River, lined by *Pandanus aquaticus* and *Terminalia*.

Fig. 5.71
Vine thickets at the Mitchell Plateau are found in small patches from a few trees to 20 hectares. These usually have some degree of topographic protection against the spread of fire, and a favourite habitat is at the edge of the plateau surface nestling against the breakaway. Another interesting feature which can be seen here is a belt of treeless grassland surrounding the thicket.

Fig. 5.72
At close quarters, the tree crowns in the vine thicket are deep green in colour and densely leafy, in contrast to the thin yellow-green crowns of the adjacent eucalypts on the plateau above. Many of the vine thicket crowns are laced with vines but these do not show clearly. The leafless tree at left is dead, not deciduous. A majority of the larger emergent trees are deciduous, however, and will appear leafless in the dry season. This particular photograph is a wet season picture.

Fig. 5.73
One of the most conspicuous deciduous trees is *Bombax ceiba* which has large red flowers in winter while the tree is out of leaf.

Fig. 5.74
The dead tree in Fig 5.72 was probably killed by fire since the annual grass fires in the savanna are capable of sweeping into the vine thickets. When they do, damage is caused since the vine thicket plants are in principle fire tender and not adapted to resist fire or to recover readily from it. However if fire is not too frequently repeated the vine thicket recovers through secondary stages of pioneer species. Damage may also be done by cyclones.

FIG. 5.75

The vine thicket and the savanna are two separate biomes in tension with one another, and fire holding the balance. Just as thickets in some areas appear to be receding, in other places the opposite can be judged to be the case. In the last generation Aborigines have ceased to visit the Bougainville Peninsula and the islands in the Admiralty Gulf. These are therefore no longer being systematically burnt off, and recovery from fire is in progress. At this site on Middle Osborn Island the savanna eucalypts are still present but the grass is being smothered by a growth of creepers, and young seedlings of vine thicket trees are establishing themselves.

FIG. 5.76

Fire scar on the butt of a vine thicket tree. This scar is more than one season old, showing that fire damage has occurred previously.

FIG. 5.77

Major scenic wonder of the North Kimberley—the falls of the Mitchell River, to the west of the Mitchell Plateau. This is a wet season picture, the falls in flood with muddy water. The main fall is in two stages, the upper partly hidden in this view. The narrow gorge to the left takes overflow at high flood. It may represent an earlier abandoned channel of the river.

FIG. 5.78

The lush tropical impression of protected parts of this country where fire does not penetrate is born out by the luxuriant growth of ferns in this cave under a waterfall. Mitchell Plateau area.

FIG. 5.79

The eastern part of the North Kimberley Region lying around Wyndham and Kununurra and the Cambridge Gulf is more varied than the Kimberley Plateau and includes rugged ranges, hilly and undulating country, sandy plains, alluvial plains and coastal mudflats. This scene is at Wyndham, looking southwest across the estuary to Mt Cockburn. Wyndham is bordered by very extensive hypersaline bare mud flats which have only narrow bands of mangrove along tidal channels.

FIG. 5.80

In this varied landscape, rugged ranges adjoin alluvial flats. In the Saw Range a granite rock rises above a black soil plain with Mitchell grass, baobabs and bauhinia *(Lysiphyllum cunninghamii)*.

FIG. 5.81

South of Kununurra, the Bungle Bungle offers fantastic mountain shapes of eroded conglomerate. Until recently this area was difficult of access and little known, but has now been included within a national park.

FIG. 5.82

The Ord River Irrigation Scheme at Kununurra was planned to utilise an extensive black soil plain laid down along a former course of the Ord River stretching for some 30 km to the north and east. The Ord River today has cut a new channel to the Cambridge Gulf and is seen flowing in a trench below the plain in the background of this photograph. The black soil plain was originally very lightly wooded with some coolabah, bauhinia and terminalias and covered with high grass as shown in Fig 1.35 of Chapter 1. Blue grass (*Dichanthium* spp.) and sorghums were the principal species.

FIG. 5.83

In contrast to the lush fertility of the black soil plains, the rocky hills around Kununurra carry only spinifex and scattered shrubs.

FIG. 5.84

With the general barren appearance of the rocky ranges, it is surprising to find large baobabs rooting in the scree slopes, an apparent change of habitat from their usual occurrence along alluvial plains. We must evidently infer that the scree below the rain-shedding rock above acts as a substantial moisture reservoir.

FIG. 5.85

The poplar gum *Eucalyptus bigalerita* is a very handsome tree found occasionally on the alluvial flats. It has large glossy poplar-shaped leaves, and a pink or salmon-coloured smooth bark. It deserves to be more widely used as an ornamental.

FIG. 5.86

Extensive sandy plains cover much of the country between Kununurra and the coast. Where poorly drained, one finds groves of teatree, *Melaleuca viridiflora* and *M. nervosa*, over a grass layer of *Plectrachne pungens* and *Sorghum* spp.

than 10 m) and sparser. *E. terminalis* and *E. argillacea* are the common dominants but sometimes are replaced by *E. pruinosa*. The grass layer changes to the 'tippera tall grass' type with *Themeda australis, Sehima nervosum* and *Chrysopogon fallax*. On sandy plains overlying limestone the vegetation tends to be a mixture of the above two types of 'basalt savanna'. Much of the plains country here however is of the 'black soil' type, and is the basis for the Ord River Irrigation Scheme at Kununurra. Tropical black earth is formed throughout the world on plains of fine-grained calcareous alluvia usually derived from basalt or dolerite. It is agriculturally fertile but extremely plastic when wet as it has a high capacity to absorb moisture, and cracks deeply on drying out. Such soil is frequently treeless or almost so, only a few specialised species being able to tolerate the rooting conditions. The Ord River at one time laid down such a plain in taking a course northward from Kununurra and flowing round to the west and north of the Pincombe Range to join the Victoria River to the east, but has subsequently abandoned this direction in favour of a westerly outlet to the Cambridge Gulf. Below Kununurra the Ord will be seen entering a deeper channel 15 m below its own alluvial plain and flowing off westward. The abandoned black soil plain was to be utilised for the full irrigation scheme extending for 30 km or more, but agricultural problems have so far prevented extension beyond the original small pilot scheme area served by the diversion weir.

As noted above, the Ord Plains were originally sparsely wooded, with scattered low trees of 3-6 m or often none. The species are locally variable and include *Eucalyptus microtheca* (coolabah), *Lysiphyllum cunninghamii* (bauhinia), *Acacia suberosa, Terminalia arostrata, T. volucris, T. platyphylla* and *Excoecaria parvifolia*. The shrubs *Acacia farnesiana* and *Carissa lanceolata* occur sparsely. The normal ground layer is a blue grass-tall grass community of perennial grasses reaching 2 m and more in height, in which the blue grasses *Dichanthium tenuiculum* and *D. fecundum*, the sorghums *S. plumosum* and *S. timorense, Eulalia fulva, Ophiuros exaltatus* and *Astrebla squarrosa* are the most common. Some tall herbs such as *Sesbania cannabina, Aeschynomene indica, Trichodesma zeylanicum* and *Hibiscus* spp. are also common.

Naturally in a brief account of this kind only the most conspicuous and/or widespread types of vegetation can be described, but it is hoped that this will be sufficient guide for the visitor and student.

BIBLIOGRAPHY

1 Petheram, R.J. & Kok, B., 1983. *Plants of the Kimberley Region of Western Australia.* Univ. W.A. Press, Nedlands.

2 Beard, J.S., 1979. The Vegetation of the Kimberley Area. Vegetation Survey of W.A., 1:1 000 000 Series, Explan. Notes to Sheet 1, *Kimberley.* Univ. W.A. Press, Nedlands.

3 McKenzie, N.L. & Kenneally, K.F., 1983. Wildlife of the Dampier Peninsula, Part I, Background and Environment. Wildlife Research Bulletin 11, Dept Fisheries and Wildlife, Perth.

4 Semeniuk, V., Kenneally, K.F. & Wilson, P.G., 1978. Mangroves of Western Australia. Handbook 12, W.A. Naturalists' Club, Perth.

5 Western Australian Museum, 1981. Biological Survey of Mitchell Plateau and Admiralty Gulf, Kimberley, Western Australia. W.A. Museum, Perth.

6 Beard, J.S., 1988. Vegetation Map of the Mitchell Plateau Area (map with explanatory notes). Dept of Conservation & Land Management, Perth.

FIG. 5.87
You have now come to the end of your botanical journey through Western Australia. We hope you have enjoyed your stay.

GLOSSARY

Archaean: Precambrian rocks exceeding 2400 million years in age

Arid zone: Area of the centre of Australia where rainfall is inadequate to give an assured growing season for plants

Basalt: Dark, dense basic igneous rock of volcanic origin

Bauxite: Ore of alumina (Al_2O_3)

Billabong: Branch of a river usually forming a cut-off lagoon

Bioclimate: Classification of climate according to qualities important for plant life

Black soil plain: Plain of dark coloured cracking clay, becoming boggy when wet

Bradyspory: Retention of seed on plant until plant dies or is killed by fire

Breakaway: Small scarp marking edge of laterite duricrust

Buckshot plain: Plain with surface covered by small round pebbles 5–10 mm in diameter

Bunch-grass: Perennial tussock grass in which the leaves die at the end of the wet season, sprouting again next season from vegetative organs

Bunch-grass savanna: Grassland formed of the above. Classified as either short or tall

Calcarenite: Rock formed by consolidation of calcareous sand

Calcrete: Massive deposit in soil of calcium carbonate

Catenary sequence: Linked topographic sequence of soil and/or vegetation types, up and down hill

Chenopodiaceous steppe: Dry vegetation dominated by members of the family Chenopodiaceae, i.e. saltbushes and bluebushes

Cladode: A leaf-like flattened branch

Coastal limestone: Limestone on the west coast of Western Australia formed of calcarenite (see above)

Curly spinifex: *Plectrachne pungens*, a perennial coarse wiry grass not adopting a hummock-grass form

Deep weathering: Chemical decomposition of surface layers of rocks to depths up to 50 m

Dolerite: Intrusive micro-crystalline igneous rock similar to basalt

Dolomite: Form of limestone compounded of both magnesium and calcium carbonate

Dominant: When used of plants, means those which by superior size or numbers affect or control the growth of others

Donga, donger: Local term on the Nullarbor Plain for a shallow round depression

Duplex soil: Soil with contrasting textural horizons, e.g. with sand over clay

Duricrust: Indurated surface layer of the ground

Ecology: Study of forms of life in relation to environment. Hence *ecologist, ecological*

Endemic: Peculiar to a particular locality

Epicormic growth: Growth originating from adventitious buds in the bark of a tree's stem or branches

Ericoid shrubs: Small shrubs with minute leaves resembling the *Erica* species of Europe

Fynbos: Local term for the mediterranean scrub of South Africa

Glaucous: Leaves of light bluish-green colour and/or with a whitish bloom

Gneiss: Metamorphic rock made up of bands which differ in colour and composition

Gravel plain: Plain strewn with stones 1–5 cm in size

Halophyte: Salt-tolerant plant

Hamada: Desert plain strewn with stones. From the Arabic; in eastern Australia but not in the west the term *gibber plain* is used

Heath: Vegetation of low evergreen ericaceous shrubs in Europe, and in Australia vegetation morphologically similar

High-grass savanna: Formation of tall bunch-grasses > 100 cm in height

Horst: Isolated upstanding block of rock surrounded by faults

311

Hummock grass: Evergreen perennial drought-resisting grasses, usually spiny, which form intertwined dome-shaped hummocks. Hence *hummock grassland*. Popular term *spinifex*

Jaspilite: Fine-grained siliceous multicoloured rock

Karri loam: Red earth which is the characteristic soil type for karri, *Eucalyptus diversicolor*

Kopi dune: Salt lake dune containing large amounts of gypsum (calcium sulphate)

Kunkar: Concretionary nodules of calcium carbonate in soil

Kwongan: Aboriginal term for the Western Australia sandplain and its vegetation which consists typically of a layer of small ericoid shrubs < 1 m. If of this layer only it is classified as *heath*, if with scattered taller shrubs as *scrub-heath*, if with dense taller shrubs as *thicket*

Laterite: Deposits of iron oxide in the soil varying from small pebbles (pea ironstone) to massive duricrusts

Limestone: Rock predominantly of calcium carbonate

Lithology: Study of the composition and fabric of sedimentary rocks

Mallee: Shrubland dominated by multistemmed small eucalypts

Mallee-heath: Heath with scattered small mallee eucalypts

Mangrove: Tropical trees adapted to growth in tidal swamps

Marlock: Single-stemmed small eucalyptus tree

Maquis: Scrub vegetation of the mediterranean region. French; also *macchia* (Italian)

Mediterranean climate: Climate with a winter rainfall maximum as experienced around the Mediterranean Sea

Metamorphic rock: Rock altered in structure or constitution by natural processes

Mulga: *Acacia aneura*, or vegetation dominated by this species

Mulga parkland: A formation in which mulga occurs irregularly as component of a mosaic

Natural region: An area possessing distinctive characteristics of physiography, geology, soil and vegetation

Obligate seeder: A plant which is incapable of resprouting from the root after a bushfire and must regenerate from seed

Ombrothermic diagram: A chart used for classifying climate in which mean monthly rainfall and temperature are plotted

Palaeoriver: Inferred course of former river no longer active

Pallid zone: Subsoil horizon in which the original material has been strongly weathered and leached of important constituents, the residue stained white by kaolin

Phanerozoic: Rock less than 570 million years old, i.e later than the Cambrian period

Phyllode: Leaf-like organ derived from a flattened leaf stalk

Pindan: Formation found on sandy soils in tropical northern Australia

Plant community: The assemblage of plants found at any given locality

Plant association: Group of communities with consistent dominants of the same or closely allied species

Plant formation: Group of associations having the same structure and life form

Plantagenet beds: Geological formation of Eocene age found on the south coast of Western Australia

Pleistocene: Division of the Cainozoic era < 2 million years old

Podzol, podzolic soil: Soil with a bleached eluviated surface horizon

Poison plant: Leguminous shrub whose leaves are poisonous to stock

Precambian: Rocks antedating the Cambrian period, 570 million years ago

Proteoid roots: Clusters of fine roots produced by members of the family Proteaceae

Proterozoic: Division of the Precambian from 570 to 2400 million years in age

Red-brown hardpan: Concretionary siliceous deposit in lower soil layers of mulga country; in place exposed on surface by deflation of topsoil

Savanna: Tropical grassland with or without scattered trees

Scald: Bare area denuded by agricultural or pastoral malpractice

Sclerophyll: Character of hard, brittle leaves. Also *sclerophylly*

Sclerophyll forest, wet or dry: Term used for eucalypt forests mainly in eastern Australia, derived incorrectly from Diels' term *Sklerophyllenwald* which actually referred to the sclerophyll understorey

Scrub: Shrubland formation, open rather than closed (see *thicket*)

Scrub heath: Formation with lower layer of small ericoid shrubs and upper emergent layer of scattered taller shrubs

Seif dune: Long linear sand dune

Silcrete: Siliceous concretionary deposit in soil

Shingle plain: Plain covered with stones > 5 cm in size

Soloth: Duplex sand-over-clay soil, the B-horizon of acid reaction

Spinifex grassland: Formation of hummock grasses of the genera *Triodia* and *Plectrachne*

Symbiosis: The living together of two species of organisms, usually in mutually beneficial relationship

Synusia: Stratum forming a division of a plant community

Systematic botany: Study of the classification of plants

Taxonomy: Classification of organisms

Tectonic inertness: Lack of movement of the earth's crust

Thicket: Plant formation comprising a closed shrubland, the shrubs > 1 m in height

Travertine: A form of limestone deposited in the soil

Vine thicket: Formation of mixed tropical deciduous orthophyll shrubs and small trees with numerous lianas

Whaleback: A hill or ridge evenly curved on all sides

Index

Abelmoschus ficulneus 285
Acacia 20, 21, 22, 24, 27, 53, 66, 98, 100, 102, 104, 109, 110, 115, 122, 123, 130, 146, 147, 156-7, 159, 160, 163, 166-7, 186, 196, 202, 204-5, 212-3, 220-1, 223-4, 226, 234, 240-1, 252, 263, 267-8, 272, 273, 288, 294
   A. aciphylla 186
   A. acuaria 160
   A. acuminata 80, 84, 87, 100, 102, 104, 105, 108, 114, 118, 121, 125
   A. alata 79, 82
   A. ampliceps 272
   A. aneura 22, 33, 35, 58, 156, 169, 171, 176, 178, 180, 185-6, 188-9, 190, 192-3, 199, 201-2, 204, 208, 213-4, 216, 223, 226-7, 233-4, 238, 244, 246, 257, 263
   A. beauverdiana 130
   A. bivenosa 208, 212, 223, 278
   A. blakelyi 294
   A. browniana 66, 77, 82
   A. burkittii 176
   A. citrinoviridis 204, 223
   A. cochlearis 90
   A. colletioides 118
   A. coriacea 189, 192, 202, 204, 208, 213
   A. craspedocarpa 189, 193
   A. cyclops 75, 77, 90, 91-2, 146
   A. cyperophylla 234
   A. delibrata 295-6
   A. dictyophleba 233
   A. dilatata 92
   A. divergens 66, 76
   A. drummondii 149
   A. eremaea 202, 209
   A. ericifolia 130
   A. eriopoda 268-9, 270
   A. extensa 77, 82
   A. farnesiana 53, 290, 309
   A. fragilis 130, 155
   A. grasbyi 189, 190, 192-3, 215, 223, 227, 233-4
   A. helmsiana 184, 233
   A. hemiteles 130, 162
   A. holosericea 54, 268, 278, 296
   A. kempeana 189
   A. lasiocalyx 125, 136
   A. lasiocarpa 90, 100-1
   A. ligulata 100, 105, 182, 184, 186, 189, 203-5, 214, 226, 233-4, 254-5, 257
   A. linophylla 186, 189, 193, 216, 233
   A. littorea 75, 76, 77
   A. longispinea 104
   A. lycopodifolia 288
   A. lysiphloia 278
   A. monticola 254, 257, 262, 268, 278, 282, 290, 294

   A. murrayana 184, 202
   A. myrtifolia 66, 69, 149
   A. nervosa 80
   A. neurophylla 121-2, 158
   A. orthocarpa 234, 237
   A. oswaldii 160, 176
   A. pachyacra 233
   A. pachycarpa 244-5, 254, 257, 262, 268, 270, 282
   A. papyrocarpa 23, 171, 175-6
   A. pentadenia 66
   A. platycarpa 269, 271
   A. pruinocarpa 186, 189, 192, 214-5, 226, 233-4, 238
   A. pulchella 66, 84, 92, 94, 149
   A. pyrifolia 208, 212-3, 243-6, 252, 278, 291
   A. quadrimarginea 156, 161, 192-3, 215
   A. ramulosa 189, 192-3, 199, 202-4, 207-9, 211, 216
   A. resinomarginea 122-3, 158
   A. rhodophloia 202
   A. rostellifera 53, 89, 90-1, 93, 100-3, 105, 109
   A. saligna 90, 94
   A. scirpifolia 100
   A. sclerosperma 189, 192, 201-2, 204, 208-9, 214, 223
   A. sericata 288, 294
   A. spathulifolia 90, 100, 103, 208, 213
   A. stenophylla 276
   A. stereophylla 104
   A. stipuligera 268
   A. subcaerulea 53, 155
   A. suberosa 276, 290, 292, 309
   A. subtessarogona 202, 208, 211
   A. tenuissima 282
   A. tetragonophylla 158, 164, 176, 186, 189, 192-3, 196, 202, 208-9, 214, 223, 234, 238
   A. translucens 213, 244, 290
   A. tumida 254, 269, 270, 282, 294
   A. urophylla 66, 82
   A. victoriae 189, 192, 196, 202, 208-9, 211, 213-4, 218, 222-3, 234, 254
   A. wanyu 222-3
   A. xanthina 100-1
   A. xiphophylla 208, 213-4, 219, 223, 246
Acanthocarpus preissii 89, 92, 206
Achyranthes aspera 276
Actinodium cunninghamii 79
Actinostrobus arenarius 94, 102, 105, 111-2, 122
Actites megalocarpa 90
Adansonia gregorii 272-3, 286, 288, 290
Adenanthos acanthophyllus 109, 114
   A. barbigerus 82
   A. cuneatus 76, 79
   A. cygnorum 92, 94, 105
   A. obovatus 69, 79

   A. sericeus 76, 155
   A. stictus 102
Aegialitis annulata 208, 242, 296
Aegiceras corniculatum 242, 296
Aerva javanica 282
Aeschynomene indica 309
Agonis 69
   A. flexuosa 66, 69, 76, 90, 92, 96, 146, 152
   A. hypericifolia 77, 79
   A. juniperina 66, 69
   A. linearifolia 82, 141, 155
   A. marginata 69, 71, 77, 79
Albany bottlebrush = *Callistemon speciosus*
Alloteropsis semialata 298
Alyogyne hakeifolia 100
   A. huegelii 100
   A. pinoniana 184, 233
Alysicarpus rugosus 280
Ammophila arenaria 90
Amphipogon caricinus 233
Anarthria 69
   A. gracilis 149
   A. prolifera 69, 79
Andersonia 79, 95
   A. caerulea 69
   A. echinocephala 149
   A. parvifolia 144
   A. simplex 76
Angianthus 176
   A. conocephalus 173
   A. cunninghamii 90
Anigozanthos 13
   A. flavidus 66, 76
   A. humilis 94
   A. manglesii 83
Anisomeles malabaricum 278
Annual sorghum = *Sorghum stipoideum* and *S. timorense*
Anthocercis genistoides 138
   A. littorea 92, 100, 105
   A. viscosa 76
Anthotroche pannosa 233
Arctotheca calendula 117
   A. populifolia 90, 91
Aristida 282, 288
   A. browniana 278
   A. contorta 183, 244, 246, 284
   A. hygrometrica 278, 288, 298
   A. inaequiglumis 278, 282, 288, 290
   A. latifolia 244, 247, 252, 273-4, 278, 285, 288
   A. pruinosa 278, 284, 288, 290
Artemisia 178
Ashburton pea = *Swainsona maccullochiana*
Astartea fascicularis 81, 82
Astrebla 238, 277
   A. elymoides 273-4, 278, 280

313

Astrebla contd.
   A. pectinata 242, 252, 263, 273–4, 278, 280
   A. squarrosa 273–4, 278, 280, 309
Astroloma microdonta 95
Astrotricha hamptonii 252
Atalaya hemiglauca 242, 290–1, 298
Atriplex 22, 100, 105, 119, 161, 164, 167, 186, 198, 201, 204, 206, 223
   A. acutibractea 176
   A. amnicola 208
   A. bunburyana 205
   A. cinerea 246, 250
   A. cryptocarpa 176
   A. eardleyae 246, 250
   A. hymenotheca 130–1, 160, 163–4, 233
   A. isatidea 90–1, 100
   A. lindleyi 214, 223
   A. nummularia 157–8, 160, 162–3, 176
   A. paludosa 122, 206
   A. vesicaria 157, 162, 176
Avena 125, 208
Avicennia marina 92, 95, 208, 242, 269, 296, 298

Baeckea 210
   B. behrii 138
   B. camphorosmae 82
   B. crispiflora 122
   B. cryptandroides 184
   B. floribunda 193
   B. grandibracteata 130
   B. ovalifolia 150
   B. pachyphylla 100, 104
   B. pentagonantha 104
   B. platystemona 130
Bald Island marlock = *Eucalyptus lehmannii*
Banksia 16, 21, 52, 67, 87, 90, 92, 98, 102, 105, 107, 109, 113, 123, 140, 142
   B. archaeocarpa 47
   B. ashbyi 102, 104, 109, 112, 114
   B. attenuata 20, 47, 79, 82, 87, 90, 92, 94, 97, 100, 102, 105, 106, 122, 141, 145, 208, 213
   B. audax 134
   B. baueri 133
   B. baxteri 144
   B. brownii 79
   B. burdettii 94, 102
   B. caleyi 144
   B. candolleana 100
   B. coccinea 79, 144–5
   B. dryandroides 143
   B. elderiana 133
   B. gardneri 135
   B. goodii 79
   B. grandis 16, 66, 75–7, 79, 80, 82, 92, 94, 149
   B. hookeriana 106
   B. ilicifolia 77, 79, 92, 97
   B. lehmanniana 150
   B. littoralis 67, 82, 92
   B. media 133, 152, 155
   B. menziesii 20, 90, 92, 97, 100, 105–6
   B. occidentalis 146
   B. petiolaris 149
   B. prionotes 90, 94, 100, 102, 105, 122
   B. quercifolia 79, 150
   B. repens 79
   B. sceptrum 102, 105, 112, 113
   B. solandri 148–9
   B. speciosa 142, 144, 146–7, 155
   B. sphaerocarpa 79, 94, 100, 149
   B. verticillata 66, 79
Baobab, Boab = *Adansonia gregorii*
Bauhinia = *Lysiphyllum cunninghamii* (formerly *Bauhinia cunninghamii*)
Baumea articulata 92
   B. juncea 92
Beaufortia anisandra 79
   B. bracteosa 90, 155
   B. cyrtodonta 149
   B. decussata 149
   B. elegans 104
   B. sparsa 69
   B. squarrosa 57, 100
Beefwood = *Grevillea striata*
Blackboy = *Xanthorrhoea* spp.
Blackbutt = *Eucalyptus patens*
Bluebush = *Maireana sedifolia*

Blue grass = *Dichanthium* spp.
Boab, Baobab = *Adansonia gregorii*
Boerhavia coccinea 285, 290
Bombax ceiba 304
Boree = *Melaleuca pauperiflora*
Boronia albiflora 155
   B. crenulata 149
   B. scabra 94
   B. spathulata 69
   B. ternata 130
Borreria 290
Borya nitida 81, 124–5, 144, 147
Bossiaea 94
   B. aquifolium 66, 82
   B. bossiaeoides 294
   B. laidlawiana 73
   B. linophylla 66, 72–3, 77, 79, 149
   B. ornata 66, 77
Bothriochloa 298
   B. bladhii 272, 291
   B. ewartiana 290
Bowgada = *Acacia ramulosa*, *A. linophylla*
Brachyachne convergens 280, 288, 290
Brachychiton 273
   B. diversifolius 294
   B. rupestris 272
   B. obtusilobus 208
Brachycome 162
   B. ciliaris 176
   B. ciliocarpa 189
   B. latisquamea 207
Brachysema aphyllum 53
   B. latifolium 141
Broombush: shrubs with ascending branches, generally *Eremophila* spp., e.g. *E. scoparia*
Brown mallet = *Eucalyptus astringens*
Bruguiera exaristata 296
   B. parviflora 296
Brunonia australis 184, 189, 197, 236
Buchanania obovata 286, 294
Bullitch = *Eucalyptus megacarpa*
Bulrush = *Typha orientalis*
Bursaria occidentalis 189
Burtonia polyzyga 233, 247, 250
   B. scabra 79
   B. villosa 149
Byblis 51

Cabbage gum = *Eucalyptus grandifolia*
Cadjuput = *Melaleuca leucadendra* or *M. cajuputi*
Cakile maritima 90, 91
Caladenia latifolia 92
Calandrinia 201, 217, 280
Calectasia cyanea 95
Californian redwood = *Sequoia sempervirens*
Callistemon phoeniceus 109, 114, 117
   C. speciosus 69
Callitris glaucophylla 164, 191, 193, 196, 202, 226, 234, 238–9, 249
   C. intratropica 288–9, 294
   C. preissii 89, 90–1, 93, 137–8, 146, 155, 182, 186
   C. roei 130
Calocephalus brownii 77, 90
Calothamnus 95, 208
   C. asper 156
   C. chrysantherus 109, 114, 165, 238
   C. gilesii 125, 137–8
   C. gracilis 149
   C. pinifolius 150, 152
   C. planifolius 80
   C. quadrifidus 82, 84, 90, 92, 94, 137–8, 154–5
   C. validus 150
Calotis breviradiata 176
   C. multicaulis 189, 250
Calycopeplus ephedroides 125
Calytrix 95, 211, 200
   C. brevifolia 111, 208–9, 213
   C. carinata 183, 186, 233
   C. exstipulata 247, 268, 284, 297
   C. leschenaultii 294
   C. strigosa 94
Camptostemon schultzii 296
Canarium australianum 296
Canthium latifolium 189, 214, 238
Carissa lanceolata 244, 288, 290–1, 309
Carpobrotus 100, 103, 118

C. aequilaterus 90
Cassia 22, 163, 186, 192–4, 202, 208, 212–3, 214–7, 222, 232, 238, 245–6, 257, 262, 282
   C. cardiosperma 160
   C. chatelainiana 278
   C. desolata 189, 238, 247, 290
   C. helmsii 194
   C. luersenii 189, 221, 223
   C. nemophila 158, 162, 176, 221, 223
   C. notabilis 233
   C. oligophylla 221, 223
   C. pilocarina 214, 219
   C. pleurocarpa 247
   C. sturtii 189, 238
   C. venusta 245, 284
Cassytha glabella 66
Casuarina 20, 21, 28, 52, 92, 98, 100, 104, 115, 122–3, 127, 133, 137, 141, 148, 156–7, 159, 166
   C. acutivalvis 104, 122, 133–4, 156, 196
   C. campestris 104, 121–2, 125, 133, 156–8, 165, 196
   C. corniculata 122, 133
   C. cristata 160, 164, 171, 176, 186, 234
   C. decaisneana 226, 233–4, 238, 260, 263
   C. decussata 66
   C. fraseriana 16, 19, 66, 72, 77, 82, 90, 92
   C. huegeliana 80, 81, 84, 102, 109, 118, 125, 136, 147, 165
   C. humilis 76, 82, 92, 94–5, 149, 150, 155
   C. lehmanniana 100–1
   C. obesa 94–5, 109, 114, 117, 146, 201
   C. pinaster 135, 138
   C. scleroclada 155
   C. trichodon 155
Caustic bush = *Grevillea dimidiata*
Caustis dioica 82, 149
Celtis philippinensis 272, 277
Cephalipterum drummondii 162, 189, 195–6, 207
Cephalotus follicularis 69
Ceriops tagal 208, 242, 296
Chamelaucium drummondii 94
Chenopodium 95
   C. auricomum 280
Chittick = *Lambertia inermis*
Chloris 288
Chorilaena quercifolia 66
Chorizema diversifolium 66
   C. ilicifolium 66
Chrysopogon 267, 271–3, 276, 285, 290–1, 292
   C. fallax 252, 269, 270, 278, 282, 284–5, 288, 296, 309
   C. latifolius 288
Chthonocephalus 219
Clematis microphylla 90, 92
   C. pubescens 66, 73, 79, 82
Cleome 290
   C. viscosa 276, 284
Clerodendrum tomentosum 231
Clianthus formosus 170, 189, 196, 215, 252
Cochlospermum fraseri 277, 286, 288, 290, 298
Codonocarpus cotinifolius 57, 124
Comesperma drummondii 155
Compass plant = *Casuarina pinaster*
Conospermum 92, 100, 106, 142
   C. caeruleum 79
   C. dorrienii 149
   C. flexuosum 79
   C. petiolare 79
   C. stoechadis 82, 94–5, 102, 110–2
Conostephium 95
Conostylis 79, 94, 113
Coolabah = *Eucalyptus microtheca*
Coopernookia strophiolata 141
Corchorus 290
   C. walcottii 223
Correa reflexa 177
Cosmelia rubra 69
Cotula coronopifolia 117
Cratystylis conocephala 161–2, 165, 167, 176, 178
Crotalaria cunninghamii 182, 186, 233, 254–5, 272
   C. dissitiflora 244
   C. medicaginea 280, 285
Crowea angustifolia 66
Cryptandra arbutiflora 82
   C. parvifolia 130
Curly spinifex = *Plectrachne pungens*
Cyathochaeta avenacea 79

Cycas armstrongii 300
Cymbopogon bombycinus 290, 300
Cypress pine = *Callitris* spp.

Dacrydium 45, 46
Dactyloctenium radulans 290
Dampiera 149, 197, 233
    D. alata 82
    D. candicans 258
    D. hederacea 66
    D. linearis 66
Danthonia caespitosa 173-5, 178
Darwinia 138
    D. collina 148
    D. diosmoides 149
    D. hypericifolia 148
    D. lejostyla 148
    D. macrostegia 148
    D. meeboldii 148
    D. squarrosa 148
    D. vestita 79
Dasypogon bromeliifolius 79, 82
Daviesia benthamii 162
    D. decurrens 82
    D. divaricata 94
    D. grahamii 193
    D. horrida 84
    D. preissii 94
    D. teretifolia 141
Desert blackboy = *Xanthorrhoea thorntonii*
Desert bloodwood = *Eucalyptus chippendalei*
Desert oak = *Casuarina decaisneana*
Desert walnut = *Owenia reticulata*
Desmodium 290
Dichanthium 267, 271, 273, 277, 291-2, 298, 307
    D. annulatum 285
    D. fecundum 276, 278, 285, 290, 309
    D. tenuiculum 290, 309
Dicliptera 290
Dicrastylis exsuccosa 184-5, 233
Dillwynia 82
    D. pungens 155
Diplolaena dampieri 100, 105, 204
    D. microcephala 82, 125
Diplopeltis huegelii 92
Disphyma crassifolium 117-9
Diuris longifolia 149
Dodonaea 163
    D. adenophora 162
    D. aptera 90, 92
    D. inaequifolia 165
    D. lobulata 158, 160
    D. microzyga 165, 189
    D. rigida 233
    D. stenozyga 162
    D. viscosa 247
Dolichandrone heterophylla 268, 270, 282, 290-1
Drosera 51, 79, 149
Drummondita ericoides 138
Dryandra 52, 84, 102, 107, 140
    D. arborea 156
    D. armata 80, 150, 154-5
    D. bipinnatifida 95
    D. carlinoides 79, 94
    D. cirsioides 84
    D. formosa 77, 148-9
    D. fraseri 102
    D. longifolia 141, 155
    D. nivea 80, 84, 92, 94, 141, 149
    D. proteoides 149
    D. pteridifolia 143
    D. quercifolia 150, 152-3, 155
    D. sessilis 76, 84, 90, 92, 94
    D. shuttleworthiana 94
    D. squarrosa 79
Dwarf blackboy = *Xanthorrhoea nana*

Ecdeiocolea monostachya 104, 111, 113, 122
Echinochloa colona 280, 290
Einadia nutans 118
Enchylaena 122
    E. tomentosa 246, 250
Enneapogon 277-8, 281, 284, 288
    E. caerulescens 201, 238
    E. polyphyllus 278, 286, 290
Eragrostis dielsii 189
    E. eriopoda 186, 189, 226, 238, 257-8
    E. falcata 271
    E. japonica 280, 290
    E. lanipes 189
    E. setifolia 242, 247, 250, 252, 290
    E. xerophila 247
Eremaea 95
    E. pauciflora 94, 104, 109, 122
Eremophila 22, 130, 160-1, 163, 166-7, 192-4, 196, 202, 208, 211-6, 222, 226, 238, 246, 250, 257, 262-3
    E. abietina 216, 221, 233
    E. alternifolia 160, 162, 176
    E. bignoniiflora 280
    E. clarkei 189
    E. cuneifolia 208, 215, 218, 221, 223
    E. decipiens 162
    E. duttonii 189
    E. elderi 234
    E. exilifolia 189
    E. foliosissima 189
    E. fraseri 186, 189, 193-4
    E. georgei 189
    E. gilesii 162, 189, 234
    E. glabra 160, 162-3
    E. granitica 189
    E. ionantha 162
    E. latrobei 186-7, 233, 238
    E. lehmanniana 130
    E. leucophylla 184, 189, 207, 233
    E. longifolia 176
    E. mackinlayi 189
    E. macmillaniana 189
    E. maculata 176
    E. 'magnifica' 252
    E. margarethae 194
    E. miniata 164, 186
    E. oldfieldii 158, 160, 162-3
    E. oppositifolia 189
    E. pachyphylla 162
    E. platycalyx 189, 194
    E. pterocarpa 201, 207, 209, 211, 250
    E. punicea 189
    E. saligna 162
    E. scoparia 160, 162-3
    E. spathulata 189
    E. spectabilis 189
    E. viscida 189
    E. weldii 162
    E. willsii 233
Eriachne ciliata 286
    E. glauca 290
    E. helmsii 189
    E. mucronata 189, 286
    E. obtusa 278, 298
    E. pulchella 189
    E. sulcata 290
Eriochiton sclerolaenoides 176
Eriostemon spicatus 95
    E. thryptomenoides 122
    E. tomentellus 193
Erodiophyllum elderi 174
Erodium cicutarium 176
    E. cygnorum 189
Erythrophleum chlorostachys 257, 268, 270, 298
Eucalyptus 153, 171, 208, 240, 252, 270, 281
    E. accedens 80, 86
    E. alba 298
    E. albida 138
    E. anceps 176
    E. angulosa 76, 146, 150
    E. annulata 130
    E. apodophylla 298
    E. argillacea 26, 278, 284, 290, 292, 309
    E. aspera 233-4, 254, 256, 262
    E. astringens 79, 80, 84
    E. beardiana 109, 114
    E. bigalerita 300
    E. bleeseri 294
    E. brevifolia 247, 249, 262, 278-9, 280, 282, 283, 285-6, 287-8, 290, 292
    E. burracoppinensis 138
    E. caesia 124-6
    E. calophylla 16, 59, 61, 66, 71, 77-87, 90, 149
    E. calycogona 165
    E. camaldulensis 102, 104, 109, 114, 197, 201, 204, 217, 223, 226, 244, 247-8, 273, 296
    E. campaspe 160-2, 165
    E. capillosa 48, 118, 120, 130
    E. carnei 193
    E. celastroides 130, 162
    E. cerasiformis 130
    E. chippendalei 226, 233, 254-5
    E. chrysantha 151
    E. clelandii 160-1
    E. collina 288-9, 290
    E. comitae-vallis 186
    E. concinna 184
    E. confertiflora 278, 294, 298, 302
    E. conglobata 130, 144, 147, 150, 165, 177-8
    E. cooperiana 130, 141, 154
    E. cornuta 61, 147, 153
    E. corrugata 157, 161
    E. crucis 125-6
    E. cylindriflora 130
    E. dampieri 268, 270
    E. decipiens 79, 90
    E. deflexa 130
    E. dichromophloia (sens. lat.) 27, 28, 213, 233-4, 238, 244-5, 254, 262, 274, 278-9, 282, 285-6, 288, 290, 294, 302
    E. dielsii 130
    E. diptera 130, 132
    E. diversicolor 58-77
    E. diversifolia 165
    E. dongarraensis 100-1
    E. doratoxylon 149
    E. drummondii 84
    E. dundasii 162, 165
    E. eremophila 127, 130, 132, 157, 165
    E. erythrocorys 100, 103
    E. erythronema 118, 130
    E. eudesmioides 104, 113, 202
    E. falcata 129, 130, 134
    E. ferruginea 285, 288
    E. ficifolia 65, 66
    E. flocktoniae 129, 130-1, 144, 158, 159, 160-2, 165, 167, 176
    E. foecunda 80, 130, 157, 166-7
    E. foelscheana 294, 298, 302
    E. forrestiana 130
    E. gamophylla 215, 233-4, 245
    E. gardneri 130, 150, 153, 156
    E. georgei 130
    E. gomphocephala 16, 87, 90, 93, 96
    E. gongylocarpa 178, 180-2, 193, 199
    E. goniantha 130, 141
    E. gracilis 118, 130, 160
    E. grandifolia 234, 237, 269, 278, 288, 291-2, 298, 302
    E. griffithsii 165
    E. grossa 130, 137
    E. guilfoylei 61, 64
    E. herbertiana 294
    E. incrassata 80, 130, 139, 141, 144, 147, 155, 164
    E. jacksonii 61, 64
    E. jensenii 298
    E. kingsmillii 189, 196, 233
    E. kondininensis 118, 130, 132
    E. kruseana 125-6
    E. laeliae 84
    E. latifolia 294, 296, 298, 300, 302
    E. lehmannii 147, 150, 153
    E. leptocalyx 130
    E. leptopoda 184
    E. lesouefii 160, 161-2, 165, 167
    E. leucophloia 170, 244, 246-9
    E. longicornis 118, 130, 157, 159, 162
    E. loxophleba 79, 80, 87, 98, 102, 108-9, 115, 118, 121, 125, 130, 153, 156-8
    E. lucasii 189
    E. macrocarpa 13
    E. marginata 16, 58-9, 61, 66, 69, 70-1, 77-87, 92, 149
    E. megacarpa 66, 72, 77
    E. melanoxylon 119
    E. merrickiae 130, 165
    E. micranthera 130, 154
    E. microtheca 226, 230, 233, 242, 246, 250, 263, 267, 269, 271, 273-4, 276, 278, 280, 290, 292, 309
    E. miniata 289, 292-5, 300-2
    E. mooreana 288

Eucalyptus contd.
    E. nesophila 300
    E. nutans 150, 153
    E. obliqua 71
    E. obtusiflora 100-1
    E. occidentalis 79, 80, 86, 130, 143, 146
    E. odontocarpa 262
    E. oldfieldii 104, 113, 202
    E. oleosa 130, 132, 157-8, 160-2, 165-7, 176-8,
        184, 186, 189, 196, 199, 202, 208, 212, 215, 240
    E. oligantha 296
    E. orbifolia 125-6
    E. ovularis 130
    E. oxymitra 234
    E. pachyphylla 262
    E. papuana 234, 242, 244, 272-3, 298-9
    E. patellaris 302
    E. patens 77, 82
    E. perfoliata 286, 288
    E. phoenicea 285, 288, 290
    E. pileata 130
    E. platypus 130, 133, 146-7
    E. polycarpa 288, 291, 294, 296
    E. preissiana 150-3
    E. prominens 208, 212
    E. pruinosa 281, 283, 288, 309
    E. pyriformis 102
    E. redunca 102, 104, 113, 117, 118, 122, 130, 132,
        139, 141, 144, 152, 156
    E. rigidula 184
    E. roycei 109, 114
    E. rudis 77, 92, 94-5, 117
    E. salmonophloia 16, 20, 58, 115, 118, 120, 130, 132,
        153, 158, 160, 162, 175
    E. salubris 118, 120, 130, 158, 162-3, 165, 167, 175
    E. sepulcralis 150-1
    E. setosa 213, 226, 245, 254, 257, 268-9, 270, 278
    E. sheathiana 130, 156-8
    E. socialis 130, 167, 177-8
    E. spathulata 130, 141, 146
    E. staeri 79
    E. stoatei 153
    E. stricklandii 160, 162
    E. tectifica 26, 269, 270, 280, 285, 288, 290, 291-2,
        298, 300, 302
    E. terminalis 278-9, 282, 284, 309
    E. tetragona 21, 139, 140-1, 144, 147-8, 149, 150,
        152
    E. tetraptera 141
    E. tetrodonta 292-6, 300-2
    E. todtiana 56, 87, 92, 94, 100
    E. torquata 160-1
    E. transcontinentalis 118, 126, 130, 158-9, 160-2,
        165-7
    E. uncinata 130, 132, 144, 154
    E. wandoo 77, 80, 84, 85, 108, 115, 118, 120
    E. websteriana 125-6
    E. youngiana 24, 178, 181-2, 184, 186
Eulalia fulva 280, 290, 309
Euphorbia 272
    E. drummondii 176
Eutaxia obovata 72
Evandra aristata 69
Excoecaria agallocha 296
    E. parvifolia 309
Exocarpos 51, 272
    E. sparteus 77

Feathertop spinifex = Plectrachne schinzii
Ficus 286
    F. coronulata 273, 296
    F. opposita 277
    F. platypoda 208, 212, 226, 238
    F. racemosa 273, 296
Fimbristylis 288
    F. dichotoma 286, 288
Flemingia 276
    F. pauciflora 285
Flinders grass = Iseilema spp.
Frankenia 100, 105, 130, 164, 186, 193, 198, 201, 206,
    215, 246, 259
    F. ambita 242
    F. magnifica 214, 218
    F. pauciflora 162
Franklandia fucifolia 79

Gahnia 92, 122
Gardenia 288
    G. keartlandii 257, 262, 268, 270
    G. pyriformis 294
    G. resinosa 286, 288
Gastrolobium 80, 86
    G. bilobum 80
    G. calycinum 82
    G. grandiflorum 247
    G. microcarpum 84, 87
    G. oxylobioides 102
    G. parvifolium 130
    G. pycnostachyum 155
    G. spinosum 80, 84, 102, 108
Geijera linearifolia 176
Ghost gum = Eucalyptus papuana
Gimlet = Eucalyptus salubris
Glischrocaryon aureum 124
Gnephosis 218
    G. brevifolia 222
    G. burkittii 189
    G. skirrophora 176
Gompholobium 95
Gomphrena 288
    G. brachystylis 282
    G. canescens 282, 288, 300
    G. cunninghamii 247
Goodenia azurea 233
    G. concinna 189
    G. forrestii 216
    G. hirsuta 189
    G. maideniana 215, 252
    G. pinnatifida 176
    G. scaevolina 231-2, 250, 290
Gossypium australe 282
    G. sturtianum 290
Grevillea 21, 52, 208, 212, 224, 226, 252
    G. agrifolia 288, 294
    G. asteriscosa 135
    G. biformis 94, 110, 111
    G. bipinnatifida 81, 82
    G. brevicuspis 69
    G. concinna 155
    G. crithmifolia 92
    G. cunninghamii 288, 294
    G. deflexa 189
    G. dielsiana 112
    G. dimidiata 278, 282, 284
    G. diversifolia 82
    G. endlicheriana 82
    G. eriostachya 94, 186, 202, 208, 210, 213, 233, 254
        var. excelsior 157, 160
    G. gordoniana 102, 104, 109, 208, 213
    G. heliosperma 288, 294
    G. hookeriana 133-4, 141, 142, 144
    G. juncifolia 183-4, 186, 233, 235, 245, 254
    G. leucopteris 13, 100, 102, 105, 113
    G. nematophylla 160, 201
    G. oncogyne 162
    G. paradoxa 130
    G. pectinata 141
    G. petrophiloides 122, 130
    G. pinaster 102
    G. pritzelii 122
    G. pteridifolia 283, 288, 294, 296
    G. pterosperma 160, 162, 184
    G. pyramidalis 213, 244, 246, 262, 268, 271, 278,
        282, 284, 288, 290-1
    G. refracta 257, 268, 288
    G. rogersoniana 109, 114
    G. sarissa 164
    G. stenobotrya 182, 186, 208, 213, 226, 254, 255-6
    G. striata 273, 288
    G. synapheae 79
    G. thelemanniana 92
    G. vestita 92
    G. wickhamii 234, 244, 247, 254, 256-7, 262, 278,
        282, 284, 288
    G. wilsonii 82
Grey box = Eucalyptus tectifica
Greybush = Cratystylis conocephala
Grewia breviflora 272
Guichenotia ledifolia 89
Gum tree, general term for any Eucalypt with smooth
    bark shed annually

Gyrocarpus americanus 27, 268-9, 270, 286, 290
Gyrostemon ramulosus 182, 186, 233

Hakea 52, 102, 107-8, 112, 133, 144, 202, 226
    H. adnata 155
    H. amplexicaulis 66, 77, 79
    H. arborescens 278, 298
    H. auriculata 102
    H. baxteri 95, 149
    H. bucculenta 114
    H. ceratophylla 81, 82
    H. cinerea 141, 142
    H. conchifolia 95
    H. corymbosa 141, 144
    H. costata 94
    H. crassifolia 144, 150
    H. cristata 84
    H. cucullata 144, 149
    H. cyclocarpa 82
    H. elliptica 76, 81
    H. falcata 130
    H. ferruginea 79
    H. florida 149
    H. incrassata 95
    H. laurina 144
    H. lehmanniana 80
    H. lissocarpha 80, 82
    H. lorea 234, 284
    H. macrocarpa 268
    H. multilineata 130, 184
    H. obliqua 94, 102, 107
    H. oleifolia 76, 77, 79, 146
    H. pandanicarpa 149, 155
    H. preissii 102, 109, 117, 118, 196, 214
    H. prostrata 76, 92, 94-5, 141
    H. pycnoneura 102, 108
    H. recurva 102, 109
    H. rhombales 226, 234
    H. ruscifolia 82, 94-5
    H. stenophylla 100, 208, 211, 213
    H. suberea 184, 186, 189, 213-4, 226, 233-4, 245-7,
        254, 257, 260, 262, 278, 291
    H. trifurcata 84, 94
    H. undulata 81
    H. varia 77, 81, 82
    H. victoria 144, 151
Halgania 185
    H. preissiana 189
    H. solanacea 233
    H. viscosa 193
Haloragis odontocarpa 189
Halosarcia 119, 146, 162, 164, 193, 201, 204, 214, 259,
    263, 271
    H. halocnemoides 117-8, 130, 198, 242, 246, 269
    H. indica 117, 242, 250, 269
    H. pergranulata 117
Hardenbergia comptoniana 66, 79, 90, 92
Helichrysum apiculatum 184
    H. cordatum 77, 92
Heliotropium 290
    H. heteranthum 219
Helipterum 173, 195
    H. charsleyae 189
    H. craspedioides 189, 195
    H. fitzgibbonii 236, 238
    H. floribundum 162, 173, 189, 236, 238
    H. haigii 176
    H. humboldtianum 158, 201
    H. splendidum 189
    H. sterilescens 197, 201
    H. stipitatum 184, 233, 235, 238
    H. venustum 189
Hemichroa diandra 242, 258-9
Heterodendrum oleaefolium 176, 178, 238
Heteropogon contortus 272, 284, 286, 290-1, 298, 300
Hibbertia 80, 143
    H. amplexicaulis 66, 77
    H. cuneiformis 66, 76-7
    H. gracilipes 155
    H. grossulariifolia 66
    H. hypericoides 84, 92, 94-5
    H. lineata 82
    H. montana 84
    H. mucronata 138
    H. polystachya 82, 83

Hibbertia contd.
   H. pungens 162
   H. racemosa 92
   H. serrata 66
   H. spicata 208, 213
Hibiscus 309
   H. panduriformis 251, 290
   H. spinulosus 287
Hovea chorizemifolia 82
   H. elliptica 66, 73, 77
Hybanthus calycinus 94
   H. floribundus 130
Hypocalymma angustifolium 77, 80–2, 94
   H. cordifolium 66
   H. myrtifolium 149
Hypolaena exsulca 80

Illyarrie = *Eucalyptus erythrocorys*
Indigofera linifolia 276
Ipomoea 285, 290
   I. pes-caprae 272
Ironwood = *Erythrophleum chlorostachys*
Iseilema 280, 282, 285, 290
   I. vaginiflorum 244
Isolepis nodosa 77, 90, 91
Isopogon 52, 92, 133–4
   I. buxifolius 140, 141, 155
   I. cuneatus 79, 149
   I. divergens 94, 111
   I. dubius 52, 77, 82, 149
   I. formosus 79, 143
   I. latifolius 148–9
   I. scabriusculus 95, 130
   I. sphaerocephalus 79
   I. teretifolius 80, 130, 150

Jacksonia argentea 294
   J. cupulifera 100, 102, 109
   J. furcellata 92
   J. horrida 76
   J. sternbergiana 92, 94
   J. thesioides 288, 294
Jacquemontia browniana 282
Jam, raspberry jam tree = *Acacia acuminata*
Jarrah = *Eucalyptus marginata*
Johnsonia lupulina 79, 149
Juncus 95
   J. kraussii 92, 95

Kangaroo grass = *Themeda australis*
Kangaroo paw = *Anigozanthos* spp.
Kapok bush = *Aerva javanica*
Kapok tree = *Cochlospermum fraseri*
Karri = *Eucalyptus diversicolor*
Kennedia coccinea 66, 72–3, 79, 82–3
   K. glabrata 72
   K. prorepens 184
   K. prostrata 79, 92
Keraudrenia integrifolia 124
Kimberley heather = *Calytrix exstipulata*
Kingia australis 77, 79, 82, 149
Kondinin blackbutt = *Eucalyptus kondininensis*
Kunzea baxteri 147, 155
   K. ericifolia 69
   K. pulchella 125, 147
   K. recurva 79, 149

Labichea 208
   L. lanceolata 138
Lachnostachys eriobotrya 13, 94
Lagenifera huegelii 92
Lamarchea hakeifolia 203, 233
Lambertia 140
   L. ericifolia 149
   L. inermis 79, 94–5, 151–2, 144
   L. uniflora 149
Lamb's tails = *Lachnostachys* spp.
Lasiopetalum floribundum 82
   L. rosmarinifolium 155
Lawrencia helmsii
   L. squamata 176
Lechenaultia 83, 197
   L. biloba 83, 94–5
   L. formosa 80, 141
   L. macrantha 123

Leschenaultia = *Lechenaultia*
Leichhardt pine = *Nauclea orientalis*
Lepidium 236
   L. oxytrichum 123
   L. platypetalum 223
   L. rotundum 173
Lepidosperma 69, 90, 122
   L. angustatum 79, 82
   L. costale 125
   L. gladiatum 90–1
   L. gracile 125
   L. longitudinale 66
   L. persecans 69
   L. tenue 125
   L. tetraquetrum 82
   L. viscidum 149
Leptocarpus scariosus 69, 81–2
   L. tenax 69, 82
Leptomeria 51
   L. cunninghamii 82
Leptosema chambersii 184
Leptospermum erubescens 80, 84, 95, 125, 130
   L. sericeum 147, 152
   L. sp. inedit. 137–8
Leucopogon 95, 143
   L. apiculatus 155
   L. capitellatus 66, 82
   L. cordatus 82
   L. obovatus 69
   L. oxycedrus 82
   L. parviflorus 76, 92
   L. propinquus 66, 77, 82
   L. pulchellus 72
   L. unilateralis 149
   L. verticillatus 66, 77, 79, 82
Livistona alfredii 247, 251
   L. eastonii 300–2
   L. loriphylla 279
   L. sp. inedit. 296
Logania vaginalis 92
Lomandra 66
   L. hastilis 94
Lophostemon grandiflorus 273
Loxocarya flexuosa 76
Lumnitzera racemosa 296
Lupin = *Lupinus* spp.
Lycium australe 176
Lyginia barbata 69
Lysinema ciliatum 76, 95, 149
Lysiphyllum cunninghamii 242, 267–8, 270, 273, 276–8, 280, 286, 288, 290–1, 307, 309

Macrozamia riedlei 52, 66, 77, 82, 84, 90, 94
Maireana 122, 186, 198
   M. brevifolia 117
   M. carnosa 186
   M. georgei 176
   M. luehmannii 250
   M. polypterygia 209
   M. pyramidata 164, 201
   M. sedifolia 22, 158, 160, 161, 162, 171–8
   M. triptera 201
Malleostemon roseus 94
Malvastrum americanum 280
Mangrove: general term for trees adapted to tropical tidal swamps
Marble gum = *Eucalyptus gongylocarpa*
Marri = *Eucalyptus calophylla*
Melaleuca 20–1, 59, 69, 87, 90, 100, 102, 104, 108–9, 114, 117, 122–3, 127, 130, 142, 146–7, 153, 167, 233, 269, 273, 294
   M. acerosa 76, 90, 92, 100–1
   M. cardiophylla 100–1, 204–5
   M. citrina 150
   M. ?cordata 194, 122
   M. cuticularis 92, 95
   M. cymbifolia 118
   M. densa 69
   M. elliptica 125, 131
   M. fulgens 137–8, 147
   M. glaberrima 155
   M. globifera 147
   M. glomerata 230, 257–8, 263
   M. hamulosa 118, 130
   M. huegelii 90, 92, 100–1

   M. lanceolata 146–7, 178
   M. lasiandra 230, 234, 244, 254, 257–8, 262–3
   M. lateriflora 130
   M. leiocarpa 184
   M. leucadendra 217, 223, 247, 296, 299
   M. megacephala 100, 102, 108, 126
   M. minutifolia 290–1
   M. nervosa 226, 271, 309
   M. pauperiflora 16, 122, 130, 157, 159, 162, 244
   M. pentagona 146, 152, 155
   M. polygaloides 149
   M. preissiana 67–9, 81–2, 92, 96, 146
   M. pungens 130
   M. quadrifaria 178
   M. rhaphiophylla 67–8, 92, 95–6
   M. scabra 102, 108, 110–1, 130
   M. spicigera 130
   M. striata 79
   M. subtrigona 146
   M. thymoides 146
   M. thyoides 79, 118, 130
   M. uncinata 20, 104, 117–8, 122, 146, 165, 196, 201
   M. undulata 126
   M. urceolaris 130
   M. viminea 130
   M. viridiflora 298, 309
Mesembryanthemum 119
Mesomelaena 69
   M. stygia 149
   M. tetragona 69, 81–2, 149
Micromyrtus flaviflora 183, 186
   M. imbricata 130
Minnieritchie = *Acacia* spp. with peeling bark, esp. *A. grasbyi*
Mirbelia ramulosa 208, 213
   M. spinosa 130
Mirret = *Eucalyptus flocktoniae*
Mistletoe = *Amyema* spp.
Mitchell grass = *Astrebla* spp.
   Barley m. grass = *A. squarrosa*
   Bull m. grass = *A. pectinata*
   Weeping m. grass = *A. elymoides*
Monachather paradoxa 189, 238
Monotoca oligarrhenoides 155
Moort = *Eucalyptus platypus*
Morrell = *Eucalyptus longicornis*
Morrison = *Verticordia* spp.
Mount House box = *Eucalyptus argillacea*
Muehlenbeckia cunninghamii 280
Muellerolimon saliconiaceum 242, 246
Mulga = *Acacia aneura*
Mullamulla = *Ptilotus* spp.
Mygum = *Eucalyptus leucophloia*
Myoporum 178
   M. insulare 90–1
   M. platycarpum 162, 164, 175–6
   M. tetrandrum 92
Myriocephalus guerinae 189

Native poplar = *Codonocarpus cotinifolius*
Nauclea orientalis 273, 275
Neptunia 276, 280, 285
Newcastelia cephalantha 184, 231, 233
Nicotiana goodspeedii 176
Nine awn grass = *Enneapogon* spp.
Nothofagus 45–6
Northern morrison = *Verticordia cunninghamii*
Nuytsia floribunda 51, 67, 77, 79, 84, 87, 92, 106, 109, 141–2, 144, 146, 149, 151

Oenothera drummondii 91
Olax phyllanthi 76, 146
Old socks = *Grevillea leucopteris*
Olearia axillaris 75, 77, 90–1, 93, 100, 105, 146
   O. muelleri 162
   O. propinqua 193
Opercularia hispidula 66
Ophiuros exaltatus 309
Orthrosanthus 94
   O. laxus 66
Osbornia octodonta 296
Owenia reticulata 213, 252, 257, 270
   O. vernicosa 294–6
Oxychloris scariosa 284
Oxylobium 77, 80, 86

Oxylobium contd.
 O. atropurpureum 148-9
 O. capitatum 80
 O. drummondii 86
 O. lanceolatum 69
 O. parviflorum 80, 87, 138

Pachycornia triandra 246, 250
Pandanus 273, 293, 299
 P. aquaticus 294, 303
Panicum decompositum 250, 285
 P. mindanaense 290
 P. whitei 247, 285
Paperbark = Melaleuca spp. with laminated bark
Parakeelya = Calandrinia spp.
Paraserianthes lophantha 66
Paspalidium rarum 290
Patersonia occidentalis 82
 P. rudis 82
 P. umbrosa 66
Paterson's curse = Echium plantagineum
Pelargonium 51, 90
 P. capitatum 90
Pemphis acidula 296
Peplidium muelleri 189
Peppermint tree = Agonis flexuosa
Pericalymma ellipticum 81-2
Perotis rara 278
Persoonia 279
 P. elliptica 77
 P. falcata 294
 P. longifolia 16, 66, 77, 79, 82
 P. teretifolia 79
Petalostigma pubescens 286, 288, 290, 294
Petalostylis cassioides 232-3
 P. labicheoides 247
Petrophile 52, 92, 133, 141, 155
 P. divaricata 149
 P. diversifolia 77
 P. ericifolia 94, 130
 P. linearis 94
 P. media 95
 P. rigida 79
 P. serruriae 77, 92, 94
 P. squamata 80
Phebalium filifolium 130
 P. tuberculosum 122
Phyllanthus calycinus 66, 82, 89, 92
Phytophthora cinnamomi 82, 84
Pigface = Disphyma crassifolium, also Carpobrotus and Mesembryanthemum
Pileanthus peduncularis 102, 104, 111, 208-9, 211
Pimelea 94, 101
 P. brevifolia 155
 P. clavata 66, 76-7
 P. ferruginea 75-6, 101, 143, 146, 152
 P. lehmanniana 79
 P. spectabilis 79
 P. suaveolens 79, 130
Pin grass = Borya nitida
Pitcher plant = Cephalotus follicularis
Pittosporum phylliraeoides 130, 164-5, 173, 176, 178, 186, 201, 226
Pityrodia axillaris 130
 P. lepidota 130
Planchonia careya 273, 288, 294
Platysace compressa 155
Platytheca galioides 149
Plectrachne 91, 170, 186, 204, 271-2
 P. bromoides 285
 P. bynoei 286
 P. danthonioides 109, 114, 285
 P. melvillei 226, 234, 238, 296
 P. pungens 27, 266, 269, 270, 278, 285-6, 288, 290, 292, 294-5, 298, 302-9
 P. rigidissima 159, 166, 193, 285
 P. schinzii 266, 269, 281, 283, 287, 292, 294, 256-7, 268, 278
 P. sp. inedit. 290, 292
Podocarpus 45-6
 P. drouynianus 66, 74, 77
Podolepis 101, 110, 173, 184
 P. canescens 80, 186, 197
 P. kendallii 189
 P. rugata 176
Polycarpaea 288, 290, 300

Polypogon monspeliensis 117
Polypompholyx 84
Poplar gum = Eucalyptus bigalerita
Portulaca 276, 290
 P. filifolia 286
 P. oleracea 284
Pouteria sericea 272, 296
Poverty bush = Eremophila spp.
Powderbark wandoo = Eucalyptus accedens
Prostanthera incurvata 162
 P. sp. nov. inedit. 234, 237
Pteridium esculentum 66, 79, 82
Pterocaulon glandulosum 282
Ptilotus 196, 208, 262, 300
 P. aervoides 189
 P. calostachyus 278, 282, 290
 P. carinatus 244, 247
 P. clementii 231, 238, 250
 P. drummondii 189, 223, 233
 P. exaltatus 160, 162-3, 184, 189, 190, 215, 238, 246, 272, 282
 P. fusiformis 282
 P. gomphrenoides 250
 P. helipteroides 189, 215, 236, 238, 250
 P. macrocephalus 189, 207, 215, 233
 P. murrayi 214
 P. obovatus 158, 162, 169, 176, 184, 186, 189, 191, 223, 236, 238
 P. polakii 204, 209
 P. polystachyus 94, 184, 189, 231, 233, 238
 P. rotundifolius 189, 218, 221, 247
 P. schwartzii 231
 P. spicatus 282, 285
Pultenaea capitata 130
 P. reticulata 66, 69

Quandong = Santalum acuminatum
Queensland bottle tree = Brachychiton rupestris
Quinine tree = Petalostigma pubescens

Ranunculus colonorum 92
Red gum (marri) = Eucalyptus calophylla
Red tingle = Eucalyptus jacksonii
Reeds = Restionaceae spp., also Typha and Cyperaceae
Regelia velutina 150-1
Restio 69, 149
 R. tremulus 69
Rhadinothamnus euphemiae 155
Rhagodia 122, 186
 R. baccata 92
 R. spinescens 118
Rhizanthella gardneri 122
Rhizophora stylosa 208, 242, 296
Rhynchosia 276
 R. minima 280, 285
Ribbon grass = Chrysopogon spp.
River gum = Eucalyptus rudis
River red gum = Eucalyptus camaldulensis
Rottnest pine = Callitris preissii
Rushes = Restionaceae spp.

Salmon gum = Eucalyptus salmonophloia
Salsola kali 91, 272
Saltbush = Atriplex spp.
Samphire = succulent Chenopodiaceous plants, generally Halosarcia
Sandalwood = Santalum spicatum
Santalum 272
 S. acuminatum 183, 208, 233-4
 S. lanceolatum 211
 S. spicatum 183
Sarcocornia blackiana 130
Sarcostemma australe 184
Scaevola 100-1, 208
 S. auriculata 66
 S. crassifolia 75-6, 77, 90-1, 100, 104-5, 116, 152, 160
 S. globulifera 208
 S. nitida 75, 77, 92
 S. spinescens 125, 158, 162
 S. striata 66
 S. thesioides 92
Schoenia cassiniana 110, 186, 189, 195, 202, 207
Sclerolaena 100, 105, 122, 176, 186, 201, 214, 246, 259, 271
 S. cuneata 219, 233

 S. diacantha 117
 S. obliquicuspis 173
 S. patenticuspis 173
Sclerostegia arbuscula 162
Screw pine = Pandanus
Scyphiphora hydrophylacea 296
Sedges = Cyperaceae spp.
Sehima 271
 S. nervosum 273, 284, 288, 290-2, 296, 309,
Senecio lautus 23, 76, 90, 186, 206
Sesbania cannabina 309
Sesuvium portulacastrum 269
Seven years wattle = Acacia eriopoda
Sida calyxhymenia 176, 189
Sida fibulifera 280, 284-5
 S. spinosa 280, 285
Smooth-barked bloodwood = Eucalyptus foelscheana
Snappy gum = Eucalyptus brevifolia
Snakewood = Acacia eremaea, A. xiphophylla
Soft spinifex = Triodia pungens
Solanum 196, 208, 214
 S. lasiophyllum 189, 217, 223
 S. orbiculatum 207
Sonneratia alba 296
Sorghum 280, 286, 288, 292, 298, 301, 309
 S. plumosum 26, 272, 284, 290-1, 294, 298, 299, 309
 S. stipoideum 26, 290-1
 S. timorense 290-1, 294, 309
Sowerbaea laxiflora 92, 94
Sphaerolobium macranthum 79, 149
 S. medium 66
Sphenotoma 148
 S. dracophylloides 149
Spinifex = hummock grasses Triodia and Plectrachne, not genus Spinifex
 Curly spinifex = Plectrachne pungens
 Feathertop = Plectrachne schinzii
 Soft spinifex = Triodia pungens
 Hard or buck spinifex = any other inedible Triodia spp.
Spinifex 100, 272
 S. hirsutus 90-1
 S. longifolius 90-1, 269, 272
Sporobolus 288
 S. australasicus 284
 S. virginicus 242, 269, 271
Spyridium globulosum 76-7
Starflower = Calytrix spp.
Sterculia viscidula 275, 277, 286, 288
Stipa elegantissima 125
 S. eremophila 173-5, 178
 S. hemipogon 125
 S. nitida 173, 175
 S. variabilis 93
Stirlingia latifolia 82, 92, 94
 S. tenuifolia 79
Streptoglossa odora 244
Sturt's desert pea = Clianthus formosus
Stypandra imbricata 125
Styphelia tenuiflora 82
Suaeda arbusculoides 242
Sundew = Drosera spp.
Swainsona 170, 196, 206-8
 S. beasleyana 189
 S. campestris 176
 S. canescens 215
 S. kingii 246
 S. maccullochiana 222-3
 S. villosa 189
Swamp banksia = Banksia littoralis
Swamp yate = Eucalyptus occidentalis
Symphyobasis sp. 189
Synaphea 79
 S. favosa 149
 S. petiolaris 82, 91-5
 S. polymorpha 149
Syzygium suborbiculare 294

Tallerack = Eucalyptus tetragona
Teatree = Melaleuca spp.
Templetonia 233
 T. retusa 92, 154
 T. sulcata 118
Tephrosia 290
 T. arenicola 258

Tephrosia contd.
   T. flammea 207
Terminalia 26, 272, 277, 281–2, 284, 288, 294, 296, 298, 303
   T. arostrata 280, 285, 309
   T. canescens 281, 286, 290–1, 298
   T. ferdinandiana 286
   T. grandiflora 296
   T. petiolaris 272
   T. platyphylla 273, 275, 296, 309
   T. volucris 285, 290, 309
Tetragonia decumbens 90–1
Tetrarrhena laevis 66
Themeda arguens 300, 309
   T. australis 252, 272, 284, 290–1, 298
   T. avenacea 278, 280
Thomasia cognata 89
   T. glutinosa 82
   T. quercifolia 66
   T. triloba 66
Thryptomene 21, 208, 226
   T. australis 122, 124–5
   T. baeckeacea 204–5, 208, 213
   T. decussata 193, 202, 205
   T. elliottii 13
   T. maisonneuvii 182–3, 186, 226, 233, 254
   T. urceolaris 162, 193
Trachyandra divaricata 90–1
Trachymene coerulea 92
   T. glaucifolia 215, 233
   T. villosa 278
Tragus australianus 278, 284
Tremandra stelligera 66
Trianthema oxycalyptra 219
Trichodesma zeylanicum 222, 233, 278, 282, 285, 309
Triodia 91, 104, 113, 170, 186, 202, 220, 233, 240–1, 272, 277, 285
   T. angusta 208, 215, 244
   T. basedowii 170, 178, 180–1, 186, 193, 199, 208, 210, 213, 224, 226, 233–4, 237, 238, 245, 254, 256–7
   T. brizoides 244
   T. concinna 266
   T. intermedia 262, 278–9, 282

   T. inutilis 278
   T. irritans 193
   T. lanigera 244
   T. longiceps 215, 244
   T. plurinervata 203, 205
   T. pungens 170, 208, 210, 213, 240, 242–4, 245–6, 252, 254, 256–8, 259, 260, 262–3, 268–9, 270, 272, 278, 282, 290, 292
   T. scariosa 54, 158–9, 164–7, 186, 240
   T. wiseana 212, 240, 242, 244, 246–8, 274, 275–6, 278, 281, 284
   T. sp. inedit. 259, 296
Trymalium floribundum 66, 82
   T. ledifolium 80, 82
Tuart = *Eucalyptus gomphocephala*
Typha orientalis 92

Ursinia anthemoides 80
Utricularia 84

Velleia macrophylla 79
   V. rosea 22, 186, 189
Ventilago viminalis 277–8, 288, 294
Verrauxia reinwardtii 94–5
Verticordia 13, 102, 123, 142
   V. acerosa 82
   V. brownii 123
   V. chrysantha 102
   V. cunninghamii 284, 294
   V. densiflora 96
   V. etheliana 104, 208, 213
   V. forrestii 208, 212
   V. grandiflora 94–5, 110
   V. habrantha 79, 144
   V. monadelpha 111, 123
   V. nitens 20, 87, 92
   V. patens 94–5
   V. picta 93, 134
   V. plumosa 72, 144
   V. polytricha 95
   V. preissii 131, 136
   V. roei 134
   V. serrata 134
   V. verticillata 284

Viminaria juncea 52
Vitex glabrata 296

Waitzia 185
   W. acuminata 125, 184, 186, 238
   W. aurea 189
   W. citrina 80, 189
Wandoo = *Eucalyptus wandoo*
Warren River cedar = *Agonis juniperina*
Waterwood = *Atalaya hemiglauca*
Wedelia asperrima 282
Wehlia thryptomenoides 181, 184
Westringia dampieri 162, 177
White grass = *Sehima nervosum*
White gum = *Eucalyptus wandoo*
Wild plum = *Terminalia platyphylla*
Wild oats = *Avena* spp.
Wiregrass = *Aristida* spp.
   Feathertop wiregrass = *A. latifolia*
Wrightia 277

Xanthorrhoea drummondii 56, 95, 102–3, 106
   X. gracilis 149, 152
   X. nana 135, 138
   X. preissii 66–7, 69, 72, 77, 79, 84, 86, 90, 94, 144
   X. thorntonii 159, 183–4, 226
Xanthosia 66
   X. rotundifolia 73, 79, 149
Xanthostemon eucalyptoides 296
   X. paradoxus 296
Xerochloa laniflora 290
Xylocarpus moluccensis 296
Xylomelum 102, 105, 109, 113, 140, 142
   X. angustifolium 94, 100, 102, 105–6, 112–3, 122–3
   X. occidentale 77, 79, 92

Yate = *Eucalyptus occidentalis*
Yellow tingle = *Eucalyptus guilfoylei*
York gum = *Eucalyptus loxophleba*

Zamia palm = *Macrozamia riedlei*
Zygophyllum 186, 201
   Z. ovatum 173, 176

2 November 1991
F & M Peering / Exhibition Meeting, Edinburgh
£22.50